HURRICANES AND FLORIDA AGRICULTURE

John A. Attaway, Ph.D.

HURRICANES AND FLORIDA AGRICULTURE

Library of Congress Cataloging-in-Publication Data

Attaway, John A., 1930—
    Hurricanes and Florida agriculture/John A. Attaway.
        p. cm.
    Includes bibliographical references (p.) and index.
    ISBN 0-944961-05-3 (alk. paper)
    1. Hurricanes–Florida. 2. Crops and climate–Florida. I. Title

    QC945 . A77 1999
    551.55'2'09759–dc21

                                                            99-40086

Current printing (last digit): 10 9 8 7 6 5 4 3 2 1
Printed in the United States of America
E. O. Painter Printing Co., DeLeon Springs, Florida

TABLE OF CONTENTS

Foreward

After a highly successful historical treatise in 1997 which chronicled freezes damaging Florida citrus crops, Dr. John Attaway has ventured into the realm of hurricanes and their impacts on Florida agriculture. The vagaries of weather have always been a bane to farmers. Until recent times, probably the least verified weather phenomena affecting agriculture have been hurricanes and tropical storms. His extensive research on the occurrence of storms in the Atlantic Basin affords the reader an impressive, comprehensive account of almost every recorded Florida storm since Columbus sailed the Caribbean in the 1490s. For the avid scholar of meteorological events, Attaway for the first time provides, in one volume, accurate and interesting data on even the most obscure storms to hit the region.

The technical information on the major storm systems is nicely balanced with folksy anecdotes, often humorous, from individuals who personally experienced these storms. Much of the book follows a consistent format including detailed data on individual storm tracks, meteorological factors, and damage to agriculture. This is reinforced with a chapter on the activities of the National Hurricane Center.

Readers of all ages can harken back and relive their most memorable hurricane experiences. Special attention is devoted to the major hurricane of recent history, Andrew, which devastated south Florida in August, 1992. Its effects on Dade county agriculture is vividly described.

As a valued friend and colleague for over 17 years, I have enjoyed many discussions with John Attaway and his insights on all aspects of citrus science and technology. His is a storehouse of citrus knowledge from high tech to practical know-how. He is widely published on subjects ranging from chemistry, citrus production and processing, citrus freezes, and now on hurricane dynamics.

A second generation Florida citrus grower with a Ph.D. in chemistry from Duke University, the author served as a scientist with the Florida

Department of Citrus for 36 years, it's Scientific Research Director from 1968 to 1995, and as Adjunct Professor with the University of Florida at Lake Alfred. He retired in 1995. Under his administration, the Florida Department of Citrus conducted major studies in advancing citrus processing and handling operations. His background in agriculture, science and research methodology well qualifies him to address and interpret the impacts of hurricanes on agricultural production. He takes us from ancient, on-the-scene, first-hand verbal reports to the use of modern Doppler radar satellite images and computer modeling to predict and characterize these fierce weather phenomena.

As a noted documentor of climatological events, Attaway's first book on citrus freezes served as a model for understanding the nature of these destructive assaults. His second book provides further testimony of the risks the Florida farmer take in helping to feed the nation.

Hurricanes and Florida Agriculture is required reading for the student of meteorology with a genuine interest in mechanisms of storm systems in the Atlantic Basin. For the rest, it is just plain fascinating.

Walter J. Kender, Ph.D.
Professor
University of Florida
Lake Alfred, Florida

Note from the Author

On a personal note, when the Attaway family moved to Haines City from Atlanta in the mid-1940s, it was in the midst of the most active hurricane decade in Florida's recent history, and I acquired an immediate interest in the ferocious storms which struck the state in 1944, 1945, 1946, 1947, 1948, 1949 and 1950. When the hurricane advisory was broadcast on WSIR (Winter Haven), WLAK (Lakeland), WDBO (Orlando) or WFLA (Tampa), a radio in the household would be tuned to the station and a notepad would be handy to write down the coordinates of the storm. In those days, we had no TV and certainly no Weather Channel and no Internet. Today—the 1990s—hurricanes have not assaulted the peninsula as frequently as they did in the 1940s, but, as demonstrated by Hurricane Andrew in 1992, the potential is always there, and as a citrus grower, the advisories from the National Hurricane Center in Miami have not lost their interest or importance.

This is the second of two books on the impact of weather disasters on Florida agriculture. The first book, *A History of Florida Citrus Freezes*, (Attaway, 1997) traced the effect of freezes on the citrus industry beginning in early colonial times and continuing through the freeze of January 1997. It is interesting that both hurricanes and freezes have tended to cluster in certain decades, for example, the freezes of the 1980s and the hurricanes of the 1940s. Over the course of time, freezes have been the most damaging, but hurricanes also have the potential to do catastrophic damage to Florida agriculture and to the Florida citrus industry in particular since the industry is presently concentrated heavily in southwest Florida counties such as Collier, Lee and Hendry, which have experienced severe hurricanes on many occasions. Hurricanes which passed through Collier County and Hendry County in the 1920s and 1940s caused only minimal damage at the time they occurred. However, a hurricane of equal strength following the same path would be capable of causing severe damage if it occurred today.

In the book on freezes, it was possible to quantify the extent of damage to the citrus industry with reasonable accuracy as freezes occur after the

Florida Agricultural Statistics Service has released its October crop estimate. However, with very few exceptions, hurricanes occur prior to the citrus crop estimate so the extent of damage cannot be estimated as closely. The only major hurricane to occur after the citrus crop estimate was the hurricane of October 18-19, 1944.

Much has been written about the effects of hurricanes on urban areas, the destruction of homes and buildings by wind and storm surge (Barnes, 1998), but little has been written on the effects on farms and groves. This book will attempt to fill the void.

Dedication

To the meteorologists of the National Hurricane Center, Miami, Florida and the National Weather Service, Ruskin, Florida for their many valuable services to Florida agriculture.

Acknowledgments

Gratitude is expressed to Dr. Will Wardowski for his editorial skills and Mrs. Peggy Hicks, word processor extraordinaire, without whom this manuscript would have never made it to the printer, Mrs. Fay Ball, who can read my handwriting, and Dr. Herman J. Reitz and Dr. Walter J. Kender for reviewing the manuscript. Ms. Sarah Hardy of Altamonte Springs, Florida did a superb job preparing the figures showing hurricane tracks through Florida using information from Hugh J. Hardy, Jr. which show historically-accurate county boundaries at the time of each hurricane. Professional and prompt expertise at E. O. Painter Printing Co. helped to turn the manuscript into a book.

I owe special thanks to many dedicated individuals who took time to reminisce about the 1920s, 30s and 40s and patiently answer my questions. I am particularly indebted to William H. Krome and Phoebe Krome of Homestead, Florida for detailed written accounts of personal experiences with hurricanes in south Dade County from the great Miami Hurricane in 1926 to Hurricane Andrew in 1992. Other major contributors in Dade County were Seymour Goldweber who gave me a guided tour of the areas ravaged by Hurricane Andrew and Barney Walden whose family was influenced by south Florida hurricanes almost back to the turn of the century.

In the Okeechobee area, so devastated by the deadly 1928 hurricane, Mrs. Zetta Hunt, Mrs. Marie Box, Mr. Clayton White and Mr. Basil Harvey shared their memories of that terrible disaster with me, and Dr. Joseph R. Orsenigo, retired Director of the Everglades Research and Education Center, pointed me to documents describing these terrible days in 1928 when over 2,000 people died along the south shore of the big lake.

On the Indian River, Reuben Carlton of Fort Pierce had a wealth of memories and observations concerning several major hurricanes which struck the citrus, cattle and vegetable industries of Martin, St. Lucie and

Indian River counties over a 60-year period. Other contributors from the River were John King, Alto "Bud" Adams, Louis Forget, Peter Spyke, Anne Wilder and Ada Coates Williams.

In southwest Florida, Norman Todd and W. T. Maddox of LaBelle, former Florida Citrus Chairman Hugh English of A. Duda and Sons in LaBelle and E. Brown of Immokalee were helpful.

At the Citrus Research and Education Center in Lake Alfred, Librarian Pam Russ, Audio Visual Specialist Terri Appelboom and Audio Visual Equipment Operator Jamie Chastain made many important contributions, and Center Director, Dr. Harold Browning, helped smooth the way.

At the National Hurricane Center in Miami, Librarian Robert Britter was especially helpful, and at the National Weather Service Office in Ruskin, Meteorologist-in-Charge Mr. Ira Brenner made me welcome during my several visits. The staff of the National Climatic Data Center in Asheville, North Carolina greatly simplified our search for Weather Bureau publications.

In Gainesville at the University of Florida, E. C. (Gene) Stivender and Mr. Dale McPherson provided important information on the IFAS response to Hurricane Andrew.

And as always, input from the staff of the Florida Agricultural Statistics Service in Orlando was indispensable, particularly Dr. John Witzig, Paul Messenger, Bob Terry and Shirley Zonner.

Champneys Tunno of Haines City loaned me his copy of Nathaniel Bowditch's classical work entitled *The American Practical Navigator* and Polk County growers and ranchers Dudley Putnam and Buck Mann shared their memories of hurricanes long past. Charlie Hendrix of Clermont, Florida added his recollections, and the Estate of Luis E. Sanchez of Mayaguez, Puerto Rico loaned me his extensive materials related to Caribbean hurricanes.

Introduction

My first book, "The History of Florida Citrus Freezes," was confined to a study of the impact of freezes on the Florida citrus industry. This book will consider the effect of hurricanes, not only on citrus but tropical fruits, vegetables and field crops as well.

As stated in my first book, I am not a climatologist or a meteorologist, just a grower with a lifelong interest in the effect of weather on Florida agriculture, and my remarks should be taken with that in mind.

While all areas of Florida are vulnerable to hurricane damage, there are two "hurricane alleys" which are most susceptible to damage. These are (1) the southern counties including the Indian River area south to Miami and Key West and around to Naples and Fort Myers, and (2) the westernmost counties of the Florida Panhandle including the Pensacola area eastward to Apalachicola. Within these areas, the greatest damage occurs along the immediate coast where the hurricane storms ashore with its full fury fueled by the warm waters of the gulf stream or the Gulf of Mexico. As the hurricane moves over land, it loses this "heat sink" and the winds diminish in velocity, but the combined winds and heavy rains can still be exceedingly dangerous as exemplified by the heavy loss of life around Lake Okeechobee in 1926 and again in 1928.

Crops most subject to hurricane damage include citrus, the tropical fruits of south Dade County and Collier County, the vegetable crops in the northern Everglades area of Palm Beach and Broward counties, and the cotton, corn, pecan, hay, soybean and pine timber crops from Escambia County east through Santa Rosa, Okaloosa and Walton counties to the Apalachicola River. Citrus is most vulnerable in two areas, the Indian River Marketing area including Brevard, Indian River, St. Lucie and Martin counties, and the southwest Florida citrus producing district which includes Hendry, Collier, Lee and Charlotte counties.

Chapter 1

Historic Hurricanes of the Atlantic Basin

"Nothing but the service of God and the extension of the monarchy would induce me to expose myself to such dangers."—Christopher Columbus, July 17, 1494 (Ludlum, 1963).

Historic hurricanes are defined in this book as those which occurred prior to 1871. The National Climatic Data Center in Asheville, North Carolina maintains records from 1871 to the present. These records define the modern era.

In the earlier book on historic citrus freezes (Attaway, 1997), efforts to grow citrus in South Carolina and Georgia during the colonial period were described to put the southward migration of citrus production in Florida into perspective. To show the effect of destructive hurricanes, this chapter will also go back in time and consider some of the early historic hurricanes in the Atlantic Basin which occurred during the colonial period prior to the development of agriculture in Florida as we know it today. The majority of these hurricanes did not strike Florida, and when they did, there was little or no agriculture to be affected. The earliest Florida citrus grower to lose a crop to a hurricane was Count Odet Phillipe whose pioneer grapefruit grove on Tampa Bay was devastated by the Tampa Bay Hurricane in 1848, (DeFoor, 1997A).

Pre-History

There is evidence that hurricanes have always been with us. Thousands of years ago, very intense hurricanes comparable to Hurricanes

3

Camille and Andrew, probably category 4 or 5 according to the Saffir-Simpson scale (Table 15-4), moved north in the Gulf of Mexico toward west Florida, southern Alabama and Mississippi at intervals of approximately every 600 years (Davis, et al., 1989). These hurricanes may have originated near the Cape Verde Islands and traveled thousands of miles across the Atlantic Ocean, or they may have developed in the Western Caribbean Sea and passed into the Gulf through the Yucatan Channel. Although their origin are uncertain, it is known that these hurricanes made landfall along the Gulf Coast and left their marks on the geography of the region much as did Hurricane Camile when it struck the Mississippi coast in 1969 (Liu and Fearn, 1993).

There is also evidence that hurricanes and tropical storms formed even during the ice age. Using computer simulations, J. S. Hobgood and R. S. Cerveny (1988) indicated that, despite the colder atmosphere, there was the potential to support tropical storms during these times although they would have been weaker than present day storms.

Early West Indian Hurricanes

Hurricanes, referred by the natives as "furacanes" (Ludlum, 1963), were reported to occur in the West Indies as early as the voyages of Christopher Columbus. A particularly severe storm struck Columbus on July 16, 1494 while he was at anchor at Cape Santa Cruz provoking the great explorer to comment that, "nothing but the service of God and the extension of the monarchy would induce me to expose myself to such dangers" (Ludlum, 1963). Columbus experienced another hurricane in 1495, believed to be in October, which struck his fleet at anchor at Santo Domingo, sinking three ships. Again on his last voyage in the mid-summer of 1502, he experienced a hurricane which while only damaging his ships, completely destroyed another fleet which had previously left Santo Domingo for Spain.

One of the Great "What Ifs of History"

Ludlum (1963) notes that Columbus did not experience a hurricane on his first voyage, and speculates that if he had encountered a hurricane in 1492, it might have altered the course of history. The unexplained loss of the Nina, the Pinta and the Santa Maria with all hands would have certainly delayed further exploration. Actually, the weather was so favorable on Columbus' first voyage that the great navigator stated that, "in the Indies the weather was always like May," but if on his first voyage he had been struck by one of the storms he encountered on his later voyages, exploration of the new world might have been postponed for many years, and towns with names such as Columbus and Columbia might be named for someone else.

According to the records of Andreas Poey (cited by Tannehill, 1952), 16 hurricanes occurred in the 16th century, presumably in the area of the West Indies, seven of which occurred at Santo Domingo. However, if the total number of storms in the Atlantic Basin, including the Gulf of Mexico, Caribbean Sea and Atlantic Ocean, could have been recorded, the number would probably have approached 600. Two notable hurricanes struck Santo Domingo in the early 16th century, one in August 1508 which destroyed every house in Bonaventura and wrecked 20 ships, and another on July 29, 1509, which destroyed almost the entire city of Santo Domingo.

Florida was certainly not exempt from hurricanes in this period. In 1528, an expedition under the leadership of the Spanish explorer Narváez visited Tampa Bay and west Florida and was shipwrecked by a September storm in Apalachee Bay. Only ten of a company of 400 survived (Ludlum, 1963). Spanish fleets were reportedly wrecked by storms along the Florida coast in 1545, 1551, 1553, 1554 and 1559. The latter storm caused severe damage to a Spanish fleet in the Bay of Santa Maria Filipina of Arellano, now called Pensacola Bay, in August and is said to be the first account of the many hurricanes which have struck Pensacola in succeeding years (Herbert Priestly, cited by Ludlum, 1963). A hurricane was said to have preserved the Spanish settlement at St. Augustine by destroying a French fleet under Jean Ribaut, allowing the Spanish to retain control of east Florida (Lowery, cited by Ludlum, 1963).

The 17th Century

There were undoubtedly several hundred hurricanes and tropical storms throughout the Atlantic Basin during the 17th century (Ludlum, 1963; Tannehill, 1952). By about 1600, the native word "furacane" had become hurricane in the English language. Among the more notable hurricanes was a 1642 storm which wrecked 23 fully-loaded ships and destroyed the cotton and tobacco crops on St. Kitts. Hurricanes struck Guadeloupe in 1656 and 1664, and an intense hurricane struck Guadeloupe and Martinique on August 4, 1666 which destroyed every ship at Guadeloupe and sank a fleet of 17 ships killing 2,000 French troops. A severe storm, said to be the most violent hurricane ever known, devastated St. Kitts on September 1, 1667. It was said that the islanders saved themselves by lying prostrate in the fields. Barbados was heavily damaged by August hurricanes in 1674 and 1675, and in 1681 both Antigua and St. Kitts experienced hurricanes of great severity. The August 1674 storm leveled 300 houses at Barbados and claimed 200 lives.

The 18th Century

In 1772, a severe hurricane hit the town of Port Royal, Jamaica on August 28, wrecking some 26 ships and causing 400 fatalities (Tannehill, 1952) and on September 23-24, a massive hurricane moved ashore in Louisiana causing damage as far east as Mobile, Alabama and Pensacola, Florida. Twin hurricanes lashed Mobile in September 1740 causing further damage in west Florida. In 1759, a storm struck the Florida Keys with such force that it was reported that, "the current in the Gulf Stream was pushed into the Gulf of Mexico by the winds, the Dry Tortugas and other islands disappeared, and the highest trees were covered on the Peninsula of Largo" (Tannehill, 1952). The gulf coast of west Florida was struck by a hurricane on October 22, 1766 which did "much damage on shore as well as in the harbor at Pensacola," and a "hurricane severely felt in west Florida" struck on August 31-September 3, 1772 and was said to have destroyed the woods for 30 miles inland from the coast (Romans, 1775). Crops of rice and corn were lost (Ludlum, 1963). A peculiar effect of this hurricane was the production of a second crop of leaves and fruit on the mulberry trees, an event not known to have happened previously. The trees budded, blossomed and bore ripe fruit within only four weeks time after the storm

(Romans, 1775). On October 9, 1778, Pensacola was destroyed by yet another hurricane described by the British Governor as the severest ever felt in this part of the world.

Hurricanes also struck the West Indies in 1768 and 1772 and three particularly infamous hurricanes occurred in October 1780, so close in time that some accounts considered them to be one storm. The first was the "Savanna-la Mar" Hurricane (October 3-7, 1780) which formed in the central Caribbean south of Jamaica and completely destroyed the settlement of Savanna-la Mar on the southwest corner of Jamaica killing several hundred persons before crossing Cuba and passing through the central Bahamas and out to sea where it affected the British fleet patrolling off the Virginia capes. The second called the "Great Hurricane" (October 10-18, 1780) crossed Barbados, St. Lucia, Martinique, Dominica, St. Eustatius, St. Vincent and Puerto Rico before moving into the north Atlantic. Finally, the third, known as "Solano's Storm" (October 16-21, 1780), formed just east of the Cayman Islands and made final landfall between the present day cities of Pensacola and Port St. Joe in Florida. It was named Solano's Storm because it scattered a Spanish fleet of 64 vessels under Admiral Solano in the Gulf of Mexico. Solano's fleet had sailed from Havana to attack the main British gulf coast base at Pensacola; but thanks to the hurricane, the British garrison was spared. The "Great Hurricane" did untold havoc on every island it passed and killed about 20,000 people; 6,000 on St. Lucia, 4,000 French soldiers on ships near Martinique and 9,000 inhabitants of Martinique. This does not count the crews of numerous vessels sunk by the hurricane. From the enormous toll of lives lost and the tremendous property destruction, the "Great Hurricane" was the most severe storm to impact a land area in the 18th century and should be considered the most destructive of all time in the West Indies (Ludlum, 1963). It also influenced the American Revolution by its effect on the British and French fleets.

The 19th Century

The first major Atlantic Basin hurricane of the 19th century was the Antigua-Charleston Hurricane in 1804. This storm battered

Antigua and St. Kitts on September 3 of that year and hit Puerto Rico on the fourth after which it marched steadily to the west northwest and made landfall on the South Atlantic coast on September 7, causing major damage in Georgia and South Carolina, particularly the city of Charleston. After the storm had stripped fruit trees of their leaves, the trees flowered again and bore a crop of fruit from the new bloom (Tannehill, 1952). At Beaufort, cotton fields were flooded to a depth of 4 to 5 feet by the storm surge and the rice crop was severely damaged (Ludlum, 1963).

In 1806, another relentless storm struck the West Indies, making landfall at Dominica on September 6, 1806 causing floods and immense wind damage and leaving 131 known dead and missing (Dominica Journal, Sept. 20, 1806—as cited in Tannehill, 1952). Dominica also suffered two severe hurricanes in 1813, another in 1834 and yet another in 1884. The 1834 storm was considered especially violent. The South Carolina coast suffered a severe hurricane on September 27, 1806 which uprooted trees and ruined much of the cotton crop. One grower lamented that he would be lucky to "make 10 bales out of 94 acres" (Ludlum, 1963).

In 1807, the Virgin Islands and Puerto Rico were devastated by an August hurricane which is also thought to have caused severe damage in the Bahamas. Cuba was victimized by major storms in 1812, 1819, and 1821. The 1812 storm struck on October 14 and seriously damaged 500 houses in Trinidad, Cuba. Charleston, South Carolina was again hit by a small but severe hurricane on August 27-28, 1813 (Ludlum, 1963).

The 1815 hurricane season was very active with storms striking the Leeward Islands August 31 and September 1, a second storm at Turks Island on September 20 and a third at Jamaica on October 18. In 1821, a notable hurricane formed near the Turks and Caicos Islands on September 1, passed east of the Bahamas, crossed the North Carolina coast near Beaufort, exited over Norfolk, Virginia, and hit Cape May, New Jersey and Long Island, New York on September 3 and finally crossed into the New England states before losing force. It came to be known as the Long Island Hurricane. Another 1821 hurricane came ashore along the Alabama-Mississippi border in September 1821 affecting the gulf coast as far as Apalachee Bay (Ludlum, 1963).

A hurricane which passed north of the Leeward Islands on August 12, 1830, savaged the Bahamas on the fourteenth, skirted the east coast of Florida on the fifteenth, and hit Wilmington, North Carolina on August 16, 1830.

19th Century Hurricanes Affecting Puerto Rico

In Puerto Rico, it has long been the custom to name hurricanes after the Saint's day on which the storm strikes the island. Six hurricanes in the 19th century and a seventh in 1928 were particularly violent and were named for the appropriate saints as shown in Table 1-1. The Santa Ana Hurricane destroyed 7,000 houses, caused 374 fatalities and injured 1,210 people in Puerto Rico.

19th Century Hurricanes Affecting Florida

The most important storm of the 1835 season is referred to by Ludlum (1963) as the South Florida Hurricane of 1835. It was first detected by ship reports near Jamaica on September 12, after which it moved on a path over central Cuba on the 14th, striking Key West on the fifteenth. It was the first major well-documented hurricane to hit Key West. It was a large storm, and though it paralleled the west coast of Florida, many vessels were washed ashore on the east coast

TABLE 1-1. SEVEN PUERTO RICO HURRICANES NAMED FOR SAINTS.

Date hurricane affected Puerto Rico	Saint's Name
July 26, 1825	Santa Ana
August 2, 1837	Los Angeles
August 18, 1851	Santa Elena
October 29, 1867	San Narciso
September 18, 1876	San Felipe (The First)
August 8, 1899	San Ciriaco
September 13, 1928[1]	San Felipe (The Second)

[1]This is the same hurricane which struck a devastating blow to southeast Florida on September 16-17, 1928.
Source: Tannehill, 1952.

as far north as Cape Canaveral. The storm was described as acting against the gulf stream, banking water to a tremendous height. At Key Biscayne, the water was four feet deep, carrying away the lighthouse keepers stock of poultry (Key West Enquirer, September 19, 1835, cited by Ludlum, 1963).

In 1837, there were four hurricanes which significantly affected Florida. The Barbados Hurricane of July 26-August 2 crossed over Barbados, Santa Domingo and Andros Island before apparently crossing the southeast Florida coastline north of Miami (near the present day city of West Palm Beach), and moving north along the coast before exiting on a northeast course at Titusville on August 1. However, references are confusing with regard to the exact path of this storm. Ludlum indicated that the storm may have crossed the peninsula and hit Pensacola.

The Antigua Hurricane (July 31-August 7, 1837) crossed the Leeward Islands and struck San Juan, Puerto Rico, where it was called the "Los Angeles" Hurricane, on August 2 after which it followed a northwesterly course through the Bahamas and crossed the Florida coastline at Titusville on August 6, less than a week after the passage of the Barbados hurricane through this area. Such a double-barreled blast today by two major hurricanes would leave great devastation in the Indian River area citrus groves and vegetable producing sections of Brevard, Broward, St. Lucie and Palm Beach counties. The northeast quadrant of this storm lashed the Florida coast near Jacksonville destroying the cotton crop in coastal areas of northeast Florida. A storm called the Western Florida Hurricane or the Apalachee Bay Storm (Ludlum, 1963), formed in the Gulf of Mexico on August 30, 1837 near 25°N, 87°W and followed a course northeast into Apalachee Bay, caused major damage at Apalachicola, St. Marks and St. Joseph, after which it crossed north Florida and entered the Atlantic Ocean near Brunswick, Georgia on August 31 (Ludlum, 1963). Severe damage to the cotton crop can be assumed.

The final 1837 hurricane was called "Racer's Storm" (September 27-October 10), famous for its long duration and immense path of destruction. It formed in the Caribbean Sea southeast of Jamaica and crossed Swan Island and the Yucatan Peninsula before hitting Brownsville, Texas on October 4. The hurricane then curved to the northeast over Corpus Christi on the fifth, Galveston on the sixth, and New Orleans on the seventh before passing between Mobile and Pensacola on October 8. It then moved across south Georgia before entering the Atlantic Ocean at Charleston, South Carolina. It was named "Racer's Storm" because a British warship, the Racer, experienced the full force of the hurricane in the Yucatan Channel.

Four hurricanes affected Florida in the 1840s. The first, called the Key West Hurricane of 1841, formed in the western Caribbean and lashed Key West on October 18 and 19. A second storm formed in the Bay of Campeche in October 1842, followed a northeast course and crossed the Florida coastline between Cedar Key and Apalachicola and moved across the state to Jacksonville on October 5-6. As described by Sprague (1848), "the wind began blowing from the south at Cedar Key on the afternoon of October 4, and continued to increase in violence until midnight of the fifth. The water rose 27 feet carrying everything before it. Tierces of clothing, barrels of pork and flour were later picked up as far as five miles inland."

A storm known as the Great Havana Hurricane of October 1846 formed in the Caribbean Sea on October 6 near 13°N, 74°W, passed over the Cayman Islands on October 9, the Isle of Pines on the tenth, devastated Havana on the eleventh, lashed Tampa with Force 8 gales on the Beaufort scale (Table 1-2) on the twelfth, and finally made landfall at Cedar Key, Florida. Only 2.55 inches of rain fell from 4:00 P.M. on the eleventh until 6:00 A.M. on the twelfth. Many large trees were uprooted. The lowest pressure reported was 27.06 inches in Cuba where hundreds of people were killed. The Key West and Sand Key, Florida lighthouses were destroyed. The hurricane was described as the most destructive of any to visit these latitudes within the memory of man (Maloney, 1876). Thousands of bushels of salt were destroyed at Key West.

The Tampa Bay Hurricane of 1848

"As the eyewall of the hurricane roared north up Old Tampa Bay, Count Odet Phillipe was horrified to see a wall of water approaching Safety Harbor. The family fled to the top of the giant Indian mound and were saved, but the grapefruit grove was destroyed."—J. Allison DeFoor II, 1997B.

TABLE 1-2. THE BEAUFORT WIND SCALE.

Beaufort number	Seaman's description of wind	Wind speed in mph	Term used in Weather Bureau forecast
0	Calm	Less than 1	Light
1	Light air	1-3	Light
2	Light breeze	4-7	Light
3	Gentle breeze	8-12	Gentle
4	Moderate breeze	13-18	Moderate
5	Fresh breeze	19-24	Fresh
6	Strong breeze	25-31	Strong
7	Moderate gale	32-38	Strong
8	Fresh gale	39-46	Gale
9	Strong gale	47-54	Gale
10	Whole gale	55-63	Whole gale
11	Storm	64-75	Whole gale
12	Hurricane	Above 75	Hurricane

Source: Bowditch, 1938.

The last storm of the decade of the 1840s was a major hurricane (Dunn and Miller, 1964) called the Tampa Bay Hurricane of 1848 (Ludlum, 1963). It approached Tampa with little advance warning on September 25, 1848, and raged from 8:00 A.M. to 4:00 P.M. with greatest force from 1:00 to 3:00 P.M. destroying all wharfs and public buildings. The barometer fell from a prestorm reading of 30.12 inches at 9:00 A.M. on the twenty-fourth to a low of 28.18 inches prior to 3:00 P.M. on the twenty-fifth. The storm approached from the south and passed to the west of Tampa Bay before recurving and moving northeast across the peninsula to Jacksonville.

The destruction of Count Odet Phillipe's grapefruit grove in Safety Harbor by this hurricane was the first recorded instance of the destruction of a commercial citrus grove in Florida by a hurricane. Count Odet Phillipe is credited with the introduction of grapefruit into Florida in the 1830s and is one of the earliest members of the Florida Citrus Hall of Fame. A hurricane following this path in the 1990s would severely damage vegetables and citrus in Manatee and Hillsborough counties and citrus along the central Florida ridge.

As the hurricane moved inland it "sowed a path of destruction through central and southern Florida (Brown, 1991). At Bradenton, homes were flattened and fields inundated. Fort Brooke (Tampa) was a shambles with officers quarters swept away by the waves. Most public buildings were a total loss. *Longboat Yesterday, Today and Tomorrow* (1984) referred to the Tampa Bay hurricane as the "grandfather of all hurricanes" which created the pass at the south end of Longboat Key." After the hurricane, William Whitaker, a fisherman living on the Key, looked for the nets which he had left on the beach. He found that both his nets and the beach had disappeared, only open water remained. He named the former beach "New Pass," and that name is still used today to identify the pass which separates Longboat Key from Lido Shores and Lido Key.

Hurricanes Affecting Florida in the 1850s

The 1850s began with a strong tropical storm, probably of gale force, which formed in the western Gulf of Mexico and pummeled Apalachicola on August 23, 1850, after which it moved on a north northeast course over Tallahassee and into south Georgia causing damage to cotton and corn crops. The second storm of the decade, referred to as the Great Middle Florida Hurricane of August 1851 (Ludlum, 1963), passed near Key West on August 20, blew for 24 hours at Tampa on the twenty-second and twenty-third, and finally moved inland just west of Apalachicola before passing into Georgia and South Carolina. No reference was made to crop losses in Florida, but corn and cotton were said to be damaged in South Carolina.

A strong storm/hurricane struck west Florida again in October 1852, but details are scanty (Ludlum, 1963). The storm crossed the Florida coast in the Apalachicola/St. Marks area on October 8, damaging the wharves at St. Marks, and producing winds probably of gale force in the Tallahassee area before moving north into southern Georgia. Damage to unharvested cotton fields was serious. A hurricane which lashed Mobile in August 1852 also did damage in the Pensacola area (Ludlum, 1963) where the monthly rainfall totaled 16.25 inches (Division of Water Survey and Research, 1948).

West Florida was struck again on August 30, 1856 when a "large and powerful hurricane" came ashore between Pensacola and Apalachicola and moved to the north northeast into Alabama and Georgia. The center of the hurricane crossed the coast near Panama City, passed to the west of Marianna with very high winds and heavy rains dealing a heavy blow to Milton, Florida. Gales of Force 8 on the Beaufort scale (Table 1-2) were felt as far west as Pensacola and Mobile. This must have been an exceptionally intense hurricane as it continued to cause major damage as it crossed Alabama and Georgia. The cotton harvest had not yet commenced and crop losses were heavy.

Storms Affecting Florida from 1860-1870

The first serious storm affecting the state in the 1860s was described by Ludlum (1963) as, "The Gale at Mobile." Apparently the storm came ashore east of Biloxi, Mississippi with its strong right side passing over Mobile on August 11, 1860. Its effects were felt at Pensacola, Florida with rain and gale winds, Force 6 on the Beaufort scale (Table 1-2) at 2:00 P.M. By morning on August 12, the winds reached Force 8. Total rainfall was a moderate 3.03 inches.

Another significant storm affected the Pensacola area in October 1860 as a backlash from a hurricane which moved inland over Plaquemines Parrish, Louisiana on October 2 devastating the sugarcane fields south and west of New Orleans before moving into the central counties of Mississippi east of Natchez and Vicksburg (Ludlum, 1963). At Fort Barrancas near Pensacola, winds of Force 6 were experienced late on October 2 with rains totaling 5.85 inches. Ludlum labeled this storm Hurricane III as it was the third hurricane to hit the middle Gulf Coast in 1860.

The final storms of the decade were the "Twin Key West Hurricanes" of October 1870 (Ludlum, 1963). The first of these storms crossed Cuba on October 7, moved slowly north northwest lashing Key West with hurricane force winds for four hours before moving up the east coast of Florida. Reports of any damage to Florida agriculture from this storm are not available. The other of the two storms also crossed Cuba before following a course west of Key West producing hurricane force winds for a brief period on October 20.

Florida Agriculture During the Historic Period

Williams (1837) provides an interesting perspective on Florida agriculture in the early nineteenth century. Prior to 1835, oranges were a staple crop around St. Augustine and down the St. John's River from near Lake George to Jacksonville. However, an extremely severe freeze in 1835 (Attaway, 1997) completely destroyed these early groves and it was many years before orange production resumed in northeast Florida only to be severely damaged again in 1886 and finally wiped out during the winter of 1894-95. Peaches and nectarines were grown successfully in north and west Florida, but did not do well along the coast. This was also true of the apple and quince. Figs and pomegranates were cultivated successfully for local consumption. Even plantains and bananas succeeded in northeast Florida prior to the 1835 freeze, but can only be grown in south Florida today.

In the early days of the century, Sea-Island or black seed cotton was the principal crop near the coast and green seed cotton on the oak ridges of the inland areas with cotton yields becoming progressively smaller farther inland (Williams, 1837). Cotton pods begin to open in August and reach their peak in September, making the crop extremely vulnerable to breakage and defoliation from the high winds and heavy rains of a hurricane or tropical storm. In addition, the price of Florida cotton was only 20 to 50¢ per pound on the London market whereas South Carolina cotton brought up to $1.00 per pound, making cotton a less desirable crop for Florida than sugar or tobacco. Cotton was described by Romans (1775) as, "an article of which we can never raise too much. It will grow in any soil."

Indigo was a very popular crop when Florida was under British control as it brought the very high price of $1.00 per pound on the London market. Rice was found to be profitable in areas when adequate fresh water was available. Favorable areas included lands along the St. Mary's and Apalachicola Rivers. However, there was not sufficient acreage above the salt water line of the St. John's River for rice cultivation to succeed. Pitch, tar and turpentine were important products in the northern part of the peninsula.

Away from the coast, sugarcane was a staple crop and could be grown on a variety of soils. Three types of cane were produced: creole, otaheita and ribbon, with creole producing the most sugar, ribbon producing earlier and better in the north, while otaheita was preferred south of 30° north latitude (Williams, 1837).

For small planters, tobacco was the second most valuable crop after sugar, but it was not suitable for large plantations. Small farmers considered it to be "a source of great riches" (Romans, 1775). However, it was very vulnerable to ants and caterpillars and required more attention than other crops. Mulberry trees for silk production was suggested for areas north of the 27th parallel and limited plantings were made near Bayane and Picolata.

Corn was the major food crop in early Florida, although it was considered by the early settlers to be south of the best corn climate (Williams, 1837). Corn was grown from Pensacola to St. Augustine, but most successfully along the Chattahoochee River where yields ranged up to 40 bushels per acre, as opposed to 12 to 15 bushels in the eastern counties. Due to the length of the season, two crops could be grown, but the late crop was subject to damage by the corn worm. The late crop was also vulnerable to hurricane damage, particularly in the panhandle. In the middle districts, corn brought about 37.5 cents a bushel, but $1.00 per bushel at Pensacola and St. Augustine. However, it was not considered a major item of commerce. Its chief use was family consumption.

Next to corn, the sweet potato was the most popular food crop. A normal crop was 200 bushels per acre and 75¢ per bushel was the common price at Pensacola and St. Augustine. Except possibly in cases of serious flooding, sweet potatoes were not subject to hurricane damage.

Other crops grown include the Irish potato, pumpkins, squash, English peas, watermelons, muskmelons, cucumbers, turnips, beets, carrots, onions, radishes, tomatoes, eggplant and peanuts. On high, well-drained land, these crops were not often subject to total destruction from hurricanes. The growing of coffee in the southern part of the peninsula was considered "worthy of our notice" (Romans, 1775).

A Mariner's Self Forecast

In the 19th century, before the invention of weather satellites, hurricane hunter airplanes, or even radio to allow the dissemination of accurate hurricane forecasts to ships at sea, the captain of a vessel was left to his own devices in avoiding sailing into the center of a hurricane. To assist ships in this regard, Nathaniel Bowditch (1773-1838) devised a simple system to use in locating the storm center.

Bowditch (1938) first described the observations which indicate the presence of a hurricane, which he calls a tropical cyclone, as follows:

During the season of tropical storms, any interruption in the regularity of the diurnal oscillation of the barometer characteristic of low latitudes should be considered as indication of a change of weather. The barometer is by no means an infallible guide as a warning much in advance, but after the beginning of a storm, it will more or less accurately indicate the rapidity of approach and distance from the center. Its indications should not be disregarded.

A long swell evidently not caused by the winds blowing at the place of observation is another warning that should never be overlooked. Frequently a swell from the direction of the storm sets in before any other indication becomes marked. Such a swell has in some instances given warning of a tropical cyclone days in advance of its arrival.

As the cyclone comes nearer, the sky becomes overcast and remains so, at first with a delicate cirrus haze, which shows no disposition to clear away at sunset, but which later becomes gradually more and more dense until the dark mass of the true hurricane cloud appears upon the horizon. From the main body of this cloud, portions are detached from time to time and drift across the sky, their progress marked by squalls of rain and wind of increasing force. Rain, indeed, forms one of the most prominent features of the storm. In the outer portions, it is fine and mistlike, with occasional showers, these later increasing in frequency and in copiousness. In the neighborhood of the center, it falls in torrents. The rain area extends farther in advance of the storm than in the rear.

Surrounding the actual storm area is a territory of large extent throughout which the barometer reads a tenth of an inch or more below the average, the pressure diminishing toward the central area, but with no such rapidity as is noted within that area itself. Throughout the outer ring, unsettled weather prevails. The sky is ordinarily covered with a light haze, which increases in density as the center of the storm approaches. Showers are frequent. Throughout the northern semicircle of this area (in the Northern Hemisphere), the wind rises to force six or eight by the Beaufort scale—the "reinforced trades"—and is accompanied by squalls; throughout the other semicircle unsettled winds, generally from a southeasterly direction, prevail. Usually after the appearance of cirrus clouds, sometimes before, the barometer shows an unmistakable although gradual decrease in pressure. As the clouds grow thicker and lower and the wind increases, the fall of the barometer usually becomes more rapid. When this stage if reached, one may confidently expect a storm, and observations to determine the location of its center and its direction of movement should be begun.

Fixing the bearing of the storm center.—It is very important to determine as early as possible the location and direction of travel of the center. While this cannot be done with absolute accuracy with one set of observations, a sufficiently close approximation can be arrived at to enable the vessel to maneuver to the best advantage.

Since the wind circulates counterclockwise in the Northern Hemisphere, the rule in that hemisphere is to face the wind, and the storm center will be on the right hand. If the wind traveled in exact circles, the center would be eight points to the right when looking directly into the wind. We have seen, however, that the wind follows more or less a spiral path inward, which brings the center from 8 to 12 points (90° to 135°) to the right of the direction of the wind. The number of points to the right may vary during the same storm, and as the wind usually shifts in squalls its direction should not be taken during a squall. Ten points (112°) to the right (left in south latitude) when facing the wind is a good average allowance to make if in front of the storm, but a larger allowance should be made when in the rear. If very near the center, the allowance should be reduced to 8 or 9 points (90° to 101°) in the front quadrants.

"Based on the average, the following rules will enable an observer to fix approximately the bearing of the storm center:

In the **Northern Hemisphere**, stand with the face to the wind; the center of the cyclone will bear approximately 10 points (112°) to the observer's right.

In the **Southern Hemisphere**, stand with the face to the wind; the center of the cyclone will bear approximately ten points (112°) to the observer's left.

It may be noted here that the storm center almost always bears very close to 8 points (90°) from the direction of movement of the lower clouds of the cyclone. Therefore, when the direction of movement of the lower clouds can be observed it may serve as a more accurate indication of the bearing of the center than does the direction of surface wind.

Further assistance in locating the approximate position of the storm center may be obtained in some instances by observations of the clouds. When the sky first becomes overcast with the characteristic veil of cirrus, the storm center will most probably lie in the direction of the greatest density of the cloud. Later when the hurricane cloud appears over the horizon, it will be densest at the storm center. The hurricane cloud, sometimes called the **"bar of the cyclone,"** is a dense mass of rain cloud formed about the center of the storm, giving the appearance of a huge bank of black clouds resting upon the horizon. It may retain its form unchanged for hours. It is usually most conspicuous about sunrise or sunset. When it is possible to observe this cloud, the changes in its position at intervals of a few hours will enable the observer to determine the direction of movement of the storm.

Although the approximate bearing of the storm center is a comparatively easy matter to determine, and the direction in which the center is moving may be estimated with fair accuracy from the charted paths of similar storms, it is by no means an easy matter for the observer to estimate his distance from the storm center. The following old table from Piddington's "Horn Book" may serve as a guide, but it can only give an imperfect estimate of the distance and too much reliance must not be placed upon it:

Average fall of barometer per hour	Distance in miles from center
From 0.02 to 0.06 inch	From 250 to 150
From .06 to .08 inch	From 150 to 100
From .08 to .12 inch	From 100 to 80
From .12 to .15 inch	From 80 to 50

This table assumes that the vessel is hove-to in front of the storm and that the latter is advancing directly toward it.

With storms of varying area and different intensities, the lines of equal barometric pressure (isobars) must lie much closed together in some cases than in others, so that it is possible only to guess at the distance of the center by the height of the mercury or its rate of fall.

A further source of error arises because storms travel at varying rates of progression. In the Tropics, this ranges from 5 to 20 miles per hour, generally decreasing as the storm track turns poleward and recurves, increasing again as it reaches higher latitudes. In the North Atlantic, its rate of progression may amount to as much as 50 miles per hour. Within the Tropics, the storm area is usually small, the region of violent winds seldom extending more than 150 miles from the center. The unsettled state of the barometer described heretofore is usually found in the area between 500 and 1,000 miles in advance of the center. This gives place at a distance of 300 or 400 miles to a slow and steady fall of the mercurial column. When the region of violent winds extending about 150 miles from the center is reached, the barometer falls rapidly as the center of the storm comes on, this decrease within the violent area sometimes amounting to 2 inches.

Because of this very steep barometric gradient, the winds blow with greater violence and are more symmetrically disposed around the center of a tropical cyclone than is the case with the less intense cyclones of higher latitudes. After a tropical cyclone has recurved, it gradually widens out and becomes less severe, and its velocity of translation increases as its rotational energy grows more moderate. Its center is no longer a well-defined area of small size marked by a patch of clear sky and near which the winds blow with the greatest violence. Out of the Tropics, the strongest winds are often found at some distance from the center.

Handling the vessel within the storm area.—If, from the weather indications given above and such others as his experience has taught him, the navigator is led to believe that a tropical cyclone is approaching, he should at once—

First. Determine the bearing of the center.

Second. Estimate its distance.

Third. Plot its apparent path.

The first two of the above determinations will locate the approximate position of the center, which should be marked on the chart. The relation between the position of the ship and the position and prospective track of the center will indicate the proper course to pursue (a) to enable the vessel to keep out of or escape from the dangerous semicircle and to avoid the center of the storm; (b) to enable the vessel to ride out the storm in safety if unable to escape from it.

Should the ship be to the westward of the storm center before the path has recurved, it may be assumed that the latter will draw nearer more or less directly. It then becomes the utmost importance to determine its path and so learn whether the vessel is in the right or left semicircle of the storm area.

The right and left semicircles lie on the right and left hands, respectively, of an observer standing on the storm track and facing in the direction the center is moving. *Prior to recurving,* the winds in that semicircle of the storm which is more remote from the equator (the right-hand semicircle in the Northern Hemisphere, the left-hand semicircle in the Southern) are liable to be more severe than those of the opposite semicircle. A vessel hove-to in the semicircle adjacent to the equator has also the advantage of immunity from becoming involved in the actual center itself, inasmuch as there is a distinct tendency of the storm to move away from the equator and to recurve. For these reasons, the more remote semicircle (the right hand in the Northern Hemisphere, the left hand in the Southern Hemisphere) has been called the **dangerous**, while that semicircle adjacent to the equator (the left hand in the Northern Hemisphere, the right hand in the Southern Hemisphere) is called the **navigable**.

In order to determine the path of the storm and consequently in which semicircle the ship finds herself, it is necessary to wait until the wind shifts. When this occurs, plot a new position of the center 10 points (112°) to the right of the new direction of the wind as before, and the line joining these two positions will be the *probable* path of the storm. If the ship has not been stationary during the time between the two sets of observations (as will indeed never be the case unless at anchor), allowance must be made for the course and distance traveled in the interim.

Two bearings of the center with an interval between of from two to three hours will, in general, be sufficient to determine the course of the storm, provided an accurate account is kept of the ship's way, but if the storm be moving slowly, a longer interval will be necessary.

Should the wind not shift, but continue to blow steadily with increasing force, and with a falling barometer, it may be assumed that the vessel is on or near the storm track. Owing to the slow advance of storms in the Tropics, a vessel might come within the disturbed area through overtaking the center. In such a case, a slight decrease in speed would probably be all that would be necessary, but it should be done in mind that the storm path is by no means constant either in speed or direction, and that it is particularly liable to recurve away from the equator.

A vessel hove-to in advance of a tropical cyclonic storm will experience a long heavy swell, a falling barometer with torrents of rain, and winds of steadily increasing force. The shifts of wind will depend upon the position of the vessel with respect to the track followed by the storm center. Immediately upon the track, the wind will hold steady in direction until the passage of the central calm, the "eye of the storm," after which the gale will renew itself, but from a direction opposite to that which it previously had. To the right of the track, or in the *right-hand semicircle* of the storm the wind, as the center advances and passes the vessel, will constantly shift to the right, the rate at which the successive shifts follow each other increasing with the proximity to the center; in this semicircle, then, in order that the wind shall draw aft with each shift, and the vessel not be taken aback, a *sailing vessel* must be hove-to on the starboard tack; similarly, in the left-hand semicircle, the wind will constantly shift to the left, and here a *sailing vessel* must be hove-to on the port tack so as not to be taken aback. These two rules hold alike for both hemispheres and for cyclonic storms in all latitudes.

Maneuvering rules.—The rules for maneuvering, so far as they may be generalized, are for the Northern Hemisphere.

Right or dangerous semicircle.—Steamers: Bring the wind on the starboard bow, make as much way as possible, and if obliged to heave to, do so head to sea. Sailing vessels: Keep close-hauled on the starboard tack, make as much way as possible, and if obliged to heave to, do so on the starboard tack.

Left or navigable semicircle.—Steam and sailing vessels: Bring the wind on the starboard quarter, note the course and hold it. If obliged to heave to, steamers may do so stern to sea; sailing vessels on the port tack.

On the storm track, in front of center.—Steam and sailing vessels: Bring the wind two points (22°) on the starboard quarter, note the course and hold it, and run for the left semicircle, and when in that semicircle maneuver as above.

On the storm track in rear of center.—Avoid the center by the best practicable route, having due regard to the tendency of cyclones to recurve to the northward and eastward.

Chapter 2

From 1871 to 1899

"When the hurricane approached Jupiter, Florida on August 11, 1899 there was an almost total lack of the usual signs. The sky took on no brilliant colored hues and did not bank up with masses of threatening clouds. The sea remained light up to the time of the increase in wind with but little swell, and no moaning sounds. The tide was not high and there was little thunder or lightening. Without warning from the central office, the storm would have found this section almost totally unprepared."—J. W. Cronk, Observer, Weather Bureau, Jupiter, Florida (Monthly Weather Review, 1899, 22:347)

At first thought, one might suppose that the tracks of 19th century hurricanes and tropical storms would be very inexact, if known at all. Radio had not been invented so direct reports from islands and ships at sea were nonexistent, and weather satellites were far into the future. Consequently, it would seem impossible for meteorologists in the late 19th century to plot the path of a tropical storm or hurricane, even after the storm had passed from the scene, particularly those storms which remained at sea, never making landfall. However, hurricanes and tropical storms have always been of major concern to shipping interests, and the logs of ships have yielded much reliable data, particularly for storms which crossed major shipping lanes. Storm tracks for the Atlantic Basin are available in publications such as "North Atlantic Tropical Cyclones" (Cry, Haggard and White, 1959); "Tropical Cyclones of the North Atlantic Ocean" (Cry, 1965); "Contributions to Meteorology" (Loomis, 1874-1889); "West Indian Hurricanes" (Garriott, 1906); "Hurricanes of the West Indies" (Fassig, 1913); two articles entitled, "West Indian Hurricanes and Other Tropical Cyclones of the North Atlantic Ocean" published in the Monthly Weather Review by Mitchell (1924 and 1932) and several other sources and compiled into a book entitled, *Tropical Cyclones of the North Atlantic Ocean, 1871-1992* (Neumann, et al., 1993) which was prepared by the National Climatic Data Center, Asheville, North Carolina in cooperation with the National Hurricane Center, Miami, Florida. This book is available from the National Climatic Data Center and is a major source for much of the information which follows. It should be noted that for the years 1871-1884, no distinction was made between hurricanes and tropical storms. Therefore, during this time period, the term tropical cyclone will be used. Beginning in 1885, the tracks distinguish between which were hurricanes and which were tropical storms, and it will be so noted in the text. (Author's Note: in plotting the course of storms across Florida, present day cities are used as reference points. Many of these cities did not exist at the time of the storm, particularly during the 1800s). While wind speeds for these early hurricanes are not available, monthly rainfall figures can be found in a book entitled, *Observed Rainfall in Florida, Monthly Totals from the Beginning of Records to 31 December 1947* (Division of Water Survey and Research, 1948). Many of the early figures were compiled by observers for the Smithsonian Institution and Signal Corps observers at forts throughout the state.

1871

There were six tropical cyclones in 1871, the first developed on June 1 and dissipated June 5. The sixth lasted from September 30 to October 6. Of these six tropical cyclones, four affected Florida. The most important storm track in terms of Florida agriculture originated near the Caicos Islands on the sixteenth of August and sideswiped the Florida east coast August 17-18 on a course which would be very harmful to the Indian River citrus producing area today. This storm contributed to the 13.70 inches of rain recorded in Jacksonville in August 1871. Another of these tropical cyclones originated just south of the Cape Verde Islands and moved inland at Fernandina, while the remaining two storms formed in the Gulf of Mexico and crossed north Florida from west to east.

1872

The 1872 season produced five tropical cyclones, only one of which crossed the Florida coastline. There were three Atlantic storms, all of which remained east of longitude 65°W, a gulf storm which struck Mississippi on the eleventh of July, and a gulf storm which passed from Cedar Key to Jacksonville on October 22-23, 1872. Jacksonville recorded a monthly total of 6.37 inches of rain.

1873

There were five tropical storms in 1873, three of which crossed Florida from the gulf to the Atlantic. The first of these originated in the central gulf on September 18 and moved northeast to make landfall at Apalachee Bay on the nineteenth. Jacksonville received 10.47 inches of rain for the month. The second formed in the Yucatan Channel on September 22 and crossed the Florida coast near Tampa on the twenty-third, and finally exited into the Atlantic between Daytona Beach and St. Augustine. Tampa received 7.40 inches of rain for the month of September. The last tropical cyclone of the season had the most interesting track. It developed east of Martinique on September 26, moved slightly north of due west over Martinique and along the south coast of Hispaniola on the twenty-eighth, north of Jamaica on the twenty-ninth, across the Yucatan Peninsula on October 1, turned north on October 3, and then due east on October 4, striking the Fort Myers area on October 6, after which it crossed central Florida and entered the Atlantic Ocean near Melbourne. A hurricane, category 3 or stronger, following such a course in 1999 would significantly damage citrus and vegetables in Lee, Charlotte, DeSoto, Hendry, Highlands, Indian River, Brevard and adjacent counties.

1874

There were seven tropical cyclones in 1874, the first from July 2-4 and the seventh from October 31-November 2. However, only the sixth storm made landfall in Florida. It developed in the southwestern Caribbean near 17°N, 86°W, just east of Belize on September

25. Its initial movement was northwest across the Yucatan peninsula on September 26, after which it turned to the northeast and crossed Florida from Cedar Key to Jacksonville September 27-28. A total of 8.10 inches of rain was recorded at Fort George in Duval County and 7.07 inches in Jacksonville.

1875

This was a relatively inactive season with four tropical cyclones, the first from September 1-9 and the last October 13-16. The third storm was the only one to make landfall in Florida. It originated in the western gulf of Mexico near 23.5°N, 93.5°W on September 24, moving first north and then east northeast striking the Florida gulf coast near Apalachicola on September 27. A monthly total of 7.69 inches of rain was recorded at St. Marks. Figures for Apalachicola are not available.

1876

The 1876 season was unusually inactive with only three tropical cyclones, of which one crossed the Florida coast. The first storm of the season occurred September 12-19, contributing 14.90 inches of rain in Dade County, and the third storm, which struck Florida, developed October 12 and hit the Sunshine State on October 19. Storm three developed near 15°N, 56°W, just east of Martinique, hit Dominica hard on October 13 and moved almost due west for four days before turning north northwest across Grand Cayman on the eighteenth and western Cuba on the nineteenth before swinging to the northeast and crossing the Florida coastline near the present day city of Naples and finally exiting over Vero Beach. A storm following this course today would adversely affect citrus in Collier and Hendry counties and vegetables in the Lake Okeechobee region. The two October storms contributed to the October rainfall of 15.30 inches in Dade County and 15.87 inches in Daytona Beach.

1877

Tropical cyclone activity increased during the 1877 Atlantic Basin season with a total of eight storms, the first in early August and the last in late November. The second, fourth and seventh storms battered the Florida panhandle. Storm number two was generated in the Bay of Campeche on September 14 and after moving north northwest for three days, it turned almost due east and crossed the coast near the present day city of Port St. Joe on September 20. The fourth storm developed east of Barbados on September 21, skirted the Venezuelan coast for three days before curving through the Yucatan Channel on September 29, turning northeast and striking the Port St. Joe area again on October 3, 13 days after storm number two had blown over. People in the area must have felt shellshocked. The seventh storm developed in the west gulf in late October and moved east to make landfall at Cedar Key. Rainfall totalling 10.61 inches was recorded in St. Marks.

1878

Activity was also at a high level in 1878 with ten tropical cyclones, three of which pummeled the state of Florida. The first storm of the season developed in the mid-gulf on July first, moved due east over the Port Charlotte area on July second, and then crossed the state and into the Atlantic Ocean near Vero Beach lashing present day Lee, Charlotte, Highlands and St. Lucie counties with rain and high winds. Punta Rassa in Lee County received 9.58 inches of rain during the month.

The third storm of the season passed through peninsular Florida on a course which would be almost exactly duplicated by Hurricane Donna 82 years later. The storm developed unusually far to the south at 10°N, 55°W, due east of Trinidad, on September 1. It moved on a steady northwest course for seven days crossing Haiti and the northern shore of Cuba before turning north and storming ashore at Marco Island on September 8, after which it moved slowly north through Arcadia and Wauchula on September 9, Sanford and DeLand on September 10, and exited into the Atlantic near St. Augustine on September 11. Such a course today would ravage the citrus and vegetable industries in Lee, Charlotte, Collier and Hendry counties through DeSoto, Hardee, Highlands, Polk, Osceola, Orange, Seminole, and Volusia counties, much as was done by the October 1944 hurricane and Hurricane Donna in 1960.

The sixth storm of the year formed in the central gulf and battered the Panama City—Port St. Joe area for the third time in two years.

Monthly rainfall totals for September 1878 included 25.12 inches in Dade County, 23.78 inches in Merritt Island, 19.45 inches in Daytona Beach and 21.12 inches in Jacksonville.

With the area around Fort Basinger almost totally under water after the September 1878 hurricane, Kissimmee Valley pioneer Shadrach Chandler set out to buy groceries at Fort Odgen south of Arcadia. He followed the edge of the river to Rainey Slough, paddled up the slough to the switch grass marsh, then down the marsh to Myrtle Slough passing Telegraph Station and on to Shell Creek. Pushing down Shell Creek he finally reached the Peace River and then paddled up to the general store in Fort Odgen and loaded his boat with groceries for the return trip to Fort Basinger. The round trip of 120 miles as the crow flies took one week! (Van Landingham, 1976).

1879

The 1879 season produced eight tropical cyclones, the first during the interval August 13-19, and the eighth from November 16-21. Three of these storms, the fourth, sixth and seventh, buffeted the gulf coast of Florida. Storm four was spawned on September 11 in the Atlantic east of Guadeloupe, near 17°N, 60°W, and moved east southeast for six days before turning northwest, passing through the Yucatan Channel and recurving northeast to pass inland over Tampa and exiting in the Atlantic north of Titusville. This was a course which would be destructive to citrus along the central Florida ridge. Merritt Island recorded 10.28 inches and Daytona 12.77 inches of rain.

The sixth storm of the year originated September 9 just east of Martinique. It moved on a west to northwest course passing over Jamaica on September 12, Western Cuba on September 13-14 and finally turning due north to strike Pensacola on September 16.

The seventh storm formed in the Yucatan Channel on October 24, moved north northwest and then northeast to batter the Cedar Key area on October 27, then moved very rapidly across the state and passed into the Atlantic Ocean south of Jacksonville contributing to the monthly rainfall of 9.45 inches.

1880

There were nine tropical cyclones recorded in 1880, two of which made landfall in Florida, one on the Atlantic Coast and one the gulf coast. The first storm of the year took place from June 21-24 and the ninth from October 21-24. The first to strike Florida was the fourth of the season. It formed in the Atlantic near 25°N, 57°W, moved west northwest and moved inland north of Titusville on August 29, after which it crossed the state and emerged into the gulf near Cedar Key. It then skirted the coast of the panhandle from Apalachicola and moved inland over Pensacola. Cedar Key received 19.45 inches and Merritt Island 15.77 inches of rain in August 1880.

The eighth storm of the season developed in the western Caribbean near 18°N, 85°W, moved north through the Yucatan Channel and curved northeast to hit the Florida gulf coast at Cedar Key on October 8, then crossed the state to emerge into the Atlantic Ocean at St. Augustine. Cedar Key received 10.37 inches of rain and St. Augustine 14.29 inches of rain in October 1880. It can be assumed that this storm contributed to these figures.

1881

There were only six tropical cyclones in 1881, but two of these affected the Florida gulf coast. The first developed in the Yucatan Channel on August 16, and moved rapidly east northeast to cross Florida on a line from Fort Myers to Vero Beach on August 17-18. The second formed in the gulf near 25°N, 85°W on October 5 and also followed an east northeast course to cross Florida on October 5-6 from Cedar Key to Jacksonville.

1882

This was a very inactive season with only three tropical cyclones recorded, but two of the three affected Florida. The first was born in the Atlantic on September 2 near 20°N, 67°W, moved west over Turks Island and northern Cuba before recurving to strike the west Florida gulf coast at Pensacola on September 9 contributing to the monthly rainfall total of 8.49 inches. The other storm affecting the state formed on October 5 in the Western Caribbean near 14°N, 82°W. It moved north very slowly and struck the western tip of Cuba on the ninth, then advanced north northeast to lash the Florida gulf coast near Cedar Key October 10-11. The approach of the storm was felt at Cedar Key on October 9 and gales began at that location about 4:00 A.M. of the tenth as the wind shifted from northeast to east and finally to southeast and south attaining a velocity of 56 mph with heavy rains. Great quantities of logs and timber were washed away (Monthly Weather Review, October 1882).

1883

The 1883 season only produced four tropical cyclones, none of which crossed the coastline of Florida or any of the gulf coast states. All four were Atlantic storms. None formed in the western Caribbean or Gulf of Mexico. The only storm to make landfall struck the Wilmington, North Carolina area.

1884

Florida also escaped the wrath of tropical cyclones in 1884 which was a benign season with only three tropical systems. All three spent their fury in the Atlantic, leaving the gulf coast states unscathed for the second consecutive year.

1885

Tropical cyclone activity increased markedly in 1885, returning to a more normal eight storms for the year, the first occurring from August 7-15, and the last storm of the season from October 8-13. Four storms battered the Florida gulf coast and one sideswiped the Florida east coast. Three storms remained in the central Atlantic Ocean. The storm which sideswiped the east coast was the second of the year, originating north of Puerto Rico, near 21°N, 67°W on August 21. It passed south and west of the Bahamas, tacking Andros Island, then passed just east of Cape Canaveral on August 24 and continued to the north.

The season's third storm developed in the central Gulf of Mexico on August 29, drifted first to the north and then to the east striking the Valparaiso/Niceville area on August 30. This area was to take a beating as the fourth storm developed just east of Tampico, Mexico on September 17, moved north to near Brownsville, Texas and then drifted east and finally gave the Valparaiso/Niceville area its second lashing of the year. Storm six formed in the gulf south of New Orleans on September 24, struck Louisiana on the twenty-fifth, then paralleled the gulf coast through Biloxi, Pensacola, inland to Tallahassee and exited into the Atlantic at Jacksonville. The monthly rainfall for Jacksonville in September 1885 totaled 19.63 inches (Division of Water Survey and Research, 1948). Finally, the eighth tropical cyclone of the year formed just west of Jamaica near 18°N, 81°W on October 8, advanced to the north, crossing Cuba on the ninth and lashing the Florida west coast from the Tampa Bay area to Cedar Key on the eleventh before proceeding north into Georgia. Cedar Key reported a maximum wind velocity of 48 mph from the southeast with a barometer reading of 29.19 inches. Jacksonville had heavy rains and winds of 36 mph (Monthly Weather Review, October 1885).

1886

This was the first year that there was a distinction between tropical storms and hurricanes (Neumann, et al., 1993), the latter characterized as having wind velocities greater than 75 mph. A total of ten were recorded, eight hurricanes and two tropical storms, of which three invaded Florida, all striking the gulf coast. The earliest was June 13-14 and the latest was October 22-24. The second storm and first hurricane of the season developed on June 18 in the western Caribbean near 19°N, 86°W and promptly moved north through the Yucatan Channel, recurved slightly to the northeast and struck the Apalachicola area on June 21.

The second hurricane formed on June 27 southwest of Jamaica at 17°N, 80°W. It moved first to the northwest, crossed the tip of the Yucatan peninsula, shifted to a northerly course and then to the northeast hitting the coast near Perry on June 30.

The third and last hurricane of the season to strike Florida also formed in the western Caribbean, very near to the point of origin of the previous two hurricanes, at 19°N, 83°W on July 14. It passed northwest through the Yucatan Channel and continued north until the seventeenth when it made a 90° turn to the east and struck the Cedar Key area on July 18. The number of June and July tropical cyclones was an unusual feature of the 1886 season. As a result, many north Florida cities recorded double digit rainfall in July 1886, including 14.97 inches at Jacksonville, 14.49 inches at Archer, and 14.30 inches in Tallahassee (Division of Water Survey and Research, 1948).

1887

The 1887 Atlantic Basin hurricane season was both long and active with ten hurricanes and seven tropical storms. The first tropical storm was unusually early, occurring May 17-21, and the last was equally late, occurring December 7-12. Despite this preponderance of activity, peninsular Florida felt only one tropical storm while the Panhandle felt one hurricane. Five systems affected the U.S. gulf coast while eleven systems wasted away in the open Atlantic.

The hurricane affecting the Panhandle formed July 20 in the Atlantic at 17°N, 57°W. It moved west through the Caribbean until curving northward through the Yucatan Channel on the twenty-fifth and striking the coast just east of Pensacola on July 27.

The tropical storm which affected peninsular Florida developed in the central gulf near 25°N, 85°W on October 29. It was the fourteenth tropical storm or hurricane of the season. It moved immediately northeast making landfall in the Sarasota area and crossing the state to Titusville which received 12.17 inches of rain during the month.

1888

Storm activity in 1888 dropped to a more normal total of nine, five hurricanes and four tropical storms. Peninsular Florida was beset with an Atlantic hurricane, a gulf hurricane and an Atlantic tropical storm. The Atlantic hurricane formed on August 14 just north of the island of Hispaniola at 21°N, 70°W. It moved straight to the northwest affecting the entire Bahama Islands chain before crossing south Florida along a Homestead/Naples line on August 16.

The gulf hurricane developed in the Bay of Campeche near 22°N, 93.5°W on October 8, moved directly to the northeast and crossed Florida from Cedar Key to Jacksonville on October 10.

The Atlantic tropical storm formed east of the Bahamas near 23°N, 72°W on September 6 and moved northwest to cross Florida from West Palm Beach to Tampa on September 7, then recurved inland over Cedar Key on September 8th producing 60 mph winds with a barometer reading of 29.50 inches (Monthly Weather Review, September 1888). Cedar Key received 12.89 inches of rain for the month (Division of Water Survey and Research, 1948).

1889

The 1889 season produced five hurricanes and four tropical storms. One hurricane affected the Pensacola area and two tropical storms crossed the peninsula from the gulf to the Atlantic. The hurricane was of Atlantic origin and was first detected September 11 east of Dominica near 15.5°N, 59°W. It moved westward across the Caribbean for six days, crossed the Yucatan peninsula into the Gulf of Campeche on September 18, then turned north for four days and finally curved to the northeast and slammed into Pensacola on September 23.

The first of the tropical storms formed in the western Caribbean on June 15, passed through the Yucatan Channel and turned northeast crossing Florida from Cedar Key to Jacksonville on the seventeenth. The second tropical storm developed on October 4 west of the Cayman Islands, crossed Cuba on a north northeast course, passed through the Florida Keys and crossed south Florida on the fifth.

1890

This was a most unusual year with only one hurricane and no tropical storms. The hurricane developed in the Atlantic near 16.5°N, 54.5°W on August 25 and passed into the north Atlantic without making landfall. The lack of tropical storm activity in 1890 may have been due to an "El Niño," which is now known to directly influence weather patterns in the Atlantic Basin.

1891

If 1890 was an "El Niño" year, it would appear that 1891 was a "La Niña" year as there were eight hurricanes and three tropical storms in the Atlantic Basin. One hurricane and two tropical storms crossed the Florida peninsula. None affected west Florida.

The 1891 hurricane formed in the Atlantic on August 18 at 14°N, 58°W, east of Martinique. It crossed Dominica and Puerto Rico on a northwesterly course, passed through the Bahama Islands striking a head-on blow to Andros Island before passing across south Florida on a line from the present day city of Homestead to Naples on September 24.

The first of the two tropical storms developed east of the Virgin Islands on October 1 skirted the south shores of Puerto Rico and Hispaniola, turned north over Cuba and crossed Florida on October 7 on a northeast path heading from Naples to Melbourne. The second tropical storm formed in the western Caribbean just east of Nicaragua on the sixth of October, advanced north through the Yucatan Channel on October 8, then northeast crossing Florida October 9-10 on a line from Sarasota to Daytona Beach.

1892

This was an average year with four hurricanes and five tropical storms. Three hurricanes remained in the open Atlantic east of longitude 70°W and one struck Mexico. Two of the tropical storms formed in the gulf and crossed Florida from west to east. One was an early storm which crossed the state along a line from Fort Myers to Vero Beach on June 10. This storm produced heavy rain and gale force winds across south Florida. "Over a narrow strip along the Atlantic Coast from Titusville south to Hypoluxo in Brevard and Dade counties, very heavy rains were reported. At Hypoluxo in Dade County, 3.60 inches fell in 2.5 hours and 12.95 inches in six days. Monthly rainfall for Tampa was 12.41 inches and Titusville 16.69 inches (Division of Water Survey and Research, 1948). The second was a late season storm which crossed from Tampa Bay to Melbourne on October 24.

1893

This was a busy season in the Atlantic Basin with ten hurricanes and two late season tropical storms. Fortunately, only one hurricane crossed Florida. It was an early season storm which formed in the Bay of Campeche on June 12 and moved northeast to cross Florida from near Perry to the Okefenokee Swamp on the fifteenth. Two hurricanes which formed near the Cape Verde Islands, one in mid-August and the other in late September passed near the Florida east coast but turned north to strike the Carolinas.

1894

The number of tropical cyclones decreased sharply from 1893 to 1894, with five hurricanes and only one tropical storm recorded. However, one hurricane which formed in the Atlantic near 12°N, 50°W struck the peninsula along a potentially very destructive path from Fort Myers through central Florida to Daytona Beach on September 25-26, while the other hurricane, of western Caribbean origin near 13°N, 80°W, passed through the Yucatan Channel, turned northeast and slammed into the Apalachicola area on October 8. Regarding the September hurricane, reports are available from observers in Tampa, Titusville, and Jacksonville (Monthly Weather Review, 1898). The Tampa report indicated that the storm passed to the northeast of Tampa, putting that city in the weaker left hand side of the storm. Sustained winds for 5 minutes were 43 mph with one minute segments at 60 mph. The lowest barometer reading was 29.48 inches at 8:00 P.M. on September 25. The rainfall total for the 54 hour period ending at 8:00 A.M. on the twenty-sixth was 13.78 inches.

At Titusville on the Florida east coast, the wind speed was 48 to 60 mph from the northeast beginning at 8:00 A.M. on September 25 and continuing until midnight. The rainfall over a 48 hour period was 7.72 inches.

At Jacksonville, a northeast gale began at midnight and the wind speed peaked at 48 mph at 10:40 A.M. on September 26. Rainfall at Jacksonville totaled 11.11 inches.

With strong gales and heavy rains on both the east and west coasts of Florida, one can assume that as the storm center moved through central Florida hurricane force winds would have been sustained in Lee, Charlotte, Hendry, DeSoto, Highlands, Polk, Osceola, and other present day counties. Many mid-Florida cities recorded double digit rainfall totals for September 1894 including 19.78 inches in Clermont, 19.91 inches in Gainesville, 18.43 inches in Plant City, 16.20 inches in Kissimmee, 17.02 inches in Ocala, 15.71 inches in Orlando, 19.19 inches in Palatka and 14.65 inches in Ft. Meade.

1895

The Atlantic Basin produced only two hurricanes and four tropical storms. Neither hurricane and only one tropical storm crossed the Florida coastline. This storm formed in the Bay of Campeche on October 13 and moved east northeast to cross Florida from Naples to West Palm Beach on October 16 accompanied by extremely heavy rains. Hypoluxo received 24.39 inches of rain during the month (Division of Water Survey and Research, 1948).

1896

The 1896 season was relatively inactive with only six hurricanes and no tropical storms. However, three of the six hurricanes made landfall on the Florida gulf coast. The first of the three was born on July 4 just south of Cuba, near 20°N, 80°W. It moved northwest across Cuba then turned north and hit the west Florida coastline at Valparaiso on July 7. Milton recorded 19.97 inches of rain during the month.

The next hurricane to affect the Sunshine State formed in the Atlantic Ocean on September 22 east of Antigua at 17°N, 61°W. It moved west and passed south of Jamaica before curving to the northwest across the western tip of Cuba, after which it recurved to the northeast and crossed the gulf coast at Cedar Key and moved northeast through Levy, Lafayette, Suwanee, Columbia, Baker and Nassau counties causing an estimated $3 million dollars in damages mostly to agriculture. Gainesville recorded 7.29 inches of rain.

Florida's final hurricane of the year originated October 7 in the western Gulf of Mexico near 22°N, 92°W. It took a direct east northeast course and crossed Florida on a line from Fort Myers to Melbourne on October 8-9. Such a course today would put a large percentage of the Indian River and ridge grapefruit crops on the ground as well as a significant number of boxes of oranges.

June 1896 was a very wet month across Florida with double digit rainfall totals at many Florida cities including Dade City, 17.92 inches; Ft. Myers, 20.90 inches; Ft. Meade, 13.30 inches; Kissimmee, 12.57 inches; Pensacola, 12.46 inches; Tampa, 13.92 inches; Archer 14.46 inches and Brooksville, 11.94 inches. However, these heavy rains were not associated with a hurricane or a tropical storm.

1897

The following year had only two hurricanes and three tropical storms. One hurricane traveled north in the Atlantic never passing west of 50°W latitude. The other hurricane struck Texas. Two tropical storms crossed peninsular Florida from the west to the east, one from Fort Myers to Titusville on September 21 depositing 10.45 inches of rain, and the other from Tampa to St. Augustine on October 19. The September storm caused considerable damage to citrus fruits, tobacco and vegetables in Polk, Orange, Osceola and Brevard counties and an influx of cold air produced light frost in Washington County in west Florida. This was the first verified frost to occur anywhere in Florida so early in the season (Climate and Crop Service, September 1897). September rainfall included Bartow, 11.94 inches; Ft. Meade, 16.36 inches; Jupiter, 18.09 inches; Orlando, 15.77 inches; Sebastain, 23.01 inches, and Daytona Beach, 14.03 inches.

1898

The season turned a bit more active with four hurricanes and five tropical storms in the Atlantic Basin. One hurricane and one tropical storm made landfall in Florida, and a second hurricane skirted the northeast Florida coast. The hurricane which crossed Florida developed on August 2 just east of Fort Pierce at 27°N, 79°W and moved rapidly across the state entering the Gulf of Mexico near Clearwater, after which it continued its rapid advance to hit Apalachicola, Panama City and Valparaiso on August third. The tropical storm developed in the Atlantic east of Barbados on October 2, traversed the Caribbean in seven days, turned north over the western tip of Cuba on October 9 and curved to the northeast to cross Florida from Naples to Melbourne on October 10-11. Rainfall from this storm was generally less than one inch.

The August hurricane brought record monthly rainfall to many areas of the state, particularly the Panhandle counties. Totals included 31.26 inches at St. Andrews in Bay County, 20.68 inches in Washington County, 22.98 inches in Walton County, 20.56 inches in Wakulla County, 19.90 inches at Carrabelle, 19.70 inches at Lake Butler, 18.58 inches at Pensacola, 18.48 inches at Bradenton, 16.46 inches in Taylor County, 15.43 inches in Tallahassee, 17.83 inches in Tampa, 15.06 inches in Plant City, 13.44 inches in Brooksville, and 13.48 inches in Clermont.

Considerable damage was done to crops in Washington, Calhoun, Bay, Jackson and Lee counties (Climate and Crop Service, August 1898). The deluge at St. Andrews in Bay County ruined most crops on flat lands and made it almost impossible to transplant sweet potatoes. The wet ground prevented the hoeing of vegetable crops. Cotton was heavily damaged in Lake Butler and at Wausau. Conversely, the Plant City area in Hillsborough County was suffering a drought with rainfall through August deficient by 10.89 inches.

The center of the hurricane which threatened the northeast Florida coast passed about 50 miles east of Jacksonville during the morning of October 2, 1898. The minimum pressure at Jacksonville was 29.07 inches at 11:00 A.M. with hurricane velocity winds prevailing from 10:10 to 11:25 A.M. The hurricane crossed the Georgia coastline between the St. Marys and Altamaha Rivers. Rainfall in north Florida was not excessive with 4.43 inches being recorded at Lake City. Agricultural damage was not significant.

1899

There were five hurricanes and one tropical storm in the Atlantic Basin season in 1899. Fortunately, only a minimal hurricane and the tropical storm crossed the Florida coastline, although a large Cape Verde storm narrowly missed the Florida east coast as it curved northward to hit North Carolina.

The hurricane which hit Florida formed on July 31 in the Gulf of Mexico west of Tampa, near 27°N, 85°W, and moved north northeast to strike the Apalachicola area on August 1. A pressure of 28.9 inches was recorded by an aneroid barometer at Carrabelle where the wind blew fiercely for 10 hours. Both cotton and corn suffered severe damage but fortunately the storm did not extend inland more than 50 miles. The tropical storm originated in the Yucatan Channel on October 2 and moved north and then northeast to cross the coastline near Cedar Key on October 5. Damage to agriculture was not significant.

The large hurricane which narrowly missed the Florida east coast formed August 3 just southeast of the Cape Verde Islands near 12°N, 35°W. It followed a west northwest course passing over Antigua on August 7, Puerto Rico on August 8, the Dominican Republic on the ninth, Great Inagua Island on the tenth, Andros Island on the twelfth and recurved to the north just east of the Fort Pierce/Vero Beach area on the thirteenth. It finally smashed into the Outer Banks of North Carolina on August 17. Florida's close brush with this storm was indicated by winds of 52 mph at Jupiter at 8:00 A.M. on August 13 with the barometer reading 29.22 inches. A slight move to the west would have meant disaster for the Indian River area. This storm crossed the entire length of Puerto Rico, where it was called the San Ciriaco Hurricane (Table 1-1), and was arguably the most destructive storm in the history of the island. The hurricane finally died near the Azores on September 7 after a lifetime of five weeks.

Chapter 3

From 1900 to 1909

"My father was born in 1865 and moved to Bartow when he was 2 to 3 years old. When he was a young man, the 1903 hurricane came and poured a foot of rain on Bartow and Fort Meade. After the storm passed, he paddled a boat from the Peace River to where the Citrus and Chemical Bank stands in Bartow today."—G. W. "Buck" Mann, Jr. to John A. Attaway, Buck Mann Ranch, Polk County, Florida, July 9, 1998.

The first decade of the twentieth century was not noteworthy for severe Florida hurricanes, but four hurricanes and about a dozen tropical storms affected either peninsular Florida or west Florida during the period with some damage to Florida agriculture.

1900

The 1900 season consisted of three hurricanes and four tropical storms. The only system to strike Florida was a tropical storm which formed in the Western Caribbean, 18°N, 87°W on October 8, crossed the Yucatan peninsula on October 9, then recurved to cross Florida from Cedar Key to Jacksonville on October 11-12. Monthly rainfall during October at selected stations included 9.89 inches at Carrabelle and 7.14 inches at Jacksonville. The tropical storm would have contributed to these totals.

June was an unusually wet month with double digit precipitation in most areas of the state. One might guess that, coming at the start of the hurricane season, such heavy rains were associated with a tropical storm or hurricane. However, this was not the case. Some representative stations included Fort Meade, 17.94 inches; Tallahassee, 16.47 inches; Port St. Joe, 17.06 inches; Archer, 14.60 inches; Dade City, 14.13 inches; Tarpon Springs, 13.49 inches; Daytona Beach, 12.95 inches; DeFuniak Springs, 12.04 inches; Pensacola, 11.79 inches; Clermont, 10.62 inches; Plant City, 10.24 inches; and Gainesville, 10.12 inches. The rains had an adverse effect on agriculture in that cotton became grassy and fruit and vegetable yields were reduced (Climate and Crop Service).

1901

The Atlantic Basin produced three hurricanes and seven tropical storms in this active year. Three of the tropical storms crossed the Florida coastline, but only one of the hurricanes affected the Sunshine State as it moved inland over Mississippi. All four systems affected west Florida, but only one tropical storm made a call on the peninsula.

The first storm to affect Florida hit the Panhandle on June 14. The second storm, which later became a hurricane after crossing the peninsula on August 11, went ashore in Mississippi on August 17, the third lashed the Pensacola area on September 17 and the final storm passed inland just east of Apalachicola on the twenty-seventh.

The second storm originated in the Atlantic on August 4 near 27°N, 48°W. It moved due west for six days, then west northwest and across the peninsula from West Palm Beach to Sarasota on the eleventh and became a hurricane in the Gulf of Mexico and made final landfall west of Mobile. Fortunately, Florida did not experience hurricane force winds from this system as its path would have been very destructive to agriculture. However, gale force winds up to 70 miles per hour were reported in the Pensacola area resulting in "much damage to the cotton and corn crops from both the high winds and the heavy rains." Pensacola reported a minimum pressure of 29.75 inches and Mobile, Alabama 29.32 inches (Climate and Crops, Florida Section, August 1901).

Rainfall amounts as the storm crossed the peninsula were Jupiter 2.76 inches, Miami 2.50 inches and Ft. Meade 1.63 inches. Mobile, Alabama recorded 5.44 inches.

The third tropical storm of Atlantic origin formed on September 9 near 17°N, 50°W. It followed a course just north of Puerto Rico, crossed Hispaniola, moved south of Cuba and then recurved through the central gulf and lashed the Pensacola area on September 17.

The fourth tropical storm of the year formed in the southwest Caribbean Sea near 11°N, 80°W on September 21. It moved slowly west northwest for five days, passed through the Yucatan Channel on September 26, then moved north northeast and passed inland just east of Apalachicola on the twenty-seventh. Rainfall data for Apalachicola is not available, but the monthly total for Tallahassee was 7.55 inches.

1902

The second year of the decade was relatively inactive with three hurricanes and two tropical storms. Both tropical storms affected north Florida. The first was a June storm which formed in the southwest Caribbean Sea near 13°N, 82°W on June 10, pushed north northwest over the western tip of Cuba on the twelfth and came ashore south of Tallahassee on June 14. Rainfall totals associated with this storm included 6.60 inches at Quincy, 6.65 inches at Tallahassee, 5.92 inches at Wewahitchka, 5.70 inches at Carrabelle and 2.05 inches at Marianna.

Rare Event

The second system followed a very unusual course in that the storm was born in the Gulf of Tehuantepec off the eastern Pacific coast of Mexico near 14°N, 93°W on October 3, moved north across the narrow waist of Mexico and into the Bay of Campeche where it briefly attained hurricane force in the Gulf of Mexico on October 7. It weakened to a tropical storm and crossed the coast near Mobile and Pensacola on October 10, depositing an inch of rain at Pensacola, 1.65 inches at Mobile and 1.80 inches at Tallahassee, before becoming extratropical over Alabama. In crossing Mexico through the provinces of Chiapas and Tabasco, the storm held together over land for some 200 miles. According to the Weather Bureau records

(Neumann, et al., 1993), only two other storms have crossed from the Pacific to the Gulf. The first formed in the Pacific near 10°N, 92°W on October 12, 1923, crossed over Mexico into the Bay of Campeche and eventually made landfall in Louisiana on October 15, 1923. The second formed off the Pacific coast of El Salvador on September 28, 1949, crossed through Guatemala and Mexico into the Bay of Campeche and finally struck the coast of Texas with hurricane force winds on October 4, 1949.

1903

The 1903 Atlantic Basin season was average in numbers with eight hurricanes and one tropical storm across a time period from July 19 to November 25. Only one system affected Florida, a September hurricane which followed a course which would be very destructive to citrus and vegetables today.

The Hurricane of September 11-13, 1903

"Much damage resulted to the orange crop—but in terms of monetary loss, cotton suffered more than any other crop."—Climate and Crop Service, September 1903.

Meteorological Considerations

The September 1903 hurricane formed as a tropical storm on September 9 north of Grand Turk Island near 22°N, 74°W, and increased to hurricane force as it moved on a northwesterly course through the central Bahamas on September 10 and slammed into the southeast Florida coast between Jupiter and Miami on the eleventh. The maximum wind velocity at Jupiter was 78 mph (Climate and Crop Service, September 1903) with a low pressure reading of 29.63 inches, but winds were believed to be higher as the storm ripped through the citrus groves of the Indian River district.

Early on September 10, the Weather Bureau had advised that a major storm was approaching from the Bahamas, and northeast storm warnings had been raised along the full length of the Florida east coast from Fernandina to Miami. On the eleventh, these warnings had been changed from tropical storm to hurricane warnings over the south half of the state.

After lashing the Indian River area on the eleventh, the hurricane moved across present day Palm Beach, Okeechobee, Hardee and Manatee counties and into the Gulf of Mexico south of Tampa on September 12 (Figure 3-1). Note that in 1903, Palm Beach, Okeechobee and Hardee counties did not exist, all being part of Dade, Brevard and DeSoto counties. At that time in Florida's history, Dade, Lee and DeSoto counties were all larger in area even than Polk County. Pinellas County was part of Hillsborough County. Wind speeds and directions and minimum pressure readings are shown in Table 3-1. Rainfall from this hurricane was very heavy as it moved across the Florida peninsula (Table 3-2).

Leaving peninsular Florida to lick its wounds, the hurricane crossed the Gulf of Mexico and stormed ashore for a second time near Apalachicola and rampaged through Franklin, Liberty, Calhoun and Jackson counties before finally departing the state on September 14. Rainfall totals for west Florida are also shown in Table 3-2.

Agricultural Damage

As the hurricane whipped its way through central Florida's orange groves, initial estimates were that fully half of the state's citrus crop was lost due to fruit dropping and thorn punctures. However, later reports reduced the amount of damage. The sugarcane crop in the Okeechobee region was described as "prostrate and broken" (Climate and Crop Service, September 1903) and damaged in magnitude to the citrus crop damage. Sweet potatoes also suffered some water damage in low lying areas, and pineapples were described as "being sanded and blown out of the ground. In west Florida, the cotton crop was the major casualty as heavy rains accompanied the high winds blowing off bolls of the valuable white fiber and whipping and tearing the branches from the cotton stalks. Early estimates predicted a 50% loss of the crop. Corn, vegetables and other field crops and timber also suffered. Cattle and other livestock were killed by fallen trees or due to inadequate protection from the winds and rain.

FIGURE 3-1. THE PATH OF THE SEPTEMBER 1903 HURRICANE AS IT PASSED FROM THE INDIAN RIVER AREA ACROSS THE PENINSULA AND ENTERED THE GULF SOUTH OF TAMPA BAY. NOTE THAT IN 1903, DADE COUNTY INCLUDED PRESENT DAY BROWARD, PALM BEACH AND MARTIN COUNTIES, LEE COUNTY INCLUDED PRESENT DAY COLLIER AND HENDRY COUNTIES, DESOTO COUNTY INCLUDED PRESENT DAY CHARLOTTE, GLADES AND HIGHLANDS COUNTIES, OSCEOLA COUNTY INCLUDED OKEECHOBEE COUNTY AND BREVARD COUNTY INCLUDED INDIAN RIVER AND ST. LUCIE COUNTIES. SOURCE: CLIMATE AND CROP SERVICE, SEPTEMBER 1903.

TABLE 3-1. METEOROLOGICAL DATA AT SELECTED CITIES DURING THE HURRICANE OF SEPTEMBER 11-13, 1903.

City	Minimum pressure (inches)	Maximum sustained winds (mph) direction
Key West	—	40 SW
Jupiter	29.63	78 NE
Tampa	29.46	48 NE
Pensacola	29.75	36 N

Source: Climate and Crop Service, September 1903.

1904

The 1904 Atlantic Basin hurricane season featured only five systems, two of which attained hurricane status and three remained tropical storms. One of the hurricanes affected south Florida and one of the tropical storms brushed the Pensacola area.

The south Florida hurricane developed as a tropical storm in the central Caribbean Sea near 15°N, 76°W on October 12, moved northwest for a day then turned due north crossing central Cuba on the fifteenth. It briefly attained minimal hurricane strength on October 16 and passed into the Miami area on the seventeenth quickly losing strength and dissipating after looping over the Everglades. The maximum wind velocity at Jupiter was 68 mph, and the heavy rains accompanying the storm forced many south Florida vegetable growers to replant their vegetable crops (Climate and Crop Service, October 1904). Jupiter and Miami received 6.03 inches and 4.02 inches of rainfall, respectively.

The west Florida tropical storm formed just off the tip of the Yucatan peninsula on October 29, moved northwest into the central Gulf of Mexico, then turned northeast over the mouth of the Mississippi River on November 2 and then across the Mobile/Pensacola area depositing 2.21 inches of rain at Pensacola and 2.22 inches at DeFuniak Springs.

1905

There was one hurricane and four tropical storms in the Atlantic Basin in 1905. Fortunately, none of these systems made landfall in

TABLE 3-2. RAINFALL TOTALS FORM THE HURRICANE OF SEP-TEMBER 11-13, 1903 WITH STATIONS LISTED FROM EAST TO WEST.

Station	Rainfall (inches)
Peninsular Florida	
Miami	5.00
Jupiter	9.02
Fort Pierce	4.77
Nocatee	9.30
Avon Park	5.60
Bartow	7.43
Fort Meade	11.40
Plant City	5.90
Tampa	5.88
Tarpon Springs	6.76
Manatee	2.00
West Florida	
Carrabelle	8.42
Tallahassee	7.83
Madison	7.20
Marianna	5.00

Source: Climate and Crop Service, September 1903.

Florida. However, a gulf disturbance at the end of September generated sufficient rainfall to damage the cotton crop in west Florida, and a moderate disturbance, also originating in the gulf, produced heavy rains across all areas of the state from November 9-11.

1906

With six hurricanes and five tropical storms, the 1906 season was the most active since 1893 which produced 12 storms and hurricanes. Of these, two hurricanes and two tropical storms crossed the coastline of Florida.

Taking them chronologically, the season's first tropical storm formed south of Cuba, near 19°N, 83°W, on June 8, crossed the western tip of Cuba on a generally northward path and made landfall east of Panama City on June 12 without reaching hurricane force. The second system also formed in June near 22°N, 75°W on the fourteenth. It initially followed a westerly course then curved sharply to the north on June 16 and reached hurricane force in the Florida straits before entering the state east of Naples and following a northeasterly course which took it back into the Atlantic north of West Palm Beach on the seventeenth. Some rainfall totals included Jupiter 5.91 inches, Hypoluxo 4.62 inches and Miami 4.75 inches. A hurricane following this course today would do extensive damage to crops in Collier, Dade and Hendry counties and around Lake Okeechobee.

The entire state of Florida was hurricane free in July and August, but the Pensacola area received an unwelcome visitor in September.

The Hurricane at Pensacola, September 26-27, 1906

"This was the most terrific storm since the village of Pensacola on Santa Rosa Island was swept away 170 years ago."—W. F. Reed, Weather Bureau, Pensacola, Florida (Climatological Data, September 1906).

Meteorological Considerations

The 1906 Pensacola hurricane was first noted on September 22 as a tropical storm forming south of Grand Cayman Island. On September 24, the Central Office of the Weather Bureau in Washington transmitted the following message:

"Cyclone disturbance off western extremity of Cuba moving northward . . . "and at 9:55 A.M. on the twenty-fifth, the advisory noted:

"Center of disturbance has entered southeastern gulf and will probably move northward . . .".

At 3:25 P.M. the same day, the advisory warned:

"Northeast storm warnings 3:00 P.M., New Orleans, Pensacola, Mobile, Apalachicola, Carrabelle, Cedar Key, Dunnellon: Tropical disturbance has moved from Yucatan Channel northward over gulf; vessels in gulf and Florida ports should remain in port."

On the afternoon of Wednesday the twenty-sixth, the northeast wind at Pensacola increased to 33 mph at 3:58 P.M., to 40 mph at

6:12 P.M., and 52 mph at 8:40 P.M. The pressure fell gradually to 29.77 inches at noon, 29.58 at 7:00 P.M. and 29.36 at midnight as the center of the storm bore down on the western Panhandle.

By the morning of Thursday the twenty-seventh, the winds had shifted to due east and maintained hurricane force from 2:00 A.M. to 8:00 A.M. At 2:46 A.M., the wind was east 80 mph, at 3:53 A.M. it was east 83 mph, then shifted to the southeast maintaining 77 to 83 mph until shifting to the south after 2:00 P.M. The pressure fell to a minimum of 29.16 inches at 5:00 A.M. after which it rose to 29.23 inches at 7:00 A.M. and 29.46 inches at noon. Rain continued all day with a peak rate of 0.91 inches per hour between 12:18 A.M. and 1:18 A.M. The total for the day in Pensacola was 3.36 inches, but to the east of the storm center, Apalachicola recorded 10.12 inches of rain in 24 hours. Molino in Escambia County received 9.10 inches.

The Damage

The total damage was estimated at 3 to 4 million dollars, mostly to shipping interests and waterfront property. Total known deaths were 29 (Climatological Data, September 1906). However, the report did not include any mention of agricultural damage. It can be assumed that the winds and rain had diminished by the time the storm reached the inland agricultural areas and no major losses were incurred.

The Hurricane of October 17-18, 1906

"If it wasn't for the 1906 hurricane, I wouldn't be here today"— Barney Walden to John A. Attaway, November 17, 1997.

The most significant storm of the 1906 season did not occur until October 17 and 18, when a hurricane of small diameter but great energy struck Monroe and Dade counties and the Indian River area.

Meteorological Considerations

The storm was first noted as a shallow depression in the eastern Caribbean near 14°N, 61°W on the eleventh of October. Its course initially was west northwest and it attained hurricane force on the twelfth. The path shifted to northwest on the fifteenth and on the sixteenth, it turned north northeast across the western tip of Cuba. On October 17, reports indicated "the presence of a well-defined cyclonic disturbance" north of Havana and by October 18 it was near Key West where the barometer read 29.30 inches at 3:00 A.M. As the hurricane swept up the Keys, the barometer touched 29.25 inches with the wind speed at 75 mph. Northeast storm warnings were displayed at Miami with the wind blowing hard until noon. The calm lasted about 30 minutes, after which the hurricane moved off the coast at Jupiter and into the Atlantic. Almost reaching the coast of South Carolina, it abruptly turned southwest as a tropical storm and took one last pass at northeast Florida before dissipating in the Gulf of Mexico. Rainfall totals were: Hypoluxo, 4.50 inches; Jupiter, 5.08 inches; Miami, 4.80 inches and Key West, 2.86 inches. The most severe consequence of this hurricane was the deaths of 124 people which resulted when houseboats quartering laborers working on the Florida East Coast Railway capsized and were blown out to sea.

Agricultural Damage

According to H. P. Hardin, Weather Bureau Observer at Jupiter, Florida (Monthly Weather Review, 1906), "planters on the larger Keys lost their orange groves, pineapples and homes, and in some cases the losses were so complete that the places were abandoned." There was no mention of agricultural losses on the peninsula.

Personal Experiences

The section on this hurricane opened with the quote from Mr. Barney Walden that, "except for the 1906 hurricane I would not be here today." Mr. Walden explained as follows:

"My father and grandfather were both working in the Keys on the extension of the Florida East Coast Railway. Their boat was blown out to sea where it was swamped and capsized. My grandfather was picked up by a freighter and ultimately put ashore at Savannah. My father was picked up by a different vessel which dropped him at Key West where he was a total stranger. Fortunately, a young man at the dock felt sorry for him and took him to his house to give him something to eat. As it turned out, my mother was this young man's sister, and that is how they met.

Without the 1906 hurricane, my father would not have met my mother and I would not have been born.

"My father and grandfather each assumed the other was dead, until by chance, they met walking along a street in Miami a year later."

The final system of the 1906 season was an October storm which formed unusually far to the north, near 33°N, 60°W, on the thirteenth and moved west southwest to cross the coast near St. Augustine on October 17. Thereafter, it quickly dissipated.

1907

This was a relatively uneventful year in the Atlantic Basin with only four tropical storms and no hurricanes. Two of the tropical storms affected north Florida. The first formed in the central Caribbean near 16°N, 78°W on June 24, moved northwest through the Yucatan Channel then abruptly turned east northeast on the twenty-seventh and crossed the coast east of Apalachicola on the twenty-eighth, producing three to four inches or rain. Representative totals were Fenholloway, 4.38 inches; Carrabelle, 3.4 inches, Monticello, 4.21 inches; and Tallahassee, 3.20 inches. The second storm formed in the western Gulf of Mexico near 23°N, 94°W on September 27, moved directly northeast and crossed the Florida coast east of Niceville on the twenty-eighth of September depositing 7.65 inches of rain at Marianna, 6.61 inches at Monticello, 4.25 inches at Carrabelle, and 3.14 inches at Apalachicola.

1908

The 1908 year produced five hurricanes and three tropical storms in the Atlantic Basin, but was kind to Florida as none of these systems entered the state.

1909

Activity increased significantly in 1909 with four hurricanes and six tropical storms. One hurricane and two tropical storms crossed the Florida coastline. The one hurricane brushed southeast Florida in October with significant damage to property including agriculture.

The Hurricane of October 11, 1909

"At Sand Key, about six miles southwest of Key West, the barometer fell to 28.36 inches, believed to be the lowest atmospheric pressure ever recorded in the United States, the previous record being 28.48 inches during the Galveston hurricane of September 1900."—Climatological Data, District No. 2, South Atlantic and East Gulf States, 1909.

Meteorological Considerations

The 1909 hurricane developed from a tropical depression in the southern Caribbean Sea near 12°N, 78°W on October 6. It moved north northwest for two days to a point just southwest of Jamaica where it turned west northwest arriving over western Cuba on Sunday morning October 10, at which point it curved to the northeast. The barometer fell slowly during the day at Key West reaching 29.80 inches at 9:00 P.M. (Monthly Weather Review, October 1909), and continued to fall very slowly during the night with rising northeast winds reaching 29.52 inches at 6:00 A.M. on the morning of October 11. However, at this point, the pressure fell rapidly and by 11:40 A.M. the minimum was 28.52 inches, a full inch lower that at 6:00 A.M., then the lowest previously recorded at the Key West station. At Sand Key, the barometer reached 28.36 inches which was believed to be a new record, lower than the 1900 Galveston hurricane of 28.48 inches. The barometric track at Key West is shown in Figure 3-2. The rainfall at Key West at this point in time was described as torrential with 6.13 inches falling in two hours and 15 minutes. Total rainfall at Key West during the storm was 11.23 inches. The wind reached gale intensity at 6:45 A.M. and continued for six and one half hours until 1:15 P.M. The maximum sustained velocity for five minutes was 83 mph from the northeast at 10:05 A.M., with a one minute sustained velocity of 94 mph at 10:07 A.M.

Finally moving away from Key West on a northeasterly path, the hurricane surged across southern Dade County and into the Atlantic Ocean where it lost intensity as it crossed the Bahamas. The maximum velocity at Miami was 60 mph and a low pressure of 29.22 inches. The lowest pressure at Nassau was 29.32 inches with winds of 50 mph. Rainfall totals for South Florida set records with 10.17

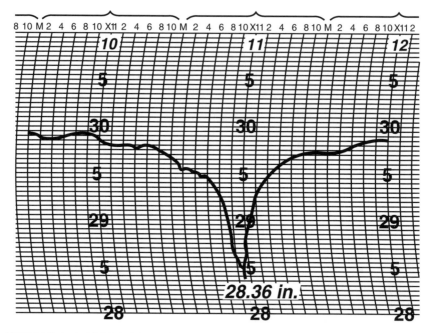

FIGURE 3-2. THE BAROMETRIC TRACE FOR THE OCTOBER 1909
HURRICANE AT KEY WEST, FLORIDA. THE LOW OF 28.36 INCHES
WAS A NEW RECORD AT THE TIME. SOURCE: CLIMATOLOGICAL
DATA, OCTOBER 1909

inches at Miami during a 24-hour period, and 21.08 inches for the
month of October. Many locations totaled 8 to 10 inches, but the
rain area did not extend into north Florida. Ft. Myers recorded 2.42
inches, but only a trace fell in Orlando and Tampa.

Agricultural Damage

In Dade County, Florida, "the principal damage was to the fruit
crops which were blown from the trees in immense quantities"
(Monthly Weather Review, October 1909). However, the flooding
must have also been damaging to vegetable crops.

Chapter 4

The Decade from 1910 to 1919

"The storm that passed over Key West on September 9 and 10, 1919 was the most violent experienced since records at this station began. The anemometer cups were shaken loose and blown away at 7:30 P.M. on the ninth and thence until 3:35 P.M. on the tenth, the wind velocity was lost."—Official in Charge, Weather Station, Key West, Florida. Climatological Data, 1919, 23(9):72.

1910

The 1910 season produced a total of only four tropical systems, three of which were hurricanes and one was a tropical storm. Unfortunately, one hurricane followed a damaging northerly path across the Keys and then through central Florida. It lost intensity as it moved inland and the winds were less severe than along the coast.

The Hurricane of October 17-18, 1910

"Some conception of the force of the wind at Sand Key may be gained from the following. The windows and doors were all kept closed, but two panes of glass were blown out on the windward side. The air that was forced in through these holes increased the pressure at the opposite side of the house about 0.05 inch. When the door was opened to let the air go through, the barometer fell about 0.05 inch and when the door was closed, it immediately rose again."—F. D. Young, Assistant Observer (Climatological Summary, 1910).

Meteorological Considerations

The October hurricane originated in the southern Caribbean Sea near 11°N, 80°W on the ninth of the month. Its initial movement was to the northwest which placed the storm center just west of the Isle of Pines on October 14 and off the north coast of Cuba on the fifteenth. At this time, hurricane warnings were issued to points in the southern United States with a special effort to warn all points in Florida of the impending danger (Monthly Weather Review, October 1910). On the sixteenth, the hurricane made an odd loop to the southeast and then turned north toward the Florida Keys where the barometer at Sand Key began to fall rapidly after midnight and reached a low of 28.62 inches at 12:20 P.M. on the seventeenth. The wind roared from the southeast until 1:05 P.M. then shifted to the south with gusts lasting several minutes at an estimated velocity of 125 mph. At 1:50 P.M., the barometer bottomed at 28.40 inches and at 3:30 P.M. began a slow rise. Telephone communication with Key West was broken off at 2:00 P.M. and did not resume until 7:00 P.M.

The barometer at Key West began to fall after 10:00 P.M. on the sixteenth but did not reach its low of 28.42 inches until 3:20 P.M. on the seventeenth. Figure 4-1 shows the barometric trace recorded at Key West on October 17. The wind attained its peak velocity at Key West between 2:30 and 4:30 P.M. on the seventeenth when it was estimated at 90 mph sustained with frequent gusts to 110 mph. The wind lessened after 5:00 P.M. but gales continued until 3:00 A.M. on the eighteenth. The hurricane had lasted 30 hours.

Continuing to the north, the storm roared ashore at Cape Romano with great waves and a storm surge reaching far inland. Survivors could only escape by climbing trees. The storm center then pushed northward through central Florida, moving the eye directly up the central Florida ridge from the Lake Placid area to Leesburg (through the heart of the state's citrus belt) and then north through Marion and Alachua counties and finally west of Jacksonville and into the Okefonokee Swamp of south Georgia on the morning of October 19 (Figure 4-2). Unfortunately, pressures and wind velocities as the eye moved up the center of the state are not available due to the lack of weather stations in that area. However, observations were made at Tampa, west of the storm center,

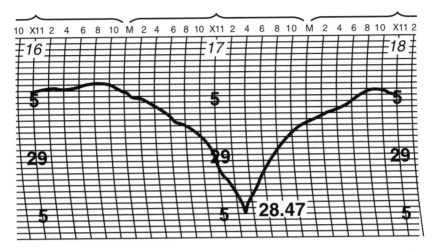

FIGURE 4-1. THE BAROMETRIC TRACE RECORDED AT KEY WEST, FLORIDA ON OCTOBER 17, 1910 DURING THE PASSAGE OF THE HURRICANE. SOURCE: CLIMATOLOGICAL DATA, OCTOBER 1910.

where wind velocities would have been much less than locations nearer to the Ridge section to the east. George B. Wurtz (Monthly Weather Review, October 1910) the local forecaster at Tampa, indicated that the barometer fell rapidly on the seventeenth with increasing wind and rain. By nightfall, northeast winds had reached gale force and by midnight the pressure had dropped to 29.30 inches and continued to fall. The early morning hours of October 18 found sustained winds at 60 mph with gusts to 70 mph. It can reasonably be assumed that winds to the east in Highlands and Polk counties must have reached hurricane force for a considerable period of time. During the nineteenth, the barometer rose steadily but did not return to normal levels until October 21.

At Jupiter on the east coast, observer H. P. Hardin noted that the storm began in earnest at 2:00 P.M. on October 17 with winds increasing to a sustained velocity of 56 mph from the southeast by 6:25 P.M., moderated for a period, then held from 50 to 58 mph from 8:20 P.M. to midnight when the wind shifted to the south and increased to 70 mph during the early morning hours of October 18 as the direction veered to the southwest and began to diminish. From October 14-18 inclusive at Jupiter, 14.27 inches of rain fell

FIGURE 4-2. THE OCTOBER 1910 HURRICANE RAVISHED FLORIDA AGRICULTURE AS IT MOVED ON A NORTHERLY COURSE FROM THE EVERGLADES THROUGH THE RIDGE SECTION TO NORTH-EAST FLORIDA. NOTE THE SIZE OF LEE, DADE, PALM BEACH, DESOTO AND BREVARD COUNTIES IN 1910. SOURCE: CLIMATO-LOGICAL DATA, OCTOBER 1910.

which was more damaging than the gale force winds. The lowest atmospheric pressure at Jupiter was 29.21 inches at 3:00 A.M. on the eighteenth.

Damage to Agriculture

Due to the high winds and excessive rainfall, damage to citrus and truck crops in central and south Florida must have been severe in some areas. However, Monthly Weather Review of October 1910 indicates that a true statement of exact losses could not be made, but that after surveying 200 localities, the Florida Citrus Exchange estimated that 10% of the citrus crop was lost, with heaviest damage in the south central counties, such as Highlands, Hardee and DeSoto but less damage further north in Lake and Marion counties.

With Jupiter recording over 14 inches of rain and rivers eight feet above the normal high water mark, flood damage in the vegetable growing areas of southeast and south central Florida was undoubtedly severe. The Jupiter observer also noted the loss of large pine trees. This must have been true elsewhere in the state as well.

Personal Observation

"My grandmother lived in a two-story house in the 10 mile section west of Fort Pierce. Her children and their families had all came to her house for safety, but half way through the hurricane the waters around the house began to rise and they feared a tidal wave. The men ripped boards from the house and made a raft, loaded the women and children, and rowed to the Indian Mound. They waited out the hurricane on the west side of the mound. When the worst of the hurricane had passed, they rowed everybody back to the house."—Reuben Carlton, 1998.

1911

The 1911 season was below average in number of storms with only three hurricanes and one tropical storm. The first hurricane of the year crossed the U.S. coastline between Pensacola, Florida and Mobile, Alabama on August 11, affecting the western-most tip of the state. Otherwise, Florida escaped damage as the second hurricane went ashore at Savannah, Georgia on August 28, the third hurricane passed into Nicaragua on September 10, and the fourth storm, which did not reach hurricane force, dissipated in the Gulf of Mexico west of Cedar Key on October 31 (Climatological Data, October 1911).

1912

The 1912 season also produced a less than average number of tropical systems with only two tropical storms and four hurricanes, none of which crossed the Florida coast. However, a tropical storm which moved ashore just west of Mobile, Alabama on September 13 did more damage to the Pensacola, Florida area than it did to Mobile (Ashenberger, 1912). High winds initially from the northeast blew from 9:45 A.M. on the thirteenth until 3:00 P.M. when the wind direction shifted to the east attaining a maximum velocity of 59 mph at 9:21 P.M. Shortly after midnight, the wind increased in velocity and reached 74 mph from the southeast at 2:00 A.M. on the fourteenth when the anemometer was blown away. The lowest pressure recorded was 29.65 inches on September 13. Rainfall associated with the storm was generally 3 to 4 inches. Damage to waterfront property was significant, but agriculture did not incur major damage.

One hurricane was noteworthy due to its lateness in the season, although it did not cross the Florida coastline. It formed as a tropical storm in the extreme southern Caribbean near 11°N, 78°W on November 11 in an area where a storm might be expected to move westward into Honduras or Nicaragua. However, after moving a short distance to the west northwest, the storm turned north on November 12 and attained hurricane force on the fifteenth as it crossed latitude 15°N. It then shifted to the northeast striking the western tip of Jamaica, turned southwest on November 19 and then resumed a northeasterly course on the twentieth and passed first through the Cayman Islands then rapidly across central Cuba and into the Bahamas on November 22. Continuing north northeast, the hurricane finally dissipated east of New Jersey, near 40°N, 67°W on November 23-24 without making landfall in the U.S.

1913

This was another below average season with three hurricanes and one tropical storm. One hurricane made landfall in Texas, one in North Carolina, one dissipated in the mid-Atlantic and the tropical storm went ashore near Charleston, South Carolina, none affecting Florida.

1914

The 1914 North Atlantic Basin season was exceedingly strange in that there were no hurricanes and only one tropical storm, which skirted the Florida coast on the sixteenth and seventeenth of September before crossing the coastline near Brunswick, Georgia. This storm produced 3.5 inches of rain at St. Augustine, 2.05 inches at Jacksonville and 2.50 inches at Fernandina. It is reasonable to assume that 1914 was an El Niño year. Gulf disturbances on the twenty-fourth and twenty-fifth and again on the twenty-eighth and twenty-ninth left 4.75 inches of rain in Pensacola and 4.30 inches in Carrabelle.

1915

After a season with only one storm, the 1915 season seemed almost active with four hurricanes and one tropical storm, still a below average season. One hurricane, of Cape Verde origin, crossed the Atlantic, traversed the Caribbean striking the western tip of Cuba and finally hit the Texas coast near Galveston on August 16. Another hurricane which formed near the Windward Islands passed through the Caribbean, the Yucatan Channel and hit New Orleans on September 29 producing tropical force winds in the Pensacola, Florida area.

The hurricane of most concern to Florida in 1915 formed on August 31 of that year in the Western Caribbean near 15°N, 79°W. It moved north northwest, crossed the Isle of Pines and Western Cuba on September 2 and hit Apalachicola, Florida with 70+ mph winds and a low pressure of 29.32 inches on September 4. The other system affecting the state was a tropical storm which made landfall north of Titusville on August 1 and moved north producing gale force winds and heavy rains as the center passed 25 miles west of Jacksonville. Heavy rains were also produced by a gulf disturbance which crossed the state August 14-17. Some 24 to 36 hour rainfall totals from the August 1 tropical storm included Avon Park, 10 inches; Fellsmere, 7.02 inches in 24 hours; Merritt Island, 10.56 inches; St. Petersburg, 15.45 inches and Tarpon Springs, 8.75 inches.

1916

Following a series of years from 1910 to 1915, in which the number of hurricanes and tropical storms was considerably below average at one to five storms per year, the 1916 season, probably a La Niña year, exploded to 14, including three tropical storms and 11 hurricanes. Both the first hurricane of the season, which began on June 29, and the last hurricane of the year, which began four and one-half months later on November 11, originated in the southwestern Caribbean. In the meantime, there were nine storms and hurricanes which formed in the Atlantic Ocean east or northeast of the West Indies, three which formed just east of the Bahamas and one additional Caribbean storm. Fortunately, none of these systems had a major effect on peninsular Florida, although the November hurricane, the fourteenth of the season, did pass through the Keys before moving northeast and dissipating in the Atlantic, and the thirteenth storm of the season hit the Pensacola area on October 18. In addition, one of the three tropical storms passed inland between Titusville and New Smyrna Beach on September 13 and broke up along the Alabama/Georgia state line on the fourteenth (Neumann, et al., 1993).

The season's first storm originated on June 29 near 11°N, 80°W in the southwest Caribbean Sea. The center actually came ashore in Mississippi, but the Florida panhandle suffered from heavy rains from July 3 to July 9 which destroyed 75% of the cotton crop west of the Apalachicola River, as well as corn, sugarcane and minor crops in low lying areas. Monthly rainfall totals in west Florida included Bonifay, 30.6 inches; DeFuniak Springs, 21.6 inches; and Pensacola, 17.9 inches.

The hurricane which passed through the Keys in November formed from a disturbance at 13°N, 77°W on November 11. It followed a steady westerly course until it curved to the north along the Honduran coast and attained hurricane force in the Yucatan Channel on the fourteenth, after which it curved to the northeast and rapidly accelerated through the Keys on November 15 and quickly lost force in the northern Bahamas.

The storm which affected the Pensacola area in October formed from an easterly wave in the Central Caribbean near 15.5°N, 75°W on October 12. It moved to the west northwest and crossed the Yucatan Peninsula with hurricane force winds on October 15-16, before curving to the north northeast and crossing the coast at Pensacola on October 18 with peak winds of 114 mph and a minimum pressure of 28.76 inches. The 24 hour rainfall total at Pensacola on the eighteenth was 4.82 inches (Climatological Data, October 1917).

1917

"Before 1917, Okeechobee County was part of Osceola County. When we went to the courthouse, we had to ride horseback from Okeechobee City to Kissimmee."—Mrs. Zetta Hunt, Okeechobee, Florida, 1998.

The number of storms dropped dramatically in 1917 to only two hurricanes and one tropical storm, only one of which affected Florida. The season's first system, a tropical storm, formed near Bermuda on August 6 and never made landfall. The second system, a hurricane, formed in the central Atlantic and curved to the north also without making landfall. The third system formed from an easterly wave over the Leeward Islands on September 21 moved east northeast across the Caribbean Sea passing over the Isle of Pines and western Cuba on September 25, after which it moved northwest until the twenty-seventh and then curved to the northeast and crossed the west Florida coastline east of Pensacola on September 29 with 103 mph winds and a minimum pressure of 28.51 inches. All crops in the northwestern counties suffered. The pecan crop was very badly damaged, particularly in Jefferson and Leon counties. Several hundred cattle were killed, and pine timber in Escambia, Santa Rosa, and neighboring counties was lost. Total damage was estimated at $1 million.

1918

This was also a relatively quiet season with three hurricanes and two tropical storms. None of these storms passed near Florida, and none formed after the fourteenth of September. The last two months of the season being exceptionally quiet (Neumann, et al., 1993).

1919

The 1919 season produced only one hurricane and two tropical storms. The hurricane went ashore in Texas on September 14 after battering Key West on the ninth and tenth. One of the tropical storms passed through Pensacola on the Fourth of July with winds peaking at 58 mph and 1.36 inches of rain, possibly dampening the holiday celebration (Climatological Data, July 1919).

The Hurricane of September 9-10 at Key West

Meteorological Considerations

The first warning was received at Key West on September 8 as follows:

> "Hoist northeast storm warning 10 a.m. Jupiter to Key West and at Ft. Myers, Florida. Disturbance near or over southwestern Bahamas apparently moving west-northwest. Strong northeast winds Monday, and will probably increase to gale force. Advise great caution until further advices later in the day."

and at 1:05 P.M., the warnings were upgraded from storm to hurricane warnings:

> "Change to hurricane warnings 1 p.m. Jupiter to Key West. Delayed report from Nassau: Barometer 29.46 and wind 56 miles from the northeast. Storm center will probably reach south Florida coast to-night attended by dangerous northeast winds. Caution all vessels to avoid the Florida Straits and the east Florida coast until further advices."

The barometer dropped steadily during the morning and afternoon of September 9 and winds reached hurricane force by 9:00 P.M. and exceeded 100 mph by midnight (Table 4-1). Hurricane force winds continued to near daybreak on September 10 accompanied by heavy rains as the hurricane moved slowly to the west and then to the northwest. The rain gauge was disabled by the winds and only a partial record was obtained, but the full amount was estimated at over 13 inches.

The Official in Charge at Key West stated that, "probably the lowest pressure readings recorded in this part of the world were noted in this storm, viz.: Key West, 28.83 inches; Sand Key, 28.36 inches; Rebecca Light, 40 miles to the westward, 27.68 inches; and Dry Tortugas, 65 miles to the westward, 27.51 inches" (Climatological Data, September 1910).

TABLE 4-1. METEOROLOGICAL DATA FOR THE SEPTEMBER 9-10, 1919 HURRICANE AT KEY WEST.

Date and time		Minimum pressure (inches)	Maximum sustained winds (mph) direction	Rainfall
Sept. 9	7 A.M.	29.61	36 NE	Light
	9	29.58	36 N	Light
	11	29.54	40 NE	Light
	1 P.M.	29.46	42 NE	Light
	3	29.31	48 NE	Moderate
	5	29.22	54 NE	Moderate
	7	29.08	58 NE	Heavy
	9	28.99	80 NE	Heavy
	11	28.93	90 NE	Heavy
	12 midnight	28.81	105 E	Heavy
Sept. 10	1 A.M.	28.90	110 E	Heavy
	3	29.02	90 E	Heavy
	5	29.13	80 SE	Heavy
	7	29.26	70 SE	Heavy
	9	29.39	60 SE	Light

Source: Climatological Data, September 1910.

The Damage

At the height of the hurricane, the walls were blown out from brick buildings and large vessels were torn from their moorings. Damage at the Key West Air Station was estimated at $800,000 and the Weather Bureau buildings suffered badly. The total property loss was near $2 million. No significant agricultural damage was reported.

Chapter 5

The 1920s, With Two of the Most Destructive Hurricanes in Florida History

"The September 1926 hurricane roared through downtown Miami. If Hurricane Andrew in 1992 had followed that course, the damage would have been incomprehensible. When the September 1928 hurricane careened across Lake Okeechobee, it was the most deadly hurricane in Florida history. Over 2,000 persons died. Those were bad days in south Florida."—Anonymous.

1920

The first year of the decade was a mild season with four storms, all of which attained hurricane force for a period of time. Only one of the four affected Florida and it had been downgraded to a tropical storm before it reached the state. This was an unusual storm in that it formed in the Gulf west of Fort Myers, near 25°N, 83°W on September 25, drifted slowly west for two days, following which it drifted slowly north for two days. Then, on September 29, it reached hurricane force briefly and turned to the northeast and crossed the Florida Gulf Coast near Cedar Key with winds of less than hurricane force. The storm accelerated across north Florida and exited into the Atlantic near Jacksonville on September 30. Maximum wind velocities recorded were Egmont Key (near Tampa), 56 mph; Key West, 50 mph; Miami, 38 mph; Pensacola, 40 mph; and Jacksonville, 39 mph.

Damage from this storm was chiefly due to the heavy rains in interior agricultural areas, particularly truck crops in counties adjacent to the southwest Florida coast. The harvesting of cotton, hay and corn was delayed in north Florida where Lake City measured eight inches of rain on September 30. The monthly rainfall total for Cedar Key was 14.60 inches.

1921

This year produced four hurricanes and two tropical storms. However, only one of these six systems, the hurricane which curved up from the western Caribbean in late October and struck the Tampa Bay area as a minimal hurricane, had an effect on Florida.

The Tampa Hurricane - October 25, 1921

"Citrus losses were 800,000 to 1,000,000 boxes and truck crops along the coast were severely damaged."—Climatological Data, October 1921.

Meteorological Considerations

The hurricane which slammed ashore just north of Tampa in late October was first detected on the twentieth of the month as a tropical disturbance in the southwestern Caribbean Sea near 13°N, 80°W. The first advisory, issued at 10 A.M. on the twenty-first noted, "disturbance appears to be forming over the western Caribbean Sea southwest of Jamaica; movement uncertain, but probably northward" (Bowie, 1921). Moving north northwest, the storm passed just to the east of Swan Island on the twenty-second, with the barometer on the island dipping to 29.20 inches between 10 A.M. and noon, and moved into the Yucatan Channel as a full-blown hurricane. Ships in the channel reported minimum barometric pressures of 27.80 and 27.84 inches on October 24, indicating a severe hurricane at this point. As the storm curved north and then northeastward, hurricane warnings were issued on the twenty-fourth for the Florida west coast from Key West to Apalachicola. The warning read:

"Change to hurricane warning at noon, Key West to Apalachicola, increasing winds and gales and hurricane velocities along the coast. Emergency - warn all interests. Tropical storm near and northwest of west end of Cuba, moving slowly northward and will change its course to north northeastward during the day."

As predicted, the storm turned to the northeast and passed over Tarpon Springs where the barometer recorded a minimum of 28.12 inches at 2:15 P.M. on October 25 with the wind at a dead calm for an hour or more. Shifting to an east northeast path, the hurricane passed between St. Leo and Brooksville, across the peninsula and exited into the Atlantic near New Smyrna Beach (Figure 5-1). The observer in Tampa (Bowie, 1921) reported:

"The wind continued from the northeast during the twenty-fourth and the early morning of the twenty-fifth, increasing to 28 mph at 3:38 A.M. It shifted between northeast and east until 7:05 A.M. with velocities between 25 and 42 mph. From 7:05 to 9:25 A.M., the wind was east with velocities occasionally reaching 40 mph. Southeast winds brought increasing velocities and when the wind shifted to south in the early afternoon, it shifted to 48 to 68 mph at 2:18 P.M. with a peak gust of 75 mph. The wind had decreased to 30 mph by 3:00 P.M. but with the rise of the barometer increased again as winds shifted to the southwest at 3:10 P.M. The highest velocity with the rising barometer was 48 mph but was only 24 mph at 8 P.M. Occasional high gusts occurred during the night."

The highest wind speeds during the hurricane occurred to the northwest of Tampa (Table 5-1) and torrential rains were felt all across the peninsula (Table 5-2). The data indicates that the heaviest rainfall occurred near the center of the hurricane and in the right hand quadrant.

The Damage

The hurricane resulted in significant losses in citrus groves of northern Pinellas, north Hillsborough and Pasco counties which felt winds of hurricane force, and across Lake, Orange, Seminole

FIGURE 5-1. COURSE FOLLOWED BY THE TAMPA HURRICANE, OCTOBER 25, 1921. FORTUNATELY, IT WAS NOT A SEVERE HURRICANE. SOURCE: CLIMATOLOGICAL DATA, OCTOBER 1921.

and Volusia counties with gale winds. Approximately 800,000 to 1,000,000 boxes of fruit were on the ground out of a total estimated crop of all varieties statewide of 15 million boxes. The monetary loss was said to exceed $1,000,000 (approximately $8,000,000 in 1998 dollars). Fortunately, the loss of trees from the minimal hurricane was only slight. Cutler (1923) noted that, "the combined forces of wind and rain stripped orange groves not only of fruit but also of leaves."

TABLE 5-1. METEOROLOGICAL DATA AT SELECTED LOCATION FOR OCTOBER 25, 1921 HURRICANE WITH STRUCK THE FLORIDA GULF COAST NORTH OF TAMPA.

Station	Minimum pressure (inches)	Maximum sustained winds (mph) direction
Key West	29.55	48 SW
Tarpon Springs, Dunedin, Safety Harbor and Egmont Key	—	80-100[1]
Tampa	28.81	68 S
Jacksonville	29.35	64 NE

[1]Estimated.
Source: Bowie, 1921.

TABLE 5-2. RAINFALL (24 HOUR) AS THE OCTOBER 25-26, 1921 HURRICANE CROSSED THE FLORIDA PENINSULA FROM WEST TO EAST.

Station	Rainfall (inches)
Fort Myers	3.30
St. Petersburg	5.05
Tarpon Springs	8.70
Pinellas Park	7.81
Tampa	5.20
St. Leo	11.73
Brooksville	9.50
Clermont	5.00
Eustis	4.46
Orlando	4.45
Sanford	3.60
DeLand	3.25
Titusville	3.02
Jacksonville	0.34

Source: Climatological Data, October 1921.

Truck crops in the counties adjacent to the coast suffered severe damage—some a total loss. The loss of fertilizer and labor added to the disaster estimated also at over $1,000,000 (Bowie, 1921), again approximately $8,000,000 in 1998 dollars.

1922

With only two hurricanes and two tropical storms, 1992 was a very inactive year in the Atlantic Basin. Of these four systems, three bypassed Florida altogether while one tropical storm struck the Pensacola area on October 16-17 and quickly dissipated in southern Alabama. Pensacola recorded less than 0.5 inches of rain (Climatological Data, October 1922).

1923

There were three hurricanes and four tropical storms during the 1923 Atlantic Basin season, but Florida was fortunate in that none of these systems crossed the state's coastline.

1924

The tropical Atlantic and Caribbean Sea produced five hurricanes and three tropical storms in 1924. Two of the hurricanes and one of the tropical storms struck Florida, all three along the gulf coast.

The West Florida Hurricane of September 15, 1924

"The torrential rains in the interior largely supplemented the wind damage, and unharvested crops on low lands suffered very much."—Climatological Data, September 1924.

The first hurricane of the year, which was not a severe storm, was first noted as a tropical disturbance the morning of September 13 in the southeastern Gulf of Mexico near 24°N, 82°W (Monthly Weather Review, September 1924). The disturbance moved northwestward for 12 hours, attained minimal hurricane intensity on the fourteenth, then recurved sharply to the east northeast and struck near Port St. Joe at about 11:00 A.M. on October 15. The highest winds at Port St. Joe were estimated at 75 to 80 mph from the northwest. At St. Andrews, the winds were estimated at 60 to 75 mph and at Carrabelle from 40 to 50 mph. At Apalachicola, the highest wind velocity was 68 mph. Minimum pressure readings were 29.10 inches at Carrabelle and 29.12 inches at Apalachicola. Rainfall totals for 24 hours at west Florida stations in inches include Apalachicola 10, Carrabelle 9.3, Madison 6.8, Monticello 6.6, Quincy 12.9, and Tallahassee 8.6

The Damage

Unharvested crops such as sugarcane, cotton, corn, truck crops and peanuts, especially in low lying fields, were severely damaged by the rains and subsequent flooding. Pecan trees in the path of the storm were sometimes pulled out of the water-soaked soil. Pine timber was lost and turpentine farms had economic losses. Tobacco sheds and barns were demolished in the Madison and Quincy areas.

The Minimal Tropical Storm of September 29, 1924

The tropical storm which affected the state formed September 27 from a low pressure area over the northwestern Caribbean Sea near 17°N, 81°W. On the morning of the twenty- ninth, "a disturbance of slight but increasing intensity" (Monthly Weather Review, September 1924) was reported on a northerly course and storm warnings were hoisted on the Florida gulf coast. The disturbance moved very rapidly first to the north and then recurved to the northeast crossing the Florida coast near Cedar Key late in the afternoon of the twenty-ninth and moving rapidly across the state. By the next morning, the storm was over the North Carolina coast. A wind speed of 69 mph and a barometric pressure of 29.60 inches were reported at Jacksonville, Florida at 11:00 P.M. on the twenty-ninth.

The Hurricane of October 20-21, 1924

"The third tropical storm to visit Florida in 1924 was attended by heavy rains and hurricane winds. The chief damage was to truck crops and citrus fruit."—Climatological Data, October 1924.

According to Mitchell (1924), this hurricane was very intense as it moved through the Yucatan Channel and poised to direct its fury toward the gulf coast of south Florida. Fortunately, it affected only the extreme western tip of Cuba and crossed south Florida through an area which was not heavily populated at the time. A hurricane following that same path today would do significant damage to the citrus groves and vegetable fields planted across Collier and Hendry counties.

Meteorological Considerations

The storm began to develop just north of the Nicaraguan coast, near 16°N, 83°W, on the fourteenth of October. For three days, it recurved slowly parallel to the Nicaraguan coast then turned due north through the Yucatan Channel brushing the western tip of Cuba on October 19. The minimum barometric pressure recorded in Cuba was 27.52 inches at Los Arroyos indicating the fierce intensity of this hurricane. The S. S. Toledo off the northwestern coast of Cuba experienced winds of hurricane force (force 12 on the Beaufort Scale, Table 1-2) from 1:00 P.M. until 7:00 P.M. on the nineteenth and recorded a minimum pressure of 27.22 inches during the lull as the hurricane center passed.

Curving to the northeast, the hurricane crossed the Florida coastline near Naples on October 20, traversed the state, and exited into the Atlantic between Fort Lauderdale and West Palm Beach on October 21. The barometer reading at Ft. Myers was 29.18 inches at 4:50 P.M. on the twentieth, with the wind speed at 46 mph. Wind gusts up to 90 mph were reported by an unofficial observer at Caxambas south of Fort Myers. Extremely heavy rains occurred along the path of the storm. Rainfall amounts at some south Florida locations are shown in Table 5-3. Minimum pressures recorded included Key West 29.44 inches, Miami 29.50 inches, Tampa 29.56 inches and an unofficial reading of 28.80 inches at Caxambas.

The Damage

The greatest damage was to truck crops which were flooded by 10 to 16 inch rains in Lee, Collier, Broward and Dade counties. Citrus losses were estimated to be near 1,000,000 boxes (Climatological Data, October 1924).

TABLE 5-3. RAINFALL TOTALS IN INCHES AT SOUTH FLORIDA LOCATIONS DURING THE HURRICANE OF OCTOBER 19-21, 1924, LISTING CITIES FROM WEST TO EAST ALONG THE PATH OF THE STORM.

Station	Rainfall (inches)
Fort Myers	15.66
Punta Gorda	8.49
Moore Haven	7.67
Lock #4	13.15
Okeechobee City	9.65
Miami	10.78
Davie	10.74
Ft. Lauderdale	16.74
Vero Beach	8.73

Source: Climatological Data, October 1924.

1925

Tropical activity was greatly diminished in the Atlantic Basin in 1925 with only two systems, a tropical storm which affected south Texas on September 6-7 and a late season hurricane of minimal strength which hit the coast just south of Tampa Bay on November 30 and exited the state near Titusville on December 1.

The November/December Hurricane of 1925

"Truck crops suffered from the heavy rains, and much citrus fruit was damaged."—Climatological Data, December 1925.

As would be expected for such a late season storm, this hurricane originated in the western Caribbean Sea.

Meteorological Considerations

The system was first noted on November 29 near 17°N, 84°W. On the thirtieth, it was already off the western tip of Cuba at which point it veered to the northeast and pointed its gale force winds toward the gulf coast of central Florida. By 8:00 P.M. on November 30, the barometer at Key West had fallen to 29.62 inches with gale force winds from the southwest. During the night of the thirtieth, the barometer at Tampa recorded a low of 29.50 inches as the storm center passed inland just south of that city. At 1:00 A.M., the wind velocity at Tampa peaked at 52 mph and by 8:00 A.M., the center was east of Titusville with the pressure at 29.50 inches. Jacksonville recorded maximum winds of 48 mph from the north, and ships in the Atlantic faced winds of hurricane force. Fortunately, no land stations in Florida recorded hurricane force winds.

The Damage

Extremely heavy rains accompanying the storm were very damaging to Florida crops, particularly in Dade County south of Miami where 15.10 inches of rain were recorded over a 24-hour period, a November record (Climatological Data, December 1925). Damage to citrus was significant, but not severe. The Citrus Exchange estimated that 300,000 boxes were lost, which would have sold for $2.00/box. However, damage to truck crops from the heavy rains in Dade County was extremely severe. Everglades area growers received less rain and losses were not heavy, but truck crops along the path of the storm across the peninsula also suffered heavy losses (Monthly Weather Review, 1925), and the planting of strawberries was delayed.

The Hurricane of July 27-28, 1926

"July is not recognized preeminently as a hurricane month."—Climatological Data, July 1926.

The hurricane of July 1926 was only a forerunner of the disaster which was to strike the state in September. However, it did result in agricultural damage along the Florida east coast and its story should be told.

Meteorological Considerations

The July 1926 hurricane originated from a tropical disturbance just east of the Leeward Islands on July 22. It gradually intensified as it moved on a west northwest course south of the Virgin Islands and Puerto Rico. San Juan reported winds of 66 mph and a low pressure of 29.70 inches as the center of the storm crossed the southwest corner of the island near Cabo Rojo. Veering to a northwesterly course, the storm center brushed the northeast coast of Santo Domingo and passed near Turks Island with 74 mph winds as it moved into the Bahamas on the twenty-fifth with winds of hurricane force. Hurricane warnings were issued for the southeast Florida coast from Miami to Jupiter on the twenty-sixth. Initial advisories indicated that the hurricane would move inland on July 27. However, it changed to a more north northwesterly course and at 8:00 A.M. on July 27, the center was still off the coast between Miami and Jupiter Inlet and hurricane warnings were extended up the coast to Jacksonville. The wind reached 50 mph on Biscayne Bay and along Miami Beach. At 2:00 P.M., the center was still offshore moving north at 10 mph with ship reports of 100 mph winds at sea. At Fort Pierce, the lowest barometer reading was 28.88 inches at 4:30 P.M. on the twenty-seventh, and 28.80 inches at Merritt Island at 11:30 P.M. on that same date. No estimate was made of the wind velocity between Miami and Jacksonville except at Titusville where sustained winds of 60 to 70 mph were recorded in the left quadrant of the hurricane. The center passed over Jacksonville in the predawn hours of the twenty-eighth, with the barometer reaching a low of 29.25 inches, after which the hurricane decreased in intensity as it moved inland into southern Georgia.

Rainfall totals along the Florida east coast are shown in Table 5-4, with wind speeds and barometer readings in Table 5-5. Hurricane force winds were confined to the open sea in the right hand quadrant of the hurricane.

The Damage

As the eye of the storm remained off the east coast of Florida, winds over the peninsula were only of tropical storm force and wind damage was not as severe as would have occurred had the hurricane passed inland. However, it was estimated that 80% of the fruit crop was lost in the Stuart area, 50% of the grapefruit and 10% of the oranges in the Fort Pierce area, and an estimated damage of

TABLE 5-4. RAINFALL TOTALS AT FLORIDA EAST COAST STATIONS DURING THE MINIMAL HURRICANE OF JULY 27-28, 1926, LISTING CITIES FROM SOUTH TO NORTH ALONG THE PATH OF THE STORM.

Station	Rainfall (inches)
Fort Pierce	5.06
Fellsmere	6.88
Malabar	6.05
Titusville	8.36
Merritt Island	10.08
New Smyrna Beach	4.04
Sanford	6.30
St. Augustine	5.19
Starke	5.10
Hastings	4.57
Jacksonville	5.66
Hilliard	6.92

Source: Climatological Data, July 1926.

TABLE 5-5. METEOROLOGICAL DATA DURING THE HURRICANE OF JULY 27-28, 1926.[1]

Station	Minimum pressure (inches)	Maximum sustained winds (mph) direction
Key West	29.78	35 NE
Miami	29.49	33 NE[2]
Titusville	29.30	60-70 ENE
Jacksonville	29.25	50 E

[1]Hurricane force winds were felt over water to the east of the center, but only gale force winds were measured over land west of the center.
[2]Approximately 50 mph over Biscayne Bay.
Source: Climatological Data, July 1926.

$150,000 ($1.2 million in 1998 dollars) to the avocado crop in Dade County. In addition, torrential rains inundated pasture land and farm acreage in Palm Beach, St. Lucie and other east coast counties (Monthly Weather Review, July 1926).

The Miami Hurricane, September 18-21, 1926

"At Moore Haven, my parents climbed into a tree and my mother held me on her lap during the storm. I could taste salt water in the rain."—Arthur Stewart, May 1951.

Meteorological Considerations

The September 1926 hurricane originated to the southwest of the Cape Verde Islands probably near or slightly east of 14-15°N, 34°W. Its initial movement was on a west northwest course at an estimated speed of 14 mph for several days to a point northeast of St. Kitts in the Leeward Islands where it was first observed on the morning of September 14, 1926 (Mitchell, 1926). As it continued its west northwest movement, the storm accelerated to almost 19 mph and covered 450 miles during the 24 hours prior to crossing the southeast Florida coast. En route to Florida, the storm had dealt a heavy blow to Grand Turk Island striking with winds estimated at 150 mph (Goodwin, 1926) and a barometer low of 29.26 inches at 1:00 P.M. on September 16 (Mitchell, 1926). Leaving Grand Turk, the storm passed through the Bahamas south of Nassau on the seventeenth where winds of 86 mph from the northeast and a low pressure reading of 29.56 were recorded. With the hurricane approaching the Florida coast from the Bahamas, tropical storm warnings were issued at noon on September 17, the Weather Bureau noting that, "this is a very severe storm" (Climatological Data, September 1926). At 11:00 P.M., the order was upgraded to hurricane warnings from Key West to Jupiter on the east coast and from Key West to Punta Gorda on the west coast. The warning was displayed from the roof of the federal building and from the warning tower at the city docks. According to the narrative of R. W. Gray, Chief Meteorologist of the Miami Station (Mitchell, 1926), "the warning was given to the long distance telephone operator who passed it to the telephone exchanges at Homestead, Dania, Hollywood and Fort Lauderdale. The warning was also telephoned to the chief dispatcher of the Florida East Coast Railroad and to the Coast Guard base at Ft. Lauderdale. Persons calling by telephone at the rate of three calls per minute were warned of the display of hurricane warnings, but telephone communications with Hollywood and Miami Beach were severed between 1:00 A.M. and 2:00 A.M. and within Miami between 2:00 A.M. and 3:00 A.M. on September 18. The Miami Daily News kept a special operator on duty to provide information to those unable to reach the Weather Bureau. The barometer fell rapidly during the early morning hours and reached a low of 27.61 inches at 6:45 A.M. with the winds gusting at the Miami station to 115 mph from the northeast as the eye of the hurricane moved directly toward the Miami Weather Bureau office. This minimum pressure was believed to be a record for the Weather Bureau (Climatological Data, September 1926). Extremely unusual pressure changes of 0.28 inch per hour occurred at Miami between 6:00 and 7:00 A.M. when the eye of the storm passed over the city. The maximum winds at Miami Beach were 128 mph from the southeast or east at 7:30 A.M. If converted to a 4-cup anemometer, the winds could have been as high as 160 mph. The Miami Beach anemometer blew away at 8:12 A.M. at which time it was recording 120 mph. The record of rainfall and temperature during the passage of the center of the storm was lost due to the recording instruments being blow away. The top of the rain gauge flew off at 3:42 A.M. in winds of 115 mph, was replaced and then was blown away again. It was found the next day on the roof of a neighboring building. Power went out at 4:00 A.M. and observations were made with a flashlight. Finally, the instrument shelter was blown away between 4:00 and 5:00 A.M. landing on a car in the street below. An hour by hour account of the wind speed at Miami Beach is shown in Table 5-6.

The storm tide ranged from 7.5 feet along the northern part of the Miami water front to 11.7 feet along the waterfront south of the Miami River. Measurements were taken from water marks inside of buildings where the action of waves was not shown.

Leaving the Miami area in shambles, the storm slowed to a forward speed of only 11.5 mph as it twisted and snarled its way across the Everglades toward a date with tragedy in Moore Haven, where over 300 residents were living out their last day on earth as the wind

TABLES 5-6. HOURLY ACCOUNTS OF WIND SPEEDS AT MIAMI BEACH DURING THE NIGHT AND MORNING OF SEPTEMBER 17-18, 1926.

Time	Maximum 5 minute sustained winds (mph)
10 P.M.	35
11 P.M.	40
12 midnight	44
1 A.M.	46
2 A.M.	57
3 A.M.	62
4 A.M.	84
5 A.M.	104
6 A.M.	104
7 A.M.	109
8 A.M.	123
8-8:12 A.M.	119

Source: Kadell, 1926.

shrieked and the water rose. Thrashing the agricultural areas of Dade, Broward, Palm Beach, Hendry, Lee and Collier counties for 10 hours (Figure 5-2), the hurricane finally exited into the Gulf of Mexico at Bonita Springs and set a course toward Pensacola.

As the storm passed south Dade County, a lull of five minutes was reported at Homestead, 28 miles south of Miami, and a calm was reported at Punta Rassa at 3:15 P.M. with a pressure low of 28.05 inches at 3:30 P.M. The highest wind velocity at Fort Myers was estimated to be 80 mph, but the heaviest rains fell on the gulf coast rather than the Atlantic coast (Table 5-7). Crossing the Gulf of Mexico, the hurricane threatened a direct hit on Pensacola on September 20 before passing south of that city and moving inland at Perdido Beach, Alabama between 3:00 and 4:00 P.M. with a minimum pressure of 28.20 inches. The minimum pressure at Pensacola as the storm passed was 28.56 inches. As Pensacola lay in the northeast quadrant of the storm, a wind velocity of 116 mph with an extreme peak velocity of 152 mph was felt even though the storm

FIGURE 5-2. THE HURRICANE OF SEPTEMBER 18, 1926 FOLLOWED A PATH THROUGH DOWNTOWN MIAMI AFTER WHICH THE STRONG RIGHT QUADRANT SWEPT OVER LAKE OKEECHOBEE MAKING IT ONE OF THE MOST DESTRUCTIVE HURRICANES OF ALL TIME. SOURCE: CLIMATOLOGICAL DATA, SEPTEMBER 1926.

center passed about 25 miles to the southwest of the city. At that time, it was described as "the most severe storm that has ever visited this locality" (Climatological Data, September 1926). Never before had the winds maintained a velocity over 100 mph for more than an hour. At Milton in Santa Rosa County, the rainfall exceeded anything previously experienced and caused one of the worst floods in history (King, 1972). Maximum wind speeds and lowest pressure readings are shown in Table 5-8.

TABLE 5-7. RAINFALL AMOUNTS AS THE SEPTEMBER 18-21,1926 HURRICANE CROSSED PENINSULAR FLORIDA AND THE FLORIDA PANHANDLE.

Station	Rainfall (inches)
Peninsular Florida	
Long Key[1]	1.88
Hypoluxo	5.60
Jupiter	5.67
Fellsmere	5.52
Belle Glade	4.26
Bradenton	8.76
Ft. Myers	8.02
St. Petersburg	6.48
Pinellas Park	6.58
Tampa	4.51
Northwest Florida	
Apalachicola	4.54
DeFuniak Springs	6.16
Quincy	6.17
St. Andrew	7.72
Bonifay	8.60
Pensacola	8.61
Bluff Springs	11.20
Blountstown	16.40

[1]Long Key is 60 miles south of Miami.
Source: Climatological Data, September 1926.

The Agricultural Damage

Hurricane force winds on the Florida peninsula were confined to the area from St. Lucie County southward, and the most concentrated damage to groves and farms was found in Dade, Broward, Lee and portions of Monroe counties. Until Hurricane Andrew in 1992, the September 1926 storm was the most destructive to strike the southeast Florida coast. In west Florida, the most severe winds were felt in Escambia and Santa Rosa counties but some damage occurred as far east as Calhoun County (Climatological Data, September

TABLE 5-8. METEOROLOGICAL DATA AT SELECTED FLORIDA STATIONS DURING THE SEPTEMBER 18-21, 1926 HURRICANE.

Station	Minimum pressure (inches)	Maximum sustained winds (mph) directions
South Florida		
Key West	29.48	60 W
Miami	27.61	115 NE
Fort Myers	28.14	80 NE
Tampa	29.36	50 NE
West Florida		
Apalachicola	29.57	64 NE
Pensacola	28.56	116 E-SE

Source: Mitchell, 1926.

1926). Total crop damage in peninsular Florida was estimated at $10 million. In west Florida, damage to sugarcane, late corn, sweet potatoes, peanuts, hay, pecans, timber, cotton and satsuma citrus fruit was considerable, but a monetary estimate was unavailable. Damage by county was furnished to the Florida Grower (1926A) by various citrus exchanges and grower associations as follows:

> Dade County: "Every packinghouse severely damaged and old citrus trees uprooted, at least 30% of trees a total loss. Do not think it will be possible to ship one box of fruit from Dade County this season."—B. E. Morrill, Manager, Dade County Citrus Sub-Exchange, Miami
>
> Lee County: "Lee County, including Fort Myers section, suffered a total crop loss with severe damage to groves."—Frank L. Skelly, Manager, American Fruit Growers, Inc., Orlando. "Reports from Fort Myers indicate 50% loss of citrus crop."—V. B. Newton, Standard Growers Exchange, Orlando.
>
> DeSoto County and Charlotte County: "Estimate citrus crop losses in these two counties at grapefruit-80%, seedling

oranges-65% and budded oranges-30%." H. L. Carlton, Manager, DeSoto Citrus Sub-Exchange, Arcadia. "Arcadia reports 30% orange loss and 60% grapefruit loss."— Archie M. Pratt, Sales Manager, Chase and Company, Orlando.

Hardee County: "Wauchula reports 30% loss of citrus crop."—Archie M. Pratt, Sales Manager, Chase and Company, Orlando.

Highlands County: "Citrus crop loss 25 to 30% in this section."—Avon Park Citrus Growers Association.

Manatee County: "Citrus crop damage from 30 to 35%, about evenly divided between oranges and grapefruit."— H. G. Gumprecht, Manager, Manatee County Citrus Sub-Exchange, Bradenton.

Volusia County to St. Lucie County: "Practically no fruit blown off from New Smyrna to Fort Pierce."—Indian River Citrus Sub-Exchange, Cocoa. "Very slight damage in Volusia. Some bruised and thorn-pricked fruit."—H. V. Pay Manager, DeLand Citrus Growers Association. "Estimate 10 to 15% loss of fruit in Fort Pierce and Wabasso sections."—Frank L. Skelly, Manager, American Fruit Growers, Inc., Orlando.

Orange County: "No material damage to citrus crop in Orange County."— L. A. Hakes, Manager, Orange County Citrus Sub-Exchange, Orlando. "Very little damage to citrus fruit in Orange County section and north of here."— W. H. Mouser President, W. H. Mouser and Company, Orlando.

Polk County: "Frostproof reports 5 to 10% damage to grapefruit, with very slight damage to orange crop."—Archie M. Pratt, Sales Manager, Chase and Company, Orlando.

Lake County: "Slight damage to citrus fruit in Eustis section of Lake County, though there is undoubtably considerable fruit thorned."—Eustis Packing Company, Eustis.

88

Other Comments: "No consequential damage in Hillsborough, Pinellas, Polk, Osceola, Orange, Lake, Volusia, Marion and Brevard counties."—Frank L. Skelly, Manager, American Fruit Growers, Inc., Orlando. "Big producing counties practically unharmed. If entire crop in Dade, St. Lucie, DeSoto and Lee counties was destroyed, it would mean only 20% of the total grapefruit crop and 7% of the orange crop. My guess would be that this storm reduces grapefruit over the state from 15 to 20% and oranges 5%."—Archie M. Pratt, Sales Manager, Chase and Company, Orlando. "From best information available it looks like an average loss for the whole state of 15% of the grapefruit crop and 10% of the orange crop."—W. H. Mouser, President, W. H. Mouser and Company, Orlando.

West Florida: "No damage whatever to Satsuma orange crop this immediate section. We had pretty high winds but the fruit stayed on the trees."—William L. Wilson, President, Florida Satsuma Fruit Growers Association, Panama City. "No damage to Satsuma crop in Jackson, Bay, Washington, Calhoun, Holmes and Walton counties."—W. A. Sessoms, Bonifay.

Citrus Crop Loss by County in Carloads

To provide the industry with a basis for planning fruit availability for shipment, the Weekly Citrus Market Summary for the week ending September 25, 1926 (Florida Grower, 1926B) estimated a state-wide loss of 5,450 cars of grapefruit and 1,850 cars of oranges. This indicated a 27% loss based on the pre-hurricane estimate of 20,000 cars of grapefruit, and a 7.5% loss on the earlier estimate of 25,000 cars of oranges. The losses by county are shown in Table 5-9 for grapefruit and Table 5-10 for oranges.

Numbers of trees in each of the important citrus producing counties are shown in Table 5-11. The 1926 tree counts for the counties hardest hit by the 1926 hurricane, and the tree counts in those counties in 1996 are shown in Table 5-12.

TABLE 5-9. ESTIMATED LOSS OF GRAPEFRUIT CROP IN COUNTIES MOST AFFECTED BY THE HURRICANE OF SEPTEMBER 18-21, 1926.

County	Pre-hurricane estimate (carloads)	% loss	Estimate loss (carloads)
Dade	1,450	90	1,350
Lee	1,000	90	900
DeSoto	600	75	450
Sarasota	140	60	80
Highlands	450	60	220
Hardee	175	50	80
Manatee	1,400	45	630
St. Lucie	1,000	75	740
Indian River	600	60	350
Polk	5,000	7	350
Hillsborough	600	9	50
Pinellas	2,200	10	200
Total		27	5,450

Source: Florida Grower, 1926B.

Damage to Other Crops

Truck Crops: String beans in south Florida were destroyed by the hurricane, greatly reducing early shipments, and peppers in Lee and Manatee counties were lost leaving Orange County the principal producer of early peppers.

West Florida: The pecan crop in west Florida was damaged by the hurricane and the 1926 cotton crop was reduced from 38,000 bales the previous year to an estimated 25,000 bales. Heavy rains from the storm caused considerable damage to peanuts in some counties (Florida Grower, 1926D).

Economic Considerations

When H. A. Marus, State Agricultural statistician for the United States Crop Reporting Service, Orlando, released his September 1, 1926 forecast for the 1926-27 season, the prediction was 9,600,000 boxes of oranges and 7,400,000 boxes of grapefruit for a total of

TABLE 5-10. ESTIMATED LOSS OF ORANGE CROP IN COUNTIES MOST AFFECTED BY THE HURRICANE OF SEPTEMBER 18-21, 1926.

County	Pre-hurricane estimate (carloads)	% loss	Estimated loss (carloads)
Dade	50	95	45
Lee	325	95	300
DeSoto	1,350	35	440
Sarasota	20	50	10
Highlands	175	40	75
Hardee	1,400	30	425
Manatee	450	20	90
St. Lucie	225	50	110
Indian River	90	50	45
Polk	4,000	5	200
Hillsborough	1,300	5	65
Pinellas	900	5	45
Totals	10,100	7.5	1,850

Source: Florida Grower, 1926B.

17,000,000 boxes (Florida Agricultural Statistics Service). This was a 2,300,000 box increase from the 8,200,000 boxes of oranges and 6,500,000 boxes of grapefruit marketed the previous year. However, after the September hurricane, it was necessary to revise the estimate to 9,000,000 boxes of oranges and 6,000,000 boxes of grapefruit, indicating a total loss of 2,000,000 boxes. The projected loss was not greater only because the hurricane, while devastating the most southern citrus producing counties, only sideswiped the major producing areas in Polk, Orange and Lake counties. The situation would be different if a hurricane followed this same course today with heavy plantings now concentrated in Hendry, DeSoto, Collier and Lee counties. Such a hurricane today would destroy far more than 2,000,000 boxes, possibly nearer to 20,000,000 boxes.

It was suggested after the hurricane that although the citrus growers in south Florida lost fruit, the total money to be returned to citrus growers statewide would be greater due to the reduced size of the crop. It was thought that packinghouses would lose money due to the

TABLE 5-11. CITRUS TREE COUNTS BY COUNTY AT THE TIME OF THE SEPTEMBER 18-21, 1926 HURRICANE.

County	Number of trees
Alachua	35,883
Brevard	414,900
Dade	474,319
DeSoto	305,125
Hardee	319,125
Hernando	70,475
Highlands	497,502
Hillsborough	617,950
Lake	624,928
Lee	400,766
Manatee	289,555
Marion	223,261
Orange	913,623
Palm Beach	128,436
Pasco	102,680
Pinellas	593,868
Polk	2,171,024
Putnam	230,725
St. Lucie	538,185
Seminole	215,120
Sumter	83,608
Volusia	583,450

Source: Walker, 1926.

smaller crop, but that increased grower returns would lead to better overall statewide returns (Florida Grower, 1926C). The prospects for a smaller crop were further enhanced by two January freezes, one occurring on January 10, 1927 and the second on January 15, 1927 when temperatures of 24 to 30 degrees were experienced across the north and central citrus growing districts. They injured an estimated 50 to 60% of the remaining orange crop, 15 to 20% of the tangerines and 80 to 90% of the remaining grapefruit. As a result of the two freezes, the estimate of the total crop was lowered an additional 1,500,000 boxes to 13,500,000 boxes. This estimate indicated 8,000,000 boxes of oranges and 5,500,000 boxes of grapefruit (Attaway, 1997).

TABLE 5-12. CITRUS TREE COUNTS IN COUNTIES HARDEST HIT BY THE 1926 HURRICANE AND TREE COUNTS IN THOSE COUNTIES IN 1996.

County	1926	1996
Charlotte	—	3,062,400
Collier	—	5,331,400
Dade	474,319	440,800[1]
DeSoto	305,125	8,541,700
Hendry	—	15,353,100
Lee	400,766	1,641,300
Manatee	289,555	2,865,500
Sarasota	—	263,400

[1]Dade lime plantings through 1994.
Source: Florida Agricultural Statistics Service, 1997.

However, as the season progressed, it became apparent that citrus production would be much greater than anticipated after the hurricane and freezes, leading Hall (1927) to comment, "it seems reasonably certain that much badly damaged fruit found its way out of the state into the terminal markets. This has unquestionably proven injurious to the citrus industry."

Fortunately, there were no fatalities in the area. Hutchinson (1975) reports that a 160 acre grove at Palm City Farms in Martin County was destroyed by the July and September hurricanes.

Effect on Fruit Prices

It is difficult to pinpoint the effect of the September 1926 hurricane on citrus fruit prices due to a wide range of factors which affected prices. At the beginning of the season, prices were depressed by shipments of fruit damaged by the high winds, the shipment of early fruit with low ratio and green flavor (Pratt, 1926), the shipment of freeze-damaged fruit after the January 1927 cold wave (Florida Grower, 1927A) the lack of a well-developed citrus processing industry and the generally chaotic marketing system (Citrus Industry, 1927 and Florida Grower, 1927B). There were ten plants canning Florida grapefruit but there was no help for the orange grower with freeze-damaged fruit. The beginnings of a cure for the chaotic marketing situation would not begin until the formation of the Florida Citrus Commission in 1935.

Hall (1927) summarized the 1926-27 season with these comments:

> The season will be long remembered for the September storm which destroyed 2,000,000 boxes of fruit with heaviest losses in Dade and Lee counties and considerable losses in DeSoto and Manatee counties, and the freezes of January 10 and January 15 which caused heavy losses.
>
> There appears to have been quite a lot of damaged fruit shipped which should have remained in the groves and this has undoubtedly proven injurious to the industry as a whole.
>
> In analyzing the situation, the fact should not be lost sight of that California had one of the finest crops in its history—and California oranges brought a substantial premium.
>
> But all things considered, the grower found himself in a better position than during certain other years. The fact that higher prices may have been hoped for or expected does not make it a bad year.

The more things change, the more they stay the same for the Florida grower, freezes, hurricanes, competition from California, and now major competition from Brazil.

The Avocado Industry

The end of the great Florida boom and the hurricane of September 1926 marked the end of the development of the avocado industry in southeast Florida (Dorn, 1928). From 1920 to 1925, it was estimated that Dade County shipped 25,000 to 40,000 boxes of avocados each year. In 1927, shipments from the county possibly totaled 4,000 to 5,000 boxes.

The hurricane broke up many trees, especially the larger and older ones. The old seedling groves of Dade County suffered most of all in broken and overturned trees, but none of the groves, even the young groves or the newer budded varieties escaped damage.

The Disaster at Moore Haven

"Late that night, in absolute darkness, it hit, the far shrieking scream, the queer rumbling of a vast and irresistible freight train."—Douglas, 1947.

Beginning in the early 1920s, engineers recognized the need for levees along the shore of Lake Okeechobee to prevent the flooding of farm fields during periods of high water. A low dike of muck and sand had been constructed along the shore of the lake. However, when the hurricane roared across the south end of Lake Okeechobee with winds from the southeast believed by residents to reach 160 mph (*Glades County Florida History*, 1985) the lake was at a record high level of 18.7 feet (Dovell, 1947) and the muck dike which ran from Sand Point to Moore Haven collapsed releasing a wall of water 10 feet deep which roared across the city inundating everything in its path. Prior to the hurricane, Moore Haven had been a thriving community with railroad service and new hotels. After the storm, it was a deserted wasteland under martial law. Over 300 Moore Haven residents had lost their lives. The total casualties for the Lake Okeechobee area were finally reported to be 392 dead, 6,281 injured and 17,784 families suffering a loss (Tebeau, 1971), but many bodies were never identified and were given a mass burial at the county cemetery in Ortona. Bodies continued to be found months after the hurricane and were buried wherever they were found. No facilities of any kind remained and survivors were evacuated to the Sebring area and the city of Moore Haven placed under martial law enforced by the National Guard. One resident who had staggered back into the city was challenged three times by a guardsman who threatened to shoot him if he didn't leave. The man replied that, "I can't find my family, my home is gone, my business is gone so what are you waiting for." The guardsman allowed him to continue his search. Pilots flying over the area after the hurricane reported seeing one huge lake where Moore Haven had been, and the Swanson family at Lakeport found snakes crawling on the walls of their house and the room full of hyacinths and dead fish (*Glades County Florida History*, 1985).

At Everglades City on the gulf coast, Brown (1993) remembers that after the hurricane, chickens were sitting on the rooftops to escape the flood, and thought that, "Barron Collier just didn't pump his city high enough."

Personal Recollections

Of the 1926 hurricane, Bill Krome (1997) recalls: "the greenish yellow sunset as we returned from a fishing trip the day before; the constant flashes of light that night which I supposed must be clouds racing across the moon (I realize now that it was lightning); my mother calling my attention to the performance of my father's barometer at about 8:00 the next morning—the needle was pumping in a most curious fashion and mother said the center of the storm must be very close; she also said we would probably never again see the pressure so low (it was 27 plus inches but I don't have a record of it). Our home was an unusually strong frame building and although rain blew in under the doors and windows, we didn't fear for our safety. I revised my opinion on this matter about mid-morning when a 20 foot branch crashed through the front door and into the living room. It gave a kind of intimacy to the elements which had heretofore escaped us.

"My recollection is that the wind subsided in late afternoon and we could open the doors and look outside. Dad's office and farm buildings (also strongly constructed) were separated from our residence by more than 40 acres of mature grove, mostly grapefruit and avocados. Ordinarily it was not possible to see through the trees more than a very short distance. Imagine my astonishment, on looking out the kitchen door, to see the office and other buildings standing alone, in the midst of a sea of overturned trees. And the smell— it was the strong, slightly acrid odor of citrus roots of hundreds of overturned grapefruit trees. The only trees left standing were in a five acre block of native pines. The color of their normally grayish-tan bark had changed to a bright red terra cotta; the wind had literally blown off some of the outer layers of bark.

"My father was in New Orleans at the time of the hurricane. Travel in 1926 was mostly by railroad, and after the storm crossed Florida and the Gulf of Mexico, it went on up through Alabama and

Mississippi, disrupting all rail travel in that area. Dad had to go north to Tennessee and down to Atlanta and Jacksonville to get home. It took three days.

"Resetting the overturned trees took several weeks. In those days, before we had equipment and techniques for trenching the soft limestone that we have in place of soil, groves were "flat planted" and shallow. Dad had all the trees put back upright, covering the roots with what loose rock and soil could be scraped up. He had a quantity of long bolts salvaged from a warehouse fire and the trees were guyed from three directions, using the bolts and some of the hundreds of miles of telephone wire that was ruined by the storm."

Cattle rancher "Buck" Mann of Winter Haven (Mann, 1998) remembers, "my daddy had put in 10 miles of new barbed wire fence in Hendry County. This was prairie country with very few trees. The ground was soft, and the hurricane blew down all 10 miles of that new fence."

W. T. Maddox of LaBelle (Maddox, 1998) remembers, "in '26, we had no warning whatsoever. We were in the woods rounding up cattle. It had rained all night, and the wind began to pick up. When 150 year old pine trees began to wring off half way up, I knew we were in trouble."

Helen Day and her family (Day, 1998) moved to Florida in 1925, living in an old Florida farmhouse. She vividly recalls that, "When the '26 hurricane slammed into south Florida, the wind blew for three days and the sky flashed from blue to green. The wallpaper drooped down, but later came back up after it dried. My mother saw that all the new grapefruit in the yard had fallen. She couldn't stand to see it wasted so she halved each fruit, squeezed out all the juice with a hand squeezer and added sugar. We drank it at every meal. My uncle was horrified. He said it would ruin our stomach as it wasn't ripe enough. However, none of us had colds for the next five years."

1927

The 1927 Atlantic Basin season can be characterized as benign in its effects on Florida. While there were four hurricanes and three tropical storms, none of these systems affected the state.

1928

Unfortunately, the 1928 season was anything but benign, even though only three tropical systems formed in the Atlantic Basin. Two storms affected the state in August. The first struck the Indian River citrus growing area on August 7-8 with hurricane force winds inflicting significant damage to groves near the coast. The second went ashore near Apalachicola as a tropical storm on August 14. However, these storms were only a prelude to the most deadly hurricane, in terms of loss of human life, which savaged the state on the sixteenth and seventeenth of September.

The Hurricane of August 7-8, 1928

"You've probably forgotten this hurricane ever happened, unless you lived in St. Lucie County"—Anonymous.

Meteorological Considerations

The August 1928 hurricane formed on August 3 near 11°N, 60°W, just east of the island of Trinidad. It moved on a steady northwest path for three days before attaining hurricane force just southeast of Andros Island in the Bahamas on the sixth, and slamming into the Vero Beach area on the seventh. The lowest barometric pressure, 28.84 inches, was recorded at the substation at Ft. Pierce at 4:00 A.M. on the eighth.

Sweeping across the state in a northwestern direction, the hurricane passed through St. Lucie, Osceola, Sumter and Citrus counties. It seemed to diminish in intensity when half-way across the state, and losses on the west coast were not so heavy as those on the east coast (Florida Grower, 1928).

The Damage

Grapefruit was damaged more heavily than oranges. Old seedling trees lost more of their crop than did budded trees. Tangerines escaped with minimal damage. Much of the fruit which was not lost by being actually blown from the trees was bruised and thorned, reducing its market grade. The loss of fruit was estimated at one million boxes (Climatological Data, 1928).

In St. Lucie County, which suffered the heaviest citrus loss, 85% of the grapefruit crop, 25% of the orange crop and 15% of the tangerine crop was destroyed. Trees in this area were defoliated, and in some instances uprooted.

Fifty percent of the grapefruit crop and 20% of the orange crop was lost in Osceola County while neighboring Polk County's loss was estimated at 8% of the grapefruit crop and 4% of the orange crop. Lake County and Orange County experienced practically no damage. Other citrus producing counties, including Hillsborough, Pinellas, Manatee, Highlands and Hardee, had no damage.

The Tropical Storm of August 14, 1928

Meteorological Considerations

This tropical storm formed on August 7 near 12°N, 80°W or only a few miles from the point of origin of the earlier storm. It moved on a west northwest course for four days, struck Cuba as a hurricane on August 11, lost force crossing the island and then moved northwest, skirting the Florida gulf coast. It passed 50 miles to the west of Tampa and finally made landfall near Apalachicola.

The Damage

Very little citrus damage resulted from this storm, and was confined principally to groves facing the gulf. The combined storms of August 7-8 and August 14 caused significant losses to corn, which was blown over and rotted in the fields, cotton blown out of the bolls and lost, and pecan crops in north Florida. Timber and turpentine farms also sustained major damage.

The Deadly Hurricane of September 16-17, 1928

"The history of this hurricane is a melancholy one associated, as it is, with the tragic ending of more than 2,000 lives on Lake Okeechobee whose waters attained a height of 10 to 15 feet as they were forced southward across the shallow rim of the lake."—Alexander J. Mitchell, Meteorologist, Jacksonville, Florida, 1928.

Meteorological Considerations

The September 1928 hurricane was a classic "Cape Verde" storm which probably originated on about September 6, 1928 from a low pressure area moving off the west African coast near 14°N, 21°W. The storm moved almost anonymously on a steady westerly course across the Atlantic Ocean until it was finally detected by three different vessels on September 10, 1928 at a point 600 miles east northeast of Bridgetown, Barbados. At the time, it was the most easterly report ever received by radio concerning a tropical cyclone (C. L. Mitchell, 1928). Having crossed the entire width of the warm ocean, the storm had attained an enormous velocity, and on the evening of September 11, an advisory was issued indicating that the center would pass over the Leeward Islands north of Martinique on September 12.

At noon on the twelfth, the hurricane passed very near Point a' Pitre, Guadeloupe where the barometer sank to the extremely low reading of 27.76 inches for over two hours. The damage was extreme in Guadeloupe, St. Kitts, Montserrat and St. Croix as the hurricane roared along a west northwesterly path toward Puerto Rico where it crossed on the thirteenth causing a widespread loss of life and destruction of property. In Puerto Rico, the hurricane is remembered as "San Felipe", for the Saint's day on which it struck, as probably the worst hurricane in the history of the island (Fassig, 1928). The hurricane stormed ashore near Guayana and passed across the island in a west northwest direction over an eight hour period at a forward speed of 13 mph before exiting between Aguadilla and Isabela. Guayana reported a barometer reading of 27.50 inches and a calm period of 20 to 30 minutes as the eye passed. At the San Juan Weather Bureau, 30 miles from the storm center, a 5-minute reading of 150 mph was recorded with an extreme gust of 160 mph. Unofficial observers believed the winds exceeded 200 mph as the eye wall came ashore at Guayana, but this cannot be confirmed. Winds of hurricane force continued at Guayana from 4:00 A.M. to 10:00 P.M., a period of 18 hours. At San Juan, hurricane force winds prevailed for 12 hours, 4:00 A.M. to 4:00 P.M. The rainfall in Puerto Rico on September 13 and 14 was the greatest recorded in the previous 30 years. In the central mountains, the rainfall exceeded 25 inches.

After leaving Puerto Rico, the path of the hurricane shifted from west northwest to nearly northwest at a steady rate of 14.5 mph toward the Bahamas and the Florida east coast. At midnight, the barometer at Grand Turks Island read 28.50 inches with winds of 120 mph as the hurricane passed 9 miles south of that island and set its sights on Nassau, capitol of the Bahamas. The following advisory was issued on Saturday night, September 15:

> "*** This hurricane is of wide extent and is attended by dangerous and destructive winds. Its center will likely pass near or slightly north of Nassau Sunday morning. Storm warnings are now displayed from Miami to Titusville, Fla. Winds of hurricane force will probably be as far west as longitude 79 degrees (60 miles off the southeast coast of Florida) by Sunday noon. Recurve of hurricane's path not yet indicated."

In addition, the warnings sent to Miami and West Palm Beach stated that, "every precaution should be taken (tonight) in case hurricane warnings should be found necessary Sunday on the east Florida coast."

Continuing its west northwest course, the hurricane center passed slightly north of Nassau on Sunday morning, September 16. The following is quoted from a report by Mr. D. Salter, the meteorological recorder in Nassau (C. L. Mitchell, 1928):

> "At 10:00 P.M. (fifteenth) the corrected barometer reading was 29.50 inches, with overcast sky, and wind northeast 40 miles and freshing rapidly. At midnight, the barometer had fallen to 29.35 inches and was still dropping rapidly, overcast sky, and wind northeast 55 miles. At 1:30 A.M. of the sixteenth the barometer was 29.22 inches, wind northeast 65 miles and increasing, accompanied by rain. Two A.M. saw the barometer down to 29.00 inches and falling every minute, wind still northeast 75 miles with rain. The 4:00 A.M. barometer was 28.25 inches, with wind shifting to northwest by west 100 miles. The wind speed recorder ceased to function at 3:30 A.M., owing to the anemometer cups blowing away while recording 96 miles sustained wind speed. At 5:00 A.M. the barometer reached its lowest point, 28.08 inches with south-

west winds estimated velocity 110 to 120 mph, and heavy rain. By 6:00 A.M. the barometer had risen to 28.50 inches and a considerable fall in the wind speed had taken place, although still blowing from the southwest, with heavy gusts, accompanied by rain. The total rainfall during the passing of the disturbance amounted to exactly 9 inches."

The Hurricane Aims at Florida

On the morning of Sunday, September 16, 1928, the following warning was issued for the Florida east coast:

"Hoist hurricane warnings 10:30 A.M. Miami to Daytona, Fla. *** No report this morning from Nassau. Indications are that hurricane center will reach the Florida coast near Jupiter early tonight. Emergency. Advise all interests. This hurricane is of wide extent and great severity. Every precaution should be taken against destructive winds and high tides on Florida east coast, especially West Palm Beach to Daytona."

At the same time, tropical storm warnings had been issued for the Florida west coast, but these were soon changed to hurricane warnings from Punta Rassa to Cedar Key. On the seventeenth, hurricane warnings were extended north of Cedar Key to Apalachicola on the west coast and north to Jacksonville on the east coast.

The center of the hurricane roared ashore just south of Jupiter with great ferocity at about 7:00 P.M. on the sixteenth. A barometric trace provided by the American Telephone and Telegraph office in West Palm Beach showed a sea level reading of 27.43 inches, 0.18 inch lower than the reading at Miami during the hurricane of September 18, 1926 (Figure 5-3). At that time, this was the lowest pressure ever recorded in the United States during a hurricane.

After moving inland, the hurricane continued to the northwest with the center passing across the northeast corner of Lake Okeechobee on the evening of the sixteenth and slightly east of Bartow the morning of the seventeenth, at which point it finally began to recurve (Figure 5-4),unfortunately about 48 hours too late for the dead and missing of southeast Florida. The hurricane had reached Lake Okeechobee with little loss in force due to the flat

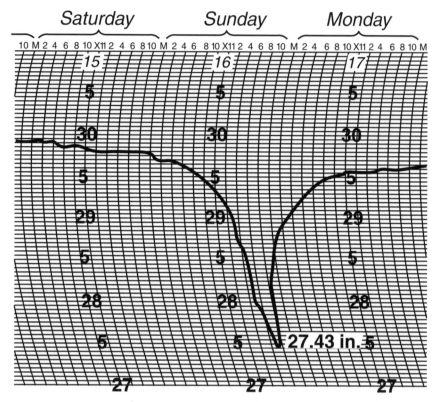

FIGURE 5-3. THE BAROMETRIC TRACE FOR THE SEPTEMBER 1928 HURRICANE AS RECORDED BY THE AMERICAN TELEPHONE AND TELEGRAPH COMPANY AT WEST PALM BEACH, FLORIDA, 7:00 P.M., SEPTEMBER 16, 1928. SOURCE: CLIMATOLOGICAL DATA, SEPTEMBER 1928.

landscape of western Palm Beach County. Moving north northwest, it passed near Ocala and finally turned north northeast to a point just west of Jacksonville about 1:00 A.M. on the morning of September 18.

According to B. A. Bourne (C. L. Mitchell, 1928) of the Bureau of Plant Industry, the barometer at Canal Point on Lake Okeechobee fell rapidly with a corresponding increase in the wind velocity after noon of the sixteenth. At 5:00 P.M., the barometer was 29.17 inches and the wind 40 mph from the north; at 7:48 P.M., the

FIGURE 5-4. THE MURDEROUS HURRICANE OF SEPTEMBER 1928 ASSAULTED FLORIDA ALONG A LINE FROM WEST PALM BEACH TO LAKE OKEECHOBEE AND NORTH THROUGH CENTRAL FLORIDA TO NEAR JACKSONVILLE. SOURCE: MONTHLY WEATHER REVIEW.

barometer was 28.54 inches and the wind 60 mph from the northwest; and at 8:15 P.M., the anemometer cups blew away after the velocity reached 75 mph from the northwest, the barometer at this time reading 28.25 inches. By 9:00 P.M., the barometer had fallen to 27.87 inches with an estimated wind velocity of 150 mph from the northwest. There was a dead calm between 9:30 and 10:00 P.M. when the center passed over the station, the lowest barometer reading being 27.82 inches at 9:45 P.M. Shortly after 10:00 P.M., the barometer began to rise and the wind immediately came with hurricane force from the southeast, reaching an estimated velocity of 160 mph about 10:45 P.M. The wind force decreased rapidly after 11:00 P.M.

The Damage

The September 1928 hurricane, leaving approximately 2,000 dead, was the second most deadly hurricane of the 20th century, surpassed only by the Galveston, Texas hurricane in 1900 which killed 6,000.

Citrus: Prior to the hurricane, C. W. Lyons of Lyons Fertilizer in Tampa had predicted that Florida's 1928-29 citrus crop would total 18,000,000 boxes, of which 11,500,000 boxes would be oranges and tangerines, and 6,500,000 boxes would be grapefruit (Florida Grower, 1928). This forecast was generally accepted although some felt that the crop could total as high as 20,000,000 boxes while others thought that it would be no more than 16,000,000 boxes. Commercial shipments the previous year had totaled only 13,600,000 boxes of all varieties indicating that a much larger crop was expected.

Due to the ferocity of the storm as it came ashore at Jupiter, in Palm Beach County, moved inland over Lake Okeechobee and recurved north through Highlands and Polk counties, heavy damage to the citrus crop would have been expected. However, citrus plantings were not heavy in the immediate area of landfall in 1928, and the prime Indian River area lay miles north of the center. By the time the hurricane reached Polk County, its velocity had diminished considerably. For this reason, the Florida Citrus Exchange (Florida Grower, 1928) estimated that the statewide loss from the hurricane would be no more than 15% of the grapefruit crop and 6% of the orange crop, with the greatest damage to Polk County. The Exchange damage estimate by area was as follows:

Polk County: Grapefruit 25%, oranges 10%

DeSoto County and Hardee County: Grapefruit 15%, oranges 7.5%

Lake County and Marion County: Grapefruit 10%, oranges none Orange County and Osceola County: About 2% of both oranges and grapefruit

Dade County: Grapefruit 2%, few oranges grown in this county

Lee County: Practically no damage

Indian River area: Practically no damage

Manatee County and Sarasota County: Practically no damage

Hillsborough County and Pinellas County: Practically no damage

Volusia County and Putnam County: Practically no damage

At season's end, the pre-season estimate of the total crop was much too low, and losses to the hurricane were evidently much too high. Final statistics from the Florida State Marketing Bureau (Citrus Industry Magazine, 1929) showed the following shipments:

Oranges - - - - 12,992,540 boxes
Grapefruit - - - - 9,335,335 boxes
Tangerines- - - - - 911,770 boxes
Total - - - - - - -23,239,645 boxes

The pre-hurricane estimates ranging from 16 million to 20 million boxes did not account for the rapid growth of the industry. The total value of the crop, based on an average price of $2.54/box was set at $49,035,795.

Damage to Citrus Crop in Puerto Rico

Prior to the 1928 hurricane (San Felipe), Puerto Rico had been shipping 750,000 boxes of early grapefruit to the New York market. However, the island's citrus crop was totally destroyed by the hurricane's 150 mph winds at an estimated loss of $10 million. Prior to the hurricane, there were approximately 500,000 grapefruit trees on the island of which 100,000 were either uprooted or broken off a foot above ground. The most damage was to the older trees with larger trunks which could not bend in the wind (Citrus Industry Magazine, 1928).

Damage to Other Crops

The Everglades were flooded and Florida snap and lima beans were lost, then replanted. Beans which survived in the north and central areas later suffered heavily from frost damage on November 22-23, 1928. Statewide fall bean production was down 63% from the previous year (Rose, 1975B). Seed beds, growing truck crops, corn, cotton, pecans and many less important crops were significantly damaged. Cane in the Everglades was prostrated by the winds but recovered. Crop losses in the Fellsmere area were estimated at 85% and at Venice 30%.

A Night of Horrors on Lake Okeechobee

Lord, hold back the waters of Lake Okeechobee,
  Lake Okeechobee water is cold,
Was in the late 20s there came a big flood,
  Drownded 4,000, their graves was the mud,
Was nothing could withstand that great tidal wave,
  Now the ghosts of the vanished cry from the grave.
Lord, hold back the waters of Lake Okeechobee,
  Lake Okeechobee water is cold,
When wild winds are blowing across Okeechobee,
  They're seeking and looking for other poor souls,
Lord, Lake Okeechobee water is cold. . . .

(Written and recorded by Will McLean, printed with permission of the Will McLean Foundation).

After the Moore Haven disaster in 1926, it was recognized by government officials, residents and engineers that something had to be done to improve the Lake Okeechobee drainage system and build a better dike to prevent future floods. However, the timing could not have been worse. The great Florida boom had run its course, and the hurricane spelled the end of the boom and spelled it emphatically. The need was obvious, but financing was difficult. Bonded indebtedness was already excessive. The area around Lake Okeechobee was very lightly populated with only a pitifully small tax base. Belle Glade and Moore Haven were the two largest towns and in 1922 the Poll List for the Moore Haven Election Districts showed only 254 voters. Meaningful tax revenue in the local area simply did not exist. Drastically complicating the problem was the fact that the death of the Florida boom had left failed real estate developments scarring the Florida landscape. The stream of money being invested in Florida had dried up. Voters in Miami, central and north Florida had their own problems. What to do about Lake Okeechobee and the Everglades had become one of the biggest issues in Florida state politics, but there was little sympathy in the more populous regions of the state for spending millions of dollars on new dikes and canals. A detailed account of the issues is available in Dovell (1947).

As a result of the financial and political problems, the summer of 1928 ended with only a muck and sand dike in place. According to Will (1968), this dike ran from Bacon Point near Pahokee along the shore and around to Moore Haven, a distance of some 47 miles. It was only six to eight feet high, and was only a patched up version of the dike which was washed away in the 1926 hurricane at Moore Haven. Under normal conditions, it might keep the lake from flooding farm fields, but it had already proven to be inadequate in a hurricane. But, before 1926, the only previous hurricane in memory had been in 1910, so there was no hurry. It probably wouldn't happen again. Particularly since the summer of 1928 had been unusually dry and the water level in Lake Okeechobee was at the lowest level ever recorded (Will, 1968). Then in late August and early September the rains came in a deluge, 14 inches, and raised the level by three feet, lapping at the top of the muck and sand dike. The canals draining the lake were brim full. The stage was set for disaster.

By early 1927, the excess water left from the 1926 hurricane had drained away, and some 3,000 farmers were cultivating an estimated 9,000 acres of land in the upper Everglades. From January to June, they had produced 3,000 carloads of fresh vegetables worth three million 1927 dollars (Dovell, 1947). Prosperity had come to the

Florida Everglades. There was money to be made and work was available for farm laborers. As a result, the numbers of migrant farmers and harvesting laborers in the communities along the south shore of the big lake had swelled to near 5,000, with no exact count known. Many lived only in tents and tar paper shacks along the primitive roads and even on the canal banks. Disaster was screaming, but few heard as the primitive communications of the day—no electricity in most homes and a paucity of battery radios—provided little news and no warning. Thousands of resident did not realize the gravity of the situation until it was too late. Attempts to escape east toward West Palm Beach would take them directly into the path of the hurricane. The road west out of Belle Glade was too primitive to support an evacuation. The muck and sand dike which people hoped would be their savior, proved instead to be their executioner. Without the dike, the water would have risen slowly, and escape might have been possible, but when the north winds piled the water into the south end of the lake and the dike broke releasing a wall of water and mud ten feet high, for many there was no escape. The final death toll has been estimated as high as 2,500. An exact count will never be known as bodies were unearthed from the muck years later. There were not enough people left to bury the dead, and hundreds of bodies were piled up and burned, creating a pall of black smoke the stench of which will never be forgotten by survivors.

At the Agricultural Experiment Station

The first effects of the hurricane were felt at the Everglades Station in Belle Glade the afternoon of September 16, with rising winds and lashing rains. The Station's anemometer was destroyed at 8:18 P.M. as it showed a sustained wind velocity of 92 miles per hour with gusts estimated to 125-135 miles per hour (Allison, 1929). The wind direction during the first half of the storm varied from northeast to north northwest, then when the eye passed between 10:00 and 11:00 P.M., the wind shifted to southwest to southeast. During the first half of the storm, the northerly winds and high water ruptured the muck dike around the southern end of the lake leading to the massive floods and heavy death toll at Belle Glade, South Bay and the neighboring communities. Then when the winds shifted and blew from the south, the water was blown north to flood Okeechobee City and

Eagle Bay. The rain gauge at the Station recorded 11.35 inches, but this figure may be deceptively low as the top of the rain gauge was blown away.

The Station property was flooded to a depth of three feet the morning of September 17, after which the waters quickly receded to a depth of 2 feet, but it was not until after December 4 that the land was again out of the water. As a result, all field work in progress was completely destroyed, and a number of experimental units were abandoned. In addition, a five-room bungalow, two labor cabins, a garage and the west section of the greenhouse were completely destroyed. The two-story boarding house building was completely wrecked and had to be torn down. The roofs of the other buildings were badly damaged and had to be replaced.

At Okeechobee City

When the eye of the hurricane passed and the winds across Lake Okeechobee began to blow toward the north, the flood waters moved toward Eagle Bay and Okeechobee City. Fortunately for the residents of Okeechobee City, they were two miles from the lake shore, but the tiny community of Eagle Bay felt the full brunt of both the wind and the rising waters. It was Belle Glade over again but on a smaller scale. There were fewer people and fewer deaths, but for those affected it was just as horrible. Present day Okeechobee residents Mr. Basil Harvey, Mrs. Marie Box, Mr. Clayton White and Mrs. Zetta Hunt tell the story in their own words.

> Basil Harvey (1998) reminds us that communications were very poor in 1928 as only a few people had electricity. A few had radios, mostly battery-powered but reception from the AM stations on the east coast was scratchy and sometimes difficult to hear. As a result, most residents along the north shore of the lake did not know a hurricane was coming. Sheriff "Pogy Bill" Collins received a telegram from Miami advising him to warn the people, and he spent all afternoon going from house to house and houseboat to houseboat along the lake shore pleading with people to get off the lake. Many ignored him and many died.
> Mrs. Marie Box (1998) remembers Lake Okeechobee in 1928 as a big saucer with only a flimsy rim. The hurricane "turned it

upside down and dumped the water out, first to the south, then after the lull, when the wind changed direction, it blew it back up on us at Okeechobee City. The water reached all the way past where the Holiday Inn sits today, and on up to the present Wal-Mart shopping center.

Clayton White (1998) still remembers Sheriff "Pogy Bill" Collins pounding on the door of their house at the north end of the lake on Eagle Bay. There were twelve or fifteen families living in houseboats as well as the houses on the lake. The White family elected to ride it out at home, but when the flood waters reached the top of the dining room table, they gathered up a few things and moved to a house on higher ground. However, as the storm got stronger and stronger Sunday night, the house began to rock and the windows blew out. Then the roof went with the tin sheets from the roof folding around the tops of pine trees where they remained for many years after the hurricane. After 10:00 A.M., the storm subsided and the family tried to go back to their house, but the Coast Guard had cordoned off the area and no one could go back until the bodies along the lake had been moved. When they finally got back to the site of their house, there was nothing left but the floor.

One of Mr. White's favorite stories concerned a Mr. L. P. Wynn and his friend Mr. Pancoast. As the hurricane raged, they became increasingly worried. Finally, Mr. Wynn suggested that Mr. Pancoast pray. At first, Mr. Pancoast demurred. He was a large and powerful man, a very self-sufficient man, but when an extreme wind gust shrieked by their refuge Mr. Pancoast prayed, "Lord, you probably wonder why I never called on you before, but I never really needed you, but if you don't do something in a hurry, this place is going to blow all to hell." It must have worked as all his people survived.

In her mind, Mrs. Zetta Hunt can still see the Coast Guard camped on the court house lawn in Okeechobee City. The court house was closed so that it could be used as an emergency morgue. Bodies were first laid in rows across the grass, then put in boxes in the court house and finally buried in trenches. When they moved back into the court house, she asked the janitor to move out a bunch of boxes which were believed to be empty. They weren't. They contained more unburied bodies which had been forgotten.

Mrs. Hunt was on the first boat to get people off the lake. They found one woman who had held on to a tree, lying on a limb, throughout the hurricane. The woman was in total shock. They took her out of the tree, pried her mouth open and fed her some soup. Miraculously, she lived.

The Coast Guard brought in numbers of people with no place to stay. The ladies worked in the kitchen night and day fixing soup and serving it in quart jars. Its been over 70 years but its taken a long time to get over it.

In Polk County

Dudley Putnam (1998) of Bartow remembers the 1928 hurricane at Hesperides in Polk County east of Lake Wales:

"We had a Model-T Ford station wagon. It sat up high and square, and the wind which had blown down my trees everywhere was ready to blow that Model-T over. The clutch had gone out, but I got it in reverse and backed down behind the railroad fill to get out of the wind. That saved the Model-T, and probably saved me too."

Buck Mann of Polk County (Mann, 1998) remembers that:

"1928 was the last year of the Model-T Ford with the high running board. In the hurricane of 1928, the rain raised the water level so high that when we crossed the Peace Creek on the old Bartow/Lake Wales Road, the water ran over the running board of that old Model-T."

The Hurricane of September 28-30, 1929

"The '29 hurricane is mostly forgotten, unless you live in south Dade."—Seymour Goldweber, 1997.

Meteorological Considerations

The 1929 Florida hurricane was first detected as a tropical disturbance east of the Bahama Islands near 24°N, 66°W on September

22. It reached hurricane force as it crossed the 70th meridian on a northwesterly course to latitude 27°N, at which point it turned abruptly to the southwest on the twenty-fourth, after which it slowed almost to a crawl as it took almost four days to pass through the Bahamas (Monthly Weather Review, September 1929). On September 27, the storm shifted to a west northwest track and struck the Florida Keys between Key West and the mainland on September 28 putting south Dade County directly in the path of the highest winds and heaviest rains in the strong northeast quadrant. A barometer low of 28.18 inches was recorded in the Keys (Climatological Data, September 1929).

Moving out to sea, the hurricane passed to the west of Tampa on September 29 and turned to the northwest striking the west Florida coastline between Pensacola and Apalachicola on September 30, where it changed course to the northeast and exited Florida in the vicinity of Thomasville, Georgia on October 1. The lowest pressure in Apalachicola was 29.06 inches on the thirtieth. Meteorological data is shown in Table 5-13.

Personal Observations

Bill Krome (1997) recalls, "The center of the hurricane passed south of Florida City and the railroad was washed out. We had spent the better part of the summer of 1929 putting back up our trees which had blown over in the 1926 hurricane. There was more grapefruit than anything else. Dade County was sixth in the state in grapefruit production in the early 1900s, particularly early grapefruit. We had a grapefruit grove where the Orange Bowl Stadium stands today. However, it was all for nothing. Grapefruit in Dade County had shallow root systems. We called it 'flat planting.' These grapefruit trees had large tops and the '29 storm blew them all over again, but this time the damage was complicated by the mess of guy wires. Most of the groves were cleared and replanted but not to grapefruit. The trees had done fine and produced high quality fruit, but they were too likely to be overturned in a hurricane. The smaller orange, tangerine and avocado trees were more wind resistant."

Barney Walden (1997) recalls, "My father was in the fertilizer business in the fall of '29. He had extended credit to the avocado growers in the Homestead area who were expecting a good season. Unfortu-

TABLE 5-13. METEOROLOGICAL DATA AT REPORTING STATIONS AFFECTED BY THE SEPTEMBER 28-30, 1929 HURRICANE.

Station[1]	Minimum pressure (inches)	Maximum sustained winds (mph) direction	Rainfall (inches)
Key West	29.21	66 W	2.87
Miami	29.41	56 E	10.58
Tampa	29.53	33 E	3.19
Apalachicola	29.06	59 SE	4.85
Pensacola	29.19	70 NE	4.84
Jacksonville	29.67	37 SE	1.91

[1]Data is not available from the areas of south Dade and Collier counties which were in the direct path of the hurricane center.
Source: Climatological Data, September 1929.

nately, the hurricane put all of the fruit on the ground and blew over most of the trees. My father had to mortgage his home, and like many others in south Dade, he eventually had to declare bankruptcy. This led to my mother's death in 1932. The stock market crash had hit almost the same day as the hurricane. Times were harder than anything endured by people today.

"The morning after the hurricane, my father and I drove into downtown Homestead in our Model-T Ford. The water was two feet deep in the streets and salt water fish were swimming up to the running board of the Model-T. There was so much salt water in the east glades west of Krome that it drowned the mules which had been pastured there to feed on the thick grass.

"The salt water had put the Homestead electric plant out of commission. My father took his steam tractor and pumped water for the City of Homestead for three or four months."

The Damage

As the southernmost counties of Florida were not major citrus producers, the Florida Citrus Exchange estimated that loss of the citrus crop would not exceed 2%. However, the loss of both truck and fruit crops was almost complete in the areas of Dade and Collier counties which experienced the center of the hurricane. Fruit loss at Homestead, south of Miami, totaled 189,000 boxes with some fruit trees uprooted. All recently planted truck and seed beds were destroyed, sugarcane was flattened and all strawberries either damaged or totally destroyed. Timber losses were serious in some areas (Climatological Data, September 1929). The dollar loss in Collier County due to fruit and truck damage was estimated at $40,000.

Chapter 6

The 1930s, The Florida Keys Hurricane and the Yankee Hurricane

The decade from 1930 to 1939 produced Florida hurricanes in only two seasons, 1933 and 1935. The remaining eight years were relatively benign with only an occasional tropical storm to worry the Florida grower; therefore, catastrophic damage to agriculture was not experienced. Two hurricanes to hit the state in 1933. Both came ashore along the Florida east coast and caused significant damage to citrus in the Indian River district. Neither of the 1935 hurricanes are known for their damage to Florida fruit and vegetable crops. On September 2-3, 1935, the Florida Keys Hurricane, a category 5 storm, resulted in a heavy loss of life and caused extreme property damage in the Florida Keys. The second hurricane in 1935 was an unusual storm called the "Yankee Hurricane" because its winds reached hurricane intensity at latitude 34°N near the island of Bermuda, from which point it moved directly south and then southwest and struck Miami on November 4. Fortunately, this was a minimal hurricane, only category 1 on the Saffir-Simpson scale (Table 15-4).

1930

This may well have been an "El Niño" year but this is not known. There were only two tropical systems. The first formed in the mid-Atlantic on August 22, 1930 and moved north without ever approaching land. The second formed near 15°N, 55°W on August

31 and attained hurricane force as it crossed the Leeward Islands on September 1. Moving west northwest, it made landfall in the Dominican Republic on September 3 after which it lost hurricane force as it crossed Haiti and Cuba. Finally recurving to the northeast on September 7, the storm moved ashore between Clearwater and Tarpon Springs before crossing the state and entering the Atlantic Ocean at St. Augustine on September 10. Approximately eight to nine inches of rain fell in southeast Hillsborough County as the weak storm passed. Crops in low lying fields received some damage, but the citrus crop was unharmed.

1931

With only two hurricanes and seven tropical storms, this was an average year in the Atlantic Basin. Fortunately, none of these storms entered Florida waters. The first of the two hurricanes formed east of Barbados but did not attain hurricane intensity until it reached the Western Caribbean, struck Belize, and dissipated as it crossed the Yucatan peninsula. The second hurricane formed as a tropical storm near 17°N, 53°W and became a hurricane as it passed through the U.S. and British Virgin Islands, skirted the northern coast of Puerto Rico, and lost intensity over the Dominican Republic. It finally passed through the Yucatan Peninsula and dissipated over Mexico.

1932

This year was more active than 1931 in that there were a total of eleven storms, six of which became hurricanes and five which remained in the tropical storm category. One storm was classified as a hurricane, but moved across Florida as a tropical storm, and did not attain hurricane status until it emerged into the Gulf of Mexico and moved away from Florida. This storm had formed just north of the Dominican Republic, near Turks Islands, on August 26 and entered Florida on a west northwest course over Key Largo late on August 24, emerging into the Gulf between Naples and Fort Myers on August 30. At Miami, northeast winds reached gale force between 6:00 and 7:00 P.M. and continued until after midnight. Excessive rainfall followed the storm across south Florida (Table 6-1). After exiting the Florida peninsula near Everglades City where the pressure dropped

TABLE 6-1. RAINFALL TOTALS AT KEY POINTS DURING THE
TROPICAL STORM OF AUGUST 29-30, 1932.

Station	Rainfall (inches)
Homestead	5.0
Miami	10.24
Hypoluxo	6.0
Indiantown	5.5
Belle Glade	2.2
Everglades City	9.5
Moore Haven	3.0
Fort Myers	7.8
Bradenton	4.0

Source: Climatological Data, August 1932.

to 29.42 inches, it intensified into a hurricane and later passed inland along the Alabama/Mississippi state line on September 1. Damage had been confined to damaged seed beds and recently planted truck crops in the extreme south where flooding occurred, and some losses to avocados and limes from the gale force winds. The second storm formed in the Gulf of Campeche on September 9, moved north and then northeast rapidly crossing Florida from Cedar Key to Jacksonville on September 14 causing some damaging rains, but without becoming a hurricane. Two of the remaining hurricanes remained over the open Atlantic, one passed over Cuba and into the north Atlantic. The other struck the Virgin Islands and Puerto Rico and continued westward to Belize and the Yucatan.

1933

"The summer of 1933 was marked by an unusual frequency of tropical disturbances. There were few days without an advisory message being received."—Climatological Data, September 1933.

The 1933 Atlantic Basin season was extremely active with ten hurricanes and eleven tropical storms. One can assume that this was not a strong El Niño year. Fortunately, only one tropical storm, on July 30-31, and a hurricane on Labor Day affected Florida.

The Tropical Storm of July 30-31, 1933

"This was a small disturbance, but winds reached near hurricane force near the center."—Climatological Data, July 1933.

Meteorological Considerations

The July tropical storm and borderline hurricane formed east of the Leeward Islands at 16°N, 57°W on July 25. It passed north of Puerto Rico on the 26th, followed a northwesterly course along the eastern edge of the Bahama Islands passing Nassau on July 29. It finally crossed the Florida coastline south of Stuart on July 30, and traversed the state exiting into the Gulf of Mexico north of Punta Gorda the evening of July 31. The wind velocity was estimated at 70 mph at Stuart with a lowest pressure reading of 29.65 inches, and 60 mph winds at Fort Pierce. Rainfall was very heavy with 12.01 inches at West Palm Beach, 12.02 inches at Indiantown, and 15.60 inches near Stuart.

The Damage

Damage to citrus, avocados and mangos in Indian River and St. Lucie counties and in the north part of Palm Beach County was significant. The estimated loss to citrus in that area averaged 10 to 20%, but some exposed groves incurred even greater losses (Climatological Data, July 1933).

Personal Observations

Reuben Carlton of Fort Pierce recalls that, "we all sat on the porch on the west side of the house watching the wind bend the trees as it blew from the east" (Carlton, 1998).

The 1933 Labor Day Hurricane

Meteorological Considerations

The Labor Day Hurricane began on August 31 as a tropical storm near 19°N, 56°W (Climatological Data, September 1933). It attained hurricane intensity quickly and passed about 170 miles northeast of San Juan, Puerto Rico on September 1 moving rapidly on a west northwesterly course. The hurricane passed northeast of Nassau in the Bahamas on Sunday morning of September 3 and tropical storm warnings were posted along the Florida east coast for points south of Fernandina. By the afternoon, the tropical storm warnings had been replaced by hurricane warnings for that part of the coast from Miami north to Melbourne as the storm directed its ferocity on a collision course with the Indian River citrus producing district, and at midnight the storm center made landfall at Jupiter thrashing the groves of Indian River, St. Lucie, Martin and Palm Beach counties with winds in excess of 100 mph. Wind velocities, barometric pressure readings and rainfall at key points are shown in Table6-2.

Leaving the east coast and moving inland (Figure 6-1), the hurricane moved west northwest toward Indiantown in Martin County where it passed at 1:45 A.M. on Labor Day Monday, September 4. It then shifted to a more northwesterly course over Okeechobee City,

TABLE 6-2. WIND VELOCITIES AND PRESSURE READINGS AND RAINFALL AT KEY POINTS DURING THE HURRICANE OF SEPTEMBER 4-5, 1933.

Station	Minimum pressure (inches)	Maximum sustained winds (mph)	Rainfall (inches)
Jupiter	27.98	100+	—
Fort Pierce	29.14	70-100	5.56
Vero Beach	—	72	—
Palm Beach	28.77	97	9.84
Bassenger	—	90	—
Bartow (Lake Garfield)	28.85	60-80	11.50
Lakeland	28.94	50-60	11.90
St. Leo	28.97	50-60	—
Tampa	29.40	46	8.14
Cedar Key	29.24	40	3.10
Jacksonville	29.63	32-43	3.53

Source: Climatological Data, September 1933.

FIGURE 6-1. THE SEPTEMBER 1933 HURRICANE STRUCK THE
INDIAN RIVER CITRUS PRODUCING AREA, THE VEGETABLE
FIELDS OF PALM BEACH COUNTY AND THE RIDGE CITRUS PRO-
DUCING SECTIONS OF HIGHLANDS COUNTY AND POLK COUNTY.
SOURCE: CLIMATOLOGICAL DATA, SEPTEMBER 1933.

and set its sights on the ridge citrus producing districts, striking
Avon Park in Highlands County, Bartow and Lakeland in Polk
County at 11:30 A.M. and 12 noon, respectively, and finally over St.
Leo in Pasco County at 2:30 P.M., after which the storm center
became diffuse and its exact course as it moved north could not be
determined. Indications are that it passed through Citrus, Levy, Ala-
chua and Columbia counties and finally broke up over eastern
Georgia on September 7.

The Damage

Damage to citrus groves in the Indian River section was severe, with losses along the immediate coastline almost complete. Damage to citrus in the ridge section was less severe, but significant. Estimated losses by county and area are shown in Table 6-3 (Climatological Data, September 1933).

Statewide, the Florida Citrus Exchange estimated the grapefruit loss at 30% or 2,400,000 1-3/5 bushel boxes, oranges at 15% or 1,950,000 boxes and tangerines at 10% or 190,000 boxes, a total statewide loss of 4,450,000 1-3/5 bushel boxes. The west coast counties had little damage (Climatological Data, September 1933).

Excessive rain from the hurricane washed out strawberry fields, and plants for resetting were scarce. A late crop resulted.

Tree damage was also a factor along the east coast which bore the brunt of the winds. Limbs were split and broken and foliage was damaged by salt spray. In vegetable growing sections, the heavy rains washed out seed beds delaying planting, and in the Plant City area some strawberry fields were flooded with resultant loss of plants.

TABLE 6-3. ESTIMATED LOSSES TO 1933-34 GRAPEFRUIT AND ORANGE PRODUCTION DUE TO THE HURRICANE OF SEPTEMBER 4-5, 1933.

County	Grapefruit	Orange
	- - - - - - - Estimated % loss - - - - - - -	
Indian River, St. Lucie, Martin and Palm Beach	60	20
Highlands	50-60	10-15
Polk:		
Frostproof section	40	15
Lake Wales section	30	10
Lakeland section	50	15
Lake and Sumter	15	5
Brevard, Seminole and Volusia	5-10	0

Source: Climatological Data, September 1933.

1934

The 1934 Atlantic Basin hurricane season was again average with 11 tropical storms and hurricanes. As in the previous year, there were six hurricanes and five tropical storms. Three tropical storms and no hurricanes crossed the Florida coastline. The first tropical storm formed just north of the western tip of Cuba, 23°N, 54°W, on May 27 prior to the official start of the season on June 1. It moved northeast and crossed Florida between Fort Myers and Daytona Beach during the day of the twenty-seventh, emerging into the Atlantic early on the twenty-eighth. The greatest rainfall from this storm was over Broward, Dade and Collier counties in the south and Seminole, Lake and Marion counties in central Florida. Crop damage, principally to celery, was estimated at $100,000 (Climatological Data, May 1934).

The second tropical storm followed an unusual course. It formed just east of Cape Hatteras, North Carolina on July 21 moved first west and then southwest to cross Florida east to west from St. Augustine to Cedar Key on July 22, after which it moved in a westerly direction across the gulf and briefly attained hurricane status before striking the Texas coast near Corpus Christi on July 25. As the storm crossed Florida, damage was generally light, but in the Monticello area, corn was "laid low" and pecans and pears blown from the trees. Rainfall in north and west Florida ranged from one to one and one-half inches (Climatological Data, July 1934).

The third storm passed inland near Pensacola on October 5 causing some rain damage, but top winds were only 35 mph. The Pensacola Weather Bureau station recorded 15.29 inches of rain.

1935

The 1935 season was much below average in numbers with only five hurricanes and one tropical storm. However, it produced two very memorable Florida hurricanes.

The Florida Keys Hurricane, September 2-4, 1935

"The area of hurricane winds was only about 30 miles in diameter, but the hurricane was probably the most intense of record over a small area to visit any part of the United States."—Climatological Data, September 1935.

Although the center of the Florida Keys Hurricane did not pass directly through any major agricultural area, gale winds and heavy rains did affect south and southwest Florida agriculture. No treatment on Florida hurricanes is complete without a description of this ferocious storm which ranks as one of the only two category 5 hurricanes to strike the United States. According to estimates by engineers (Dunn and Miller, 1964), winds of 200 to 250 mph would have been required for some of the damage done by this extremely severe hurricane.

Meteorological Considerations

The early stages of the hurricane first appeared on August 19, 1935 as an area of occasional squalls east of the central Bahamas near 24°N, 69°W. For the next two days, the incipient storm moved on a west to west southwest course through the Bahamas. Maximum winds at 34 mph were less than tropical storm force as it crossed the island chain, and the barometric pressure observed at Long Island was 29.48 inches on August 31. At this point, there was no hint of the momentous disaster which was only two days from the Florida Keys. Except for the extreme southern tip of Andros Island, no damage was reported from the Bahamas. It is known that the storm was definitely not a hurricane as it passed through the islands. However, when it reached the warm waters of the gulf stream, all the factors conducive to hurricane formation must have come together as it rapidly developed into one of the most intense hurricanes ever recorded.

As the hurricane gradually began to recurve to the west northwest, hurricane force winds began to be felt at approximately 5:00 P.M. on Alligator Reef. The diameter of the area of hurricane force winds was only 30 miles, but in intensity it exceeded any previous hurricane to strike the United States. The barometer reading at Long Key was 26.98 inches and at Lignumvita Key, 26.75 inches. The latter reading was more than a half inch lower than any previous pressure record in the U.S., and one of the lowest ever recorded in the world (Climatological Data, 1935). The pressure fell 1.16 inches at Alligator Reef in only 30 minutes, and it is estimated that winds of 150 to 200 mph must have occurred with gusts exceeding 200 mph. Such a narrow storm, with such great intensity, would not be felt again in south Florida until Hurricane Andrew in 1992. The track and cross ties were washed from a concrete railroad viaduct 30 feet above the normal water level indicating a massive storm surge. As the eye passed over Long Key and Lower Matecumbe Key during the evening of September 2, there was a calm period of 55 minutes when stars were visible.

Leaving death and destruction behind in the Keys, the hurricane curved to a more northwesterly course and paralleled the Florida coastline, 30 to 50 miles offshore from the beaches (Figure 6-2). It gradually increased in diameter and lessened in intensity as it moved north. Some low pressure readings were 29.29 inches at Everglades City at 6:00 A.M. on September 3, 28.94 inches at Egmont Key at 11:00 P.M. and 29.08 inches at Cedar Key at 1:50 P.M. on September 4. The maximum wind velocity at Tampa was 70 mph.

The storm crossed the north Florida coastline during the afternoon of September 4 and moved north through Taylor and Dixie counties, with wind gusts still exceeding hurricane force, and finally passed through Madison County and into south Georgia. Rainfall totals along the storms path are shown in Table 6-4.

The Damage

The hurricane caused a measurable loss to citrus in the Gulf Coast area as it paralleled Lee, Charlotte, Sarasota, Manatee and Pinellas counties, and the heavy rains damaged cotton and corn and washed out seed beds as it slashed through Dixie, Lafayette, Taylor, Suwanee, Madison and Hamilton counties in the north. Continued wet weather delayed harvesting and much late cotton and corn deteriorated in the fields.

Personal Observations

Bill Krome remembers the 1935 hurricane: "The 1935 hurricane that did so much damage in the Florida Keys missed us on the mainland. I was in Homestead with my mother the afternoon before the storm and saw the passenger train that was on its way to evacuate the veterans from their camp in Lower Matecumbe. "A lot of those Democrats will be Republicans by tomorrow," I joked. I had no

FIGURE 6-2. THE CATEGORY 5 LABOR DAY HURRICANE, SEPTEM-
BER 2-4, 1935, MADE A MURDEROUS RAMPAGE ACROSS THE
FLORIDA KEYS BEFORE MOVING UP THE FLORIDA WEST COAST.
SOURCE: CLIMATOLOGICAL DATA, SEPTEMBER 1935.

inkling that, Democrat or Republican, many of them would be
dead. We had a house at Islamorada, close to where the rescue train
was washed off the tracks. The house next to ours was destroyed by
the big timbers intended for bridge construction, which were
washed across the island and into the bay but our house survived.
One of the locations on the rocky shore where the corpses of the

**TABLE 6-4. RAINFALL IN INCHES ALONG THE PATH OF THE FLOR-
IDA KEYS HURRICANE, SEPTEMBER 2-4, 1935. STATIONS LISTED
FROM SOUTH TO NORTH.**

Station	Rainfall (inches)
Key West	2.23
Homestead	10.83
Everglades City	12.70
Fort Myers	9.80
Belle Glade	4.13
LaBelle	6.58
Punta Gorda	13.25
Sarasota	10.41
Bradenton	8.90
Tampa	8.20
St. Petersburg	8.71
St. Leo	10.35
Bartow	4.58
Lakeland	4.34
Tallahassee	4.94
Madison	5.59

Source: Climatological Data, September 1935.

victims from the veterans camp were cremated was right in front of
our house. It was marked by a small cross that remained there for
several years" (Krome, 1997).

The Yankee Hurricane, November 4, 1935

"Because of the sudden change in direction of the movement
of the storm, warnings could not be issued very far in
advance—but pupils who had not received the warnings were
sent home from school in ample time to reach home
safely."—Climatological Data, November 1935.

This was an unusual hurricane in two ways, first by striking so late
in the season (November hurricanes being rare) and secondly by
forming far north of the area where hurricanes normally spawn and
then moving on a southwesterly course.

Meteorological Considerations

The storm was first noted about 300 miles east of Bermuda at about 33°N, 62°W on October 30, 1935 (Byers, 1935). Its initial movement was to the west northwest, toward Cape Hatteras, N.C., a course which was maintained for two days as it increased in size and intensity. Then on November 2, it made an abrupt turn to the south, crossing latitude 30°N at approximately 4:00 A.M. on November 3 when the S. S. Davenger reported hurricane force winds and a barometer reading of 28.98 inches. During the next few hours, the hurricane increased its forward speed and gradually curved to the southwest and by 7:00 A.M. it was located about 100 miles north northwest of Great Abaco Island in the Bahamas (Climatological Data, November 1935). Gradually decreasing its forward speed the hurricane passed a short distance north of Hopetown, Great Abaco, where hurricane force winds persisted for one to three hours.

After passing over Great Abaco, the storm shifted to a west southwest course directly toward Miami. Storm warnings were raised from Jupiter to Key Largo at 5:15 A.M. on November 4, 1935. At 7:30 A.M., hurricane warnings were issued from Miami north to Stuart, and at about 1:30 P.M., the storm center passed over North Miami with the lowest barometric pressure reading of 28.73 recorded at the Weather Bureau Office. As the eye passed, Miami experienced a lull of one hour and five minutes at the Weather Bureau Office in the city and also at the airport. Rainfall came to a stop during the lull, but the sky remained cloudy. The diameter of the eye was estimated to be 15 miles. The wind velocity ahead of the eye was 72 mph from the west northwest. After the eye passed, the wind was 75 mph from the southeast at 2:07 P.M. The most extreme velocity of 94 mph was recorded at 2:17 P.M. The hurricane force winds were felt from just south of Miami to Fort Lauderdale, a diameter of only about 30 miles. Gale force winds extended further north of West Palm Beach.

After passing over Miami, the hurricane continued its west southwest course across the Everglades and emerged into the Gulf of Mexico near Cape Sable (Figure 6-3). Ships continued to report winds of hurricane force on November 5, after which the hurricane gradually lost intensity and died out in the Gulf on November 8. Rainfall at key stations affected by the storm is shown in Table 6-5.

FIGURE 6-3. THE NOVEMBER 4-5 HURRICANE IN 1935 FOLLOWED AN UNUSUAL COURSE MOVING FROM THE NORTHEAST TOWARD THE SOUTHWEST, AND WAS APTLY CHRISTENED THE "YANKEE HURRICANE." SOURCE: CLIMATOLOGICAL DATA, NOVEMBER 1935.

The Damage

Vegetable crop losses were heavy in southeast Florida, particularly beans, peppers and tomatoes. Damage to trees and tropical shrubbery was estimated at $150,000 (Climatological Data, November 1935). Avocados, which had been damaged in the December 1934 freeze (Att-

**TABLE 6-5. RAINFALL IN INCHES ALONG THE PATH OF THE YAN-
KEE HURRICANE, NOVEMBER 4-5, 1935. STATIONS LISTED FROM
EAST TO WEST.**

Station	Rainfall (inches)
Stuart	0.31
Fort Lauderdale	1.82
Miami	4.20
Chapman Field	6.25
Homestead	5.35
Long Key	11.80
Key West	4.60
Moore Haven	0.24

Source: Climatological Data, November 1935.

away, 1997), suffered further heavy losses to the fruit crop as a result
of the hurricane (Table 6-6). Fortunately, tree damage was not
severe.

Personal Observations

Bill Krome of Homestead, Florida (Krome, 1997) remembers the
November 1935 hurricane very clearly:

I was at the University of Florida in November 1935 when the
"Yankee" hurricane hit south Florida. At that time, I was in the
habit of writing my mother every week or two, and I would do
this by jotting down items of interest as they occurred. I
remember very vividly taking note about a week before
November 4 that a hurricane was reported in such an unusual
place, at such an unusually late date. Hurricanes anywhere, at
any time, were of interest to mother and me. As the days
passed, I noted the strange course of the storm and considered it a fit subject to joke about. I ended up trying to contact
my mother by telephone and being informed that phone service to Miami was disrupted due to the hurricane. I cut classes
and got myself down to Homestead as quickly as I could. In

TABLE 6-6. EFFECTS OF THE DECEMBER 1934 FREEZE AND NOVEMBER 1935 HURRICANE ON THE AVOCADO CROP.

Season	Crop (tons)	Bushels (50lb.)	Value per bushel	Gross FOB value
1934	2,000	80,000	$1.88	$150,000
1935	1,000[1]	40,000	2.38	95,000
1936	600[2]	24,000	3.00	72,000
1937	2,100	84,000	2.38	199,500
1938	2,200	88,000	1.60	140,000
1939	2,500	100,000	1.70	170,000

[1]Lower yield due to December 1934 freeze.
[2]Lower yield due to November 1935 hurricane.
Source: Florida Agricultural Statistics Service, 1997.

truth, however, damage to the groves was less that I had feared. Most of the crop had been harvested, and it wasn't necessary for me to stay to help clean up.

Helen Day (1998) writes: "In the 1935 hurricane, the rain blew straight across for three days and came around the window panes of the second floor bedrooms. We were up all night mopping window sills in the dark by candlelight—the electricity was out for a week. We had one kerosene lamp for the entire house. The yards were swamped and cars couldn't be moved. A group of my brother's friends came in one night. We had a player piano with "Stormy Weather" and "Singing in the Rain." It was fun when we were in our 20s, but then we weren't down in the Keys where it was so terrible."

1936

The state was visited by three tropical storms in 1936, on June 15, July 27-31 and August 21-22. The July storm became a hurricane after passing through south Florida, and later struck the Pensacola area with hurricane force winds. The June and July storms were accompanied by significant rainfall, but rainfall in the August storm was light.

The Tropical Storm of June 15, 1936

Meteorological Considerations

The storm formed on June 12 near 17°N, 87°W just east of Belize. It moved due north through the Yucatan Channel on the 13th reaching latitude 25°N on June 14, after which it curved to the northeast and then due east crossing the Florida coastline about 20 miles south of Fort Myers early on the fifteenth (Climatological Data, 1936). The barometric pressure at Fort Myers was 29.50 inches. Moving east southeast, the storm passed over Miami and out to sea between 8:00 and 8:30 A.M. with a pressure of 29.59 inches. No hurricane force winds or serious gales accompanied the storm, but rainfall was very heavy with more than eight inches in some places (Table 6-7).

The Damage

Wind damage from this storm was insignificant, but serious flooding resulted from the heavy rainfall. LaBelle, which received 12.47 inches of rain, was badly flooded with many forced to evacuate to the second story of homes and businesses. There was considerable damage to fields and gardens, and bridges near LaBelle were down. There was some loss of livestock due to drowning.

TABLE 6-7. RAINFALL FROM THE TROPICAL STORM OF JUNE 15, 1936. TOTALS FOR JUNE 14 AND 15.

Station	Rainfall (inches)
Fort Myers	9.61
Everglades City	7.78
LaBelle	12.47
Moore Haven	9.33
Belle Glade	7.05
Miami	5.90
Homestead	5.32
Fort Lauderdale	9.62
West Palm Beach	9.16

Source: Climatological Data, June 1936.

The Tropical Storm/Hurricane of July 27-31, 1936

This system was a tropical storm when it crossed south Florida, but a hurricane when it struck west Florida. It crossed the southern tip of Florida as a tropical storm during the twenty-eighth and twenty-ninth of July, then intensified in the Gulf of Mexico and moved into the Pensacola area as a minimal hurricane on July 31.

Meteorological Considerations

This storm began as a low pressure area just east of the Leeward Islands near 23.5°N, 74°W on July 23. It drifted slowly to the west northwest and developed a center of circulation near Cat Island in the Bahamas on the morning of July 27. Tropical storm warnings were posted along the Florida east coast from Fort Pierce to Key West. On the morning of July 28, the storm was gaining strength about 110 miles east southeast of Miami, and in the afternoon it was only 75 miles from Miami and storm warnings were extended up the west coast of Florida as far as Punta Gorda. After passing Carysfort Reef Light, with a low pressure of 29.36 inches, the storm crossed the Florida coastline at Key Largo early in the evening, and continued west northwest, passing just north of Homestead and finally exiting into the Gulf of Mexico near Everglades City. The barometer reading at Everglades City was 29.51 inches and winds were estimated at 55 mph (Climatological Data, July 1936), and shortly after entering the gulf, the storm's winds increased to hurricane intensity. Tropical storm warnings were extended north of Punta Gorda early on July 29, and later in the day hurricane warnings were posted first from Cedar Key to Apalachicola and on the morning of July 30 extended to Pensacola as the hurricane moved on a steady northwest course. The hurricane made landfall near Niceville, 40 miles east of Pensacola, on the morning of July 31 with sustained winds of 80 mph and gusts to 125 mph. The lowest barometer reading was 28.46 inches with a lull lasting from 7:25 to 8:20 A.M. Hurricane force winds were experienced from the Niceville area to east of Panama City. After making landfall, the hurricane continued to the northwest and dissipated across central Alabama on August first.

The Damage

In south Florida, neither the winds nor the rainfall was damaging, with most communities in the storm path receiving only a little over two inches of rain (Table 6-8). However, in west Florida, where hurricane force winds were experienced, crop damage was significant with pears and pecans blown from the trees (Climatological Data, July 1936).

The Tropical Storm of August 21-22, 1936

Meteorological Considerations

This storm developed near 25°N, 75°W on August 20, moved northwest, and made landfall between Titusville and Daytona Beach in the late afternoon of August 21. The barometric pressure at Titusville was 29.60 inches with winds of 40 to 55 mph. The storm moved across north Florida and dissipated near Apalachicola on the twenty-second. Rainfall along the storms's path was generally less than one inch.

TABLE 6-8. RAINFALL AMOUNTS IN SOUTH FLORIDA AND WEST FLORIDA FROM THE TROPICAL STORM AND HURRICANE OF JULY 28-31, 1936.

Station	Rainfall (inches)
South Florida	
Homestead	2.38
Miami	2.19
Everglades City	2.32
Fort Myers	2.38
Punta Gorda	2.25
West Florida	
Panama City	4.40
Apalachicola	4.96
Niceville	2.90
Pensacola	0.35

Source: Climatological Data, July 1936.

The Damage

Damage, if any, was very light. The gale force winds were not suf-ficiently strong to result in fruit loss or damage.

1937

The 1937 Atlantic Basin hurricane season was average in num-ber of storms with three hurricanes and six tropical storms. Of the nine systems, only three of the tropical storms affected Florida. The dates of these three storms were July 29-30, August 30-31 and September 20-21. The September storm was only a minor distur-bance.

The Tropical Storm of July 29-30, 1937

Meteorological Considerations

The storm developed from a low pressure area near 25°N, 84°W and moved directly northeast, made landfall over Pinellas County in the early afternoon of July 29, and moved across northern Hillsbor-ough County. The barometric pressure at both Clearwater and Tampa was 29.80 inches with a maximum wind speed of 51 mph recorded at Tampa. Unofficial estimates up to 60 mph were reported around the area. The storm was small and it passed quickly. However, it did produce heavy precipitation, Clearwater reporting 8.88 inches (Table 6-9). Winds in excess of 30 mph con-tinued in the Tampa area from 5:00 to 7:00 P.M. as the storm moved slowly northeast and lost intensity. It exited the peninsula near Day-tona Beach with 20 mph winds.

The Damage

The only citrus fruit loss was near the point of landfall in north-ern Pinellas County and was considered minor. No damage was reported from the rainfall.

TABLE 6-9. RAINFALL FROM THE TROPICAL STORM OF JULY 29-30, 1937.

Station	Rainfall (inches)
Clearwater	8.88
Tarpon Springs	5.15
St. Petersburg	1.72
Tampa	1.97
St. Leo	2.96
Lakeland	1.03
Lake Alfred	0.88
Kissimmee	0.60
Orlando	0.91
Sanford	0.86
DeLand	5.25
Daytona Beach	2.17

Source: Climatological Data, August 1937.

The Tropical Storm of August 30-31, 1937

Meteorological Considerations

The storm formed near Andros Island in the Bahamas on August 28, moved north northwest and then northwest to cross the Florida coastline near Ormond Beach on August 30. Gale winds of 50 to 60 mph were reported between New Smyrna Beach and St. Augustine, but no hurricane force winds were reported (Climatological Data, August 1937). The storm continued to the northwest producing heavy rains and squalls across northwest Florida (Table 6-10).

The Damage

Cotton was damaged in many area by the heavy rains which struck near the peak of the picking and ginning season (Climatological Data, 1937). Other crops were largely unaffected by the storm.

TABLE 6-10. RAINFALL FROM THE TROPICAL STORM OF AUGUST 30-31, 1937.

Station	Rainfall (inches)
New Smyrna Beach	4.89
Daytona Beach	3.47
St. Augustine	1.05
Crescent City	3.17
Palatka	3.63
Gainesville	5.77
Lake City	3.90
Perry	3.90
Cedar Key	2.75
Carrabelle	4.90
Tallahassee	2.60
Madison	2.98
Mariana	1.50

Source: Climatological Data, August 1937.

The Tropical Disturbance of September 20-21, 1937

Meteorological Considerations

The disturbance formed in the Bay of Campeche near 20°N, 93°W on September 16, 1937, moved north, then northeast and finally due east to cross the Florida coastline near Steinhatchee late on the twentieth. Winds on shore were less than tropical storm force with Apalachicola reporting a maximum of 30 mph with a low pressure reading of 29.79 inches (Climatological Data, September 1937).

The Damage

A tornado formed in St. Johns County on the twentieth causing some damage to fruit trees. Some damage was caused in northwest Florida by locally heavy rains. Carrabelle recorded 4.49 inches of rain on September 19 and eight inches on September 20 for a two-day total of 12.49 inches.

1938

The 1938 hurricane season in the Atlantic Basin was average in numbers with three hurricanes and five tropical storms, but below average in effect on Florida as only one very weak tropical storm affected Florida agriculture. This storm had formed off the west coast of Mexico early on the twenty-third of October, moved rapidly northeast across the gulf and crossed the Florida coastline near Perry early that evening. There was no wind damage reported, but several stations reported 24-hour rainfall totals of four inches or more (Table 6-11) (Climatological Data, October 1938). The disturbance exited Florida across the Okefenokee Swamp into south Georgia.

The Damage

Fall vegetables, truck crops and strawberries in the west coast and north central counties were damaged by the heavy rains requiring some replanting (Climatological Data, 1938).

TABLE 6-11. RAINFALL FROM TROPICAL DISTURBANCE CROSSING NORTH FLORIDA OCTOBER 23-24, 1938.

Station	Rainfall (inches)
Perry	3.70
Cedar Key	9.22
Tarpon Springs	2.22
Tallahassee	4.54
Raiford	6.80
Gainesville	5.80
Jacksonville	5.80
Glen St. Mary	5.91
Hilliard	5.10
Fernandina	5.12

Source: Climatological Data, October 1938.

1939

The 1939 season was noteworthy for the few storms produced, only three hurricanes and two tropical storms, but only a tropical storm in August crossed the Florida coastline.

The Tropical Storm of August 11-13, 1939

Meteorological Considerations

The storm formed in the western Atlantic near 19°N, 65°W, just northeast of Puerto Rico on August 7. It moved northwest on an unwavering line until it crossed the Florida coast near Stuart on the afternoon of August 11, and exited into the gulf near Tarpon Springs on the morning of the twelfth (Climatological Data, August 1939). It then crossed the gulf and made landfall near Valparaiso during the afternoon and evening of the twelfth and moved on into southern Alabama. The lowest barometer reading on the Florida east coast was 29.22 inches with a wind speed of 54 mph at Fort Pierce. Fifty mph winds were reported all across Florida, and Tampa recorded a one-minute velocity of 62 mph from the south southwest at 4:30 A.M. on August 12. When the storm crossed the west Florida coast at 3:20 P.M. on the twelfth, Apalachicola reported a low pressure of 29.26 inches with the wind at 59 mph from the northeast. Rain was heavy in both the southern portion of the peninsula and also in west Florida (Table 6-12).

The Damage

Damage to citrus was limited to the Indian River area and was minor. However, there was considerable wind damage to pecans in the Florida Panhandle, as well as rain damage to cotton, corn, peanuts, and sweet potatoes. Some panhandle areas reported a total loss of cotton and late corn.

TABLE 6-12. RAINFALL FROM THE TROPICAL STORM OF AUGUST 11-13, 1939.

Station	Rainfall (inches)
South Florida	
West Palm Beach	3.35
Stuart	3.95
Fort Pierce	1.39
Moore Haven	1.91
Lake Alfred	2.47
Lakeland	2.53
Tampa	3.54
St. Petersburg	7.06
Tarpon Springs	3.40
West Florida	
Apalachicola	6.01
Blountstown	4.76
Carrabelle	4.60
Marianna	7.10
Panama City	8.27
Pensacola	5.44

Source: Climatological Data, August 1939.

Chapter 7

The 1940s, the Period When Floridahurricane Became One Word

"I was indignant a couple of years ago when I was on a trip up east and a hurricane which hadn't come within 400 miles of Florida struck the Jersey shore causing a lot of damage, and I read in the Jersey newspaper headlines 'Florida Hurricane Strikes Jersey Coast.'"—Episcopal Bishop Wing quoted in the Citrus Industry magazine, October 1935.

Just as the 1980s are remembered as the decade of the freezes, the 1940s will be remembered as the decade of the hurricanes. During the period from 1940 through 1949, no fewer than nine hurricanes and eight tropical storms crossed the Florida coastline, with activity concentrated from 1944 to 1949. While later decades have produced major hurricanes such as Donna in 1960, Betsy in 1965, Frederic in 1979 and Andrew in 1992, and earlier decades produced the great hurricanes of 1926 and 1928 and the Florida Keys hurricane of 1935, no comparable period in Florida history has produced as many severe hurricanes in so short a time as did the 1940s.

1940

The 1940 season produced four hurricanes and four tropical storms. Only one tropical storm affected Florida. It formed in the Atlantic east of Jacksonville in early August and moved southwest to cross the state from St. Augustine to near Tarpon Springs from which point it crossed the Gulf of Mexico and moved inland along the Texas-Louisiana border. Rainfall from this disturbance was not excessive in Florida and did not result in significant damage.

1941

The 1941 North Atlantic hurricane and tropical storm season was average in the number of disturbances - six - but the entire output was confined to only 37 days, from September 11 to October 19. Only four of the season's storms reached hurricane intensity, with one hurricane and one tropical storm crossing the Florida coastline (Sumner, 1941A).

The Hurricane of October 6-7, 1941

Meteorological Considerations

The storm was first discovered on October 3-4 about 300 miles north of the Virgin Islands near 23°N, 67°W. From this point, it moved west northwest through the Bahamas, passing a short distance south of Nassau on the evening of October 5, at which time it was determined to be a small but highly developed hurricane. Winds at Nassau averaged 70 to 75 mph with gusts exceeding 100 mph. At 5:30 A.M. October 6, 1941, the hurricane moved inland about 13 miles south of Miami and at 5:45 A.M. Goulds, Florida was within the calm eye (Sumner, 1941B). The pressure reading at Fowey Rock Lighthouse was 29.28 inches at 5:00 A.M. with a wind velocity of 70 mph. At Pan American Dinner Key, a sustained velocity of 90 mph with a gust to 123 mph was recorded over a brief period (Table 7-1). The rainfall was unusually light with Miami and Homestead recording two-day totals of only 0.51 and 0.62 inches, respectively.

The hurricane continued on its west northwest course and crossed the Everglades south of Lake Okeechobee passing into the Gulf of Mexico between Everglades City and Fort Myers at 11:00 A.M. on October 6. Gusts of 60 mph were recorded along the south shore of Lake Okeechobee. A low pressure of 29.40 inches was reported at Everglades City. Very little damage was reported in the Fort Myers area.

TABLE 7-1. METEOROLOGICAL DATA FROM THE HURRICANE OF OCTOBER 6-7, 1941.

Station	Minimum pressure (inches)	Maximum Sustained winds (mph) directions
Nassau, Bahamas	29.12	75-104 E
Fowey Rock	29.28	70
Miami WB	29.38	68 NE
Pan Am Dinner Key	29.39	90[1]
Everglades City	29.40	65+ SW
Sanibel Light	29.48	60 N
Gasparilla Light	29.52	44 ENE
Egmont Key	29.69	60
Carrabelle	29.00	65-75
Tallahassee	29.36	65-75 SE
Albany, GA	29.35	65-75 NW

[1]Gusts reported to be 123 mph.
Source: Climatological Data, October 1941.

The storm curved more to the north over the open waters of the gulf passing west of the Tampa Bay area where winds of 60 mph were reported at Egmont Key. Final landfall was made at Carrabelle between 3:30 and 4:30 P.M. on October 7 with winds of 65 to 75 mph and a barometric pressure of 29.00 inches. Carrabelle received 2.50 inches of rain. Gale force winds continued across northwest Florida with the storm passing Tallahassee at 8:45 A.M. and into south Georgia.

The Damage

Dangerous winds occurred in south Florida along a path about 60 miles in width from Homestead to Fort Lauderdale on the east coast and from Punta Rassa south of Everglades City on the west coast (Climatological Data, October 1941). Over north Florida, damage extended from about the Apalachicola River eastward 80 miles to near Perry and Madison. Property and crop loss in south Florida was estimated at $325,000. In Florida, the damage over the area from the Gulf of Mexico to the Georgia state line was about $350,000. The greatest agricultural losses were to timber and pecan orchards and extended as far north as Albany, Georgia. In south Florida, salt spray damaged crops and shrubbery partially due to the lack of rainfall.

The Tropical Storm of October 18-19, 1941

A minimal tropical storm formed over the Gulf of Mexico on the night of October 18-19, and moved inland over Cedar Key, Florida on the 20th (Climatological Data, October 1941) with 40 to 50 mph winds. The barometric reading at Cedar Key was 29.70 inches as the center passed.

The storm moved across north Florida between Cross City and Gainesville and dissipated. However, it produced very heavy rains, 10 to 20 inches, extending from Cedar Key and Cross City all the way to Jacksonville and St. Augustine and south to Titusville. Extensive damage was done to recently planted truck crops and seed beds. This damage was estimated at several hundred thousand dollars (Climatological Data, October 1941).

1942

The 1942 Atlantic Basin season produced four hurricanes and six tropical storms. Fortunately, none of these ten systems threatened either peninsular Florida or the Florida panhandle.

Personal Recollections

"In 1942, I took a vacation with my family to south Florida. We stopped on the Tamiami Trail to visit an Indian village. My father asked one of the Indian men how they knew when hurricanes were approaching so that they could move to high ground. He answered that they watched the animals, particularly the birds, who were very sensitive to the approach of a severe storm" (C. Hendrix, 1998).

Dr. Roger Reep and Dr. Farol Tomson in the College of Veterinary Medicine, University of Florida, Gainesville indicated that while this had not been proven in a scientific study, they suspected that the animals responded to sounds and changes in air pressure which are beyond human perception—a keener nervous system (Reep, 1998; Tomson, 1998).

1943

Florida was again fortunate in 1943 as there were five hurricanes and five tropical storms in the Atlantic Basin, but not one of these systems affected the state.

1944

Florida suffered only one hurricane in 1944, but no hurricane before or since has been as destructive to the ridge citrus producing area of central Florida as the October 1944 storm.

The Hurricane of October 18-19, 1944

"The only hurricane which ever blew down my wind gauge was the hurricane in October 1944."—Thelma Raley, Winter Haven, Florida, November 14, 1997.

Meteorological Considerations

The 1944 North Atlantic Basin hurricane season featured seven hurricanes and four tropical storms, but only the mid-October hurricane which struck the night and morning of October 18-19 was of major consequence to Florida agriculture. This storm was first noted by ship reports between latitude 17-18°N, longitude 81°W, or about 200 miles east of Swan Island on October 12 (Sumner, 1944B). Subsequently, heavy rains and high winds buffeted Grand Cayman Island on the 13th and on the 14th, heavy individual squalls moved over the island with winds changing suddenly from moderate NNE to strong SE and the heaviest rainfall occurred with a record 24-hour rainfall for the island of 16.04 inches (Table 7-2), part of a total for the storm of 31.29 inches. The highest wind gust reported at Grand Cayman was an extreme gust of 118 mph from the east which coincided with the lowest pressure recorded of 29.06 inches.

The hurricane center made an abrupt 90° turn to the west just south of Grand Cayman, but upon reaching longitude 83°W, it made a second 90° turn, this time to the north and passed over the

TABLE 7-2. METEOROLOGICAL DATA FOR SELECTED STATIONS DURING THE OCTOBER 1944 HURRICANE.

Station	Minimum pressure (inches)	Maximum sustained winds (mph) direction	Wind gust (mph) direction	Rainfall (inches)
Grand Cayman Island	29.06	95	118 E	31.29
Havana, Cuba	28.50	140	163 SSE	—
Dry Tortugas	28.02	120 E	—	—
Sombrero Light	29.25	115 SE	—	—
Fort Myers	29.05	65 ESE	—	5.04
Tampa	28.55	68 NE	100[1]	5.49
Lakeland	28.62	81 E	86	6.83
Orlando	28.94	82 SSE	108 ESE	7.49
Jacksonville	28.94	46 NE	60	9.70

[1]Estimated.
Source: Sumner, 1944B and Climatological Data, October 1944.

Isle of Pines late on the seventeenth and crossed Cuba 10 to 15 miles west of Havana. The highest winds, 140 mph with gusts to 163 mph, were recorded by the national observatory in Havana at 10:00 A.M. on October 18 with a lowest pressure of 28.50 inches (Table 7-2). Continuing north, the calm eye of the hurricane was observed over the Dry Tortugas from 3:00 to 5:00 P.M. on the eighteenth, where the lowest pressure for the storm, 28.02 inches, was recorded.

Maximum sustained winds for the Dry Tortugas were 120 mph from the east between 1:00 and 2:00 P.M., just before the anemometer was blown away.

Leaving the Dry Tortugas, the storm shifted to the north northeast and aimed its fierce 120 mph winds at the Florida gulf coast. Coastal residents, farmers and growers from Naples to Tampa and points north feared the prospects of the worst October hurricane in the state's history. Growers who had firepots and wood piles to protect against freezing temperatures had no defense against the fierce winds. In fact, wood and firepots still in the groves were a liability as the winds turned them into projectiles.

The center of the storm screamed ashore near Nokomis, just south of Sarasota, at approximately 3:00 P.M. EST, October 19, with a low pressure reading of 28.42 inches of mercury and an eye 40 miles in diameter. Fruit trees thrashed and limbs twisted and turned in the wind and many trees were uprooted. At Pine Island in Lee County, it was reported (Citrus Industry, 1944A) that 100% of both fruit and foliage was lost plus considerable damage from salt spray. As the storm progressed to the north from Lee through Polk, Lake, Orange, Marion and Volusia counties (Figure 7-1), grapefruit flew through the air like cannonballs as tops of grapefruit and seedling orange trees were snapped by the ferocious wind gusts. As the storm moved just east of Tampa and Dade City at a forward speed of 20 mph and pointed its fury at Marion County, the Clemmons family hurried their children to an old bank vault for safe keeping. The children were more precious than bank notes (H. L. Clemmons, Jr., 1997).

The hurricane was preceded by several tornadoes as it passed through DeSoto, Hardee and Polk counties. At this point, the central core of the storm was 40 to 60 miles in diameter and required 2 to 4 hours to pass over towns directly in its path (Climatological Data, October 1944). Dangerous winds extended 200 miles to the east of the center and 100 miles to the west, thus encompassing the entire Florida peninsula. Wind gusts of 100 mph or more were felt in Tampa to the west and Orlando to the east. The heaviest wind damage occurred across a 30 mile wide strip through the truck crop and citrus producing areas of the state. The highest winds recorded in the citrus belt were 82 mph sustained and 108 mph gusts at Orlando in Orange County. The heaviest rains were 7.49 inches in Orlando and 9.70 inches in Jacksonville (Table 7-2).

The Damage

"Forrest, you won't like what you will see in your grapefruit grove. You won't be able to put your foot on the ground. You'll have to walk across a carpet of grapefruit."—John T. Lesley to Forrest F. Attaway, Sr., Haines City, Florida, October 21, 1944.

As the strong northeast quadrant of the hurricane passed inland over the citrus and vegetable growing areas of southwest Florida, it then moved north-northeast through the agricultural heartland of the state. Heavy damage was predicted and heavy damage occurred.

FIGURE 7-1. PATH OF THE OCTOBER 18-19, 1944 HURRICANE AS IT PASSED UP THE CENTER OF THE FLORIDA PENINSULA FROM NOKOMIS THROUGH THE RIDGE AREA AND NORTH TO JACKSONVILLE. SOURCE: CLIMATOLOGICAL DATA, OCTOBER 1944.

Florida counties on the stronger east side of the storm were important in Florida agriculture, including from south to north Lee, Charlotte, Hendry, DeSoto, Sarasota, Manatee, Hardee, Polk, Osceola, Lake, Orange, Seminole, Marion, Volusia, Flagler, St.

Johns and Duval counties. Heavy winds and rains caused major damage to the truck crops in the area around Lake Okeechobee, while the high winds whipped central Florida citrus groves unmercifully, with greatest damage to grapefruit and tall seedling orange trees. The greatest loss of citrus was in Sarasota, DeSoto, Hardee, Polk, Orange and Lake counties, with the total loss of all varieties predicted to be 23.7 million boxes out of a total estimated crop of 92.7 million boxes (Table 7-3). The statewide losses estimated by Warren O. Johnson (1944) of the Federal State Frost Warning Service are shown in Table 7-4. In addition, some sections suffered severe tree losses with statewide estimates at 15% for seedling trees but only 1% for budded trees.

Estimates of losses in different sections of the citrus belt were made at the request of the Citrus Industry Magazine by several knowledgeable industry leaders (Citrus Industry, 1944). The earliest official report of damage was a statement issued by the Florida Citrus Commission indicating that Florida had lost approximately 50% of its grapefruit and from 20 to 25% of its oranges. The report further indicated that additional losses might result as weakened fruit dropped to the ground later.

C. C. Commander, General Manager of the Florida Citrus Exchange, indicated that after an extensive trip around the citrus producing areas of the state, he anticipated a loss of 40 to 50% of the grapefruit crop and 15 to 25% of the orange crop, and that it

TABLE 7-3. CITRUS FRUIT LOSSES FROM THE HURRICANE OF OCTOBER 18-19, 1944 ACCORDING TO OFFICIAL GOVERNMENT FIGURES.

Fruit type	October estimate in boxes	Actual production in boxes	Loss in millions of boxes	Percent loss
------------- millions -------------				
Oranges	52	42.8	9.2	17.7
Grapefruit	36	22.3	13.7	38.1
Tangerines	4.7	3.9	0.8	17.0
TOTAL	92.7	69.0	23.7	25.6

Source: Florida Agricultural Statistics Service, 1997.

TABLE 7-4. FRUIT LOSSES TO THE FLORIDA CITRUS INDUSTRY FROM THE HURRICANE OF OCTOBER 18-10, 1944 AS ESTIMATED BY WARREN O. JOHNSON, U.S. WEATHER BUREAU, LAKELAND, FL.

Fruit type	Loss in boxes[1]	Percent loss
Grapefruit	14 million	40
Early and midseason oranges	4.5-5 million	15-20
Valencia oranges	1-1.5 million	5
Tangerines	600,000	—

[1]In 1944, boxes were oranges 90 lb, grapefruit 80 lb, and tangerines 90 lb (Trovillion, 1997).
Source: Climatological Data, October 1944 and Johnson, 1944.

would be necessary to change grading specifications because of scars and blemishes resulting from the wind.W. F. Miller of the War Food Administration (WFA) stated on October 31, 1944 that approximately 25 million boxes of fruit were blown from the trees, representing 43% of the grapefruit crop and 19% of the orange crop. W. L. Waring, Jr., President of Lyons Fertilizer Company in Tampa, Florida rendered a detailed report of its agents in various counties (Table 7-5). In later years, reports from Lyons would prove to be very helpful in describing hurricane damage in the inland and west coast counties.

TABLE 7-5. CITRUS FRUIT LOSSES BY COUNTY FROM THE HURRICANE OF OCTOBER 18-19, 1944 AS ESTIMATED BY W. L. WARING, JR. OF LYONS FERTILIZER COMPANY.

County	Grapefruit	Orange
	----------- % -----------	
Lee, Charlotte and Hendry	40-50	15-20
Sarasota, Manatee, Polk, Hardee, DeSoto, Highlands and Orange	75	25
Hillsborough	25-50	10-15
Pinellas	40-50	15-20
Pasco, Hernando and Citrus	20-25	10
Lake and Marion	25-30	10-15

Source: Citrus Industry Magazine, 1944A.

G. D. Sloan, President of Superior Fertilizer Company in Tampa, Florida reported that, "the most severe losses, with the exception of Pine Island in Lee County, were to be found in Polk County from Frostproof north." His county by county assay of grapefruit losses is found in Table 7-6. He placed the loss of early and midseason oranges at 25%, Valencias 5% and tangerines negligible.

W. H. Klee, Manager of Naco Fertilizer Company, Jacksonville, said that, "55% of grapefruit is on the ground and 20% of oranges, and this loss will be further augmented by approximately an additional 20% of grapefruit and oranges due to fruit now on twisted limbs which will eventually drop, and also due to rots and spoilage." Based on the preseason estimate of 38.1% of the grapefruit and 17.7% of the oranges (Table 7-3), the industry prognostications were not too far off.

Vegetable Damage

With the main force of the storm exerting itself across the principal vegetable areas of Lee, Hardee, Sarasota, Manatee, Polk, Orange, Seminole, Marion and St. Johns counties, and to a lesser extent in the vegetable producing counties well to the east of the storm center such as Dade, Broward, Palm Beach, Martin and St.

TABLE 7-6. GRAPEFRUIT LOSSES BY COUNTY FORM THE HURRICANE OF OCTOBER 18-19, 1944 AS ESTIMATED BY G. D. SLOAN OF SUPERIOR FERTILIZER COMPANY.

County	Percent loss
Manatee	40
Hillsborough, East and North Pasco	negligible
West Pasco	20-25
Polk, Highlands and Hardee	75
Pinellas	35-40
West Lake	25
East Lake, Orange and Volusia	75
St. Lucie, Indian River and Brevard	25

Source: Citrus Industry Magazine, 1944A.

Lucie counties, Florida's vegetable production was severely damaged (Scruggs, 1944). Crops of winter vegetables which would have been shipped from October 20 through January were reduced from 14,474 carloads in 1943-44 to an estimated total of 9,200 carloads (1,200 carloads before November 30, 3,000 carloads in December and no more than 5,000 carloads in January). The heavy fall and early winter plantings in Palm Beach County sustained losses of 75% of the bean crop and 25% of the cabbage crop. In Broward County, young beans, eggplant, peppers and tomatoes suffered a 60 to 75% loss. Vegetable losses were also heavy in Manatee, Sarasota, Hardee, Orange, Alachua, St. Johns, Flagler, and Putnam counties. The overall loss to the vegetable industry was estimated at $15 million. Combined with the loss of approximately $25 million worth of citrus fruit, plus the cost of cleaning up groves, repairing buildings and fences, it was further estimated that the total loss to fruit, vegetable and commercial flower growers might reach $50 million.

W. L. Waring of Lyons Fertilizer, reporting in the Citrus Industry Magazine (1944B) indicated that, "the west coast vegetable crop, except for celery, cabbage and escarole, was practically a total loss. Losses of celery, cabbage and escarole would not run over 25 to 30%. However, around Lake Okeechobee, the loss of all vegetables would be 75 to 80%, except celery at 25%." F. W. Scott, Lyons Fertilizer representative in southwest Florida, stated that, "vegetable growers in this territory suffered extremely heavy losses during the hurricane and many of them will not get started again until they put in their spring crops."

Economic Considerations

The Citrus Industry

As a result of the hurricane and an end-of-season drought, the 1944-45 Florida citrus season had the earliest windup of shipping activities in 15 years and prices for Florida fresh citrus fruit were the highest in 15 years. Florida's 1944-45 weighted auction averages for the standard 1-3/5 bushel box for all districts for the 1944-45 season were oranges $4.48, grapefruit $4.24 and tangerines $4.82. Tables 7-7 through 7-9 show the major increase for oranges, grapefruit and tangerines as a result of the hurricane and drought. Thanks to the

TABLE 7-7. TOTAL SALES OF FLORIDA ORANGES, ALL VARIETIES, IN THE YEARS BEFORE AND AFTER THE HURRICANE OF OCTOBER 18-19, 1944.

Season	Carloads	Weighted average price at auction per 1-3/5 bushel box
1940-41	18,971	$2.35
1941-42	17,133	$2.83
1942-43	13,606	$3.79
1943-44	12,849	$3.89
1944-45[1]	9,022	$4.48
1945-46	8,268	$4.58

[1]Hurricane year.
Source: Wilson, 1941-46.

efforts of Senator Claude Pepper, the price ceilings on Florida tangerines and Temple oranges were raised 53 cents per standard 1-3/5 bushel container to compensate for hurricane losses. The increase was effective November 10, 1944.

Grapefruit, which showed the greatest hurricane losses, felt the greatest reduction in carloads shipped (minus 1,132 cars) and corresponding price increase, from $3.56 to $4.24 per standard box of 1-3/5 bushel. The following year, there was a gain of 1,254 cars of grapefruit shipped and a price decrease from $4.24 to $3.82 per

TABLE 7-8. TOTAL SALES OF FLORIDA GRAPEFRUIT, ALL VARIETIES, IN THE YEARS BEFORE AND AFTER THE HURRICANE OF OCTOBER 18-19, 1944.

Season	Carloads	Weighted average price at auction per 1-3/5 bushel box
1940-41	8,823	$1.91
1941-42	7,325	$2.52
1942-43	5,604	$3.11
1943-44	4,264	$3.56
1944-45[1]	3,132	$4.24
1945-46	4,386	$3.82

[1]Hurricane year.
Source: Wilson, 1941-46.

TABLE 7-9. TOTAL SALES OF FLORIDA TANGERINES IN THE YEARS BEFORE AND AFTER THE HURRICANE OF OCTOBER 18-19, 1944.

Season	Carload	Weighted average price at auction per 1-3/5 bushel box
1940-41	2,717	$2.38
1941-42	2,085	$3.58
1942-43	3,399	$3.60
1943-44	644	$4.46
1944-45[1]	1,465	$4.82
1945-46	2,201	$5.22

[1]Hurricane year.
Source: Wilson, 1941-46.

standard box. Oranges and tangerines were able to hold their price gains. The smaller production caused by the hurricane and drought, coupled with higher prices for both fresh and processed citrus products, resulted in a new record for gross returns to the industry. Without an increase in estimated production of California Valencia oranges from 31 million to 37 million boxes and an increase in Texas grapefruit from 17.7 to 22.4 million boxes, the impact on fruit prices would have been even more dramatic. Before the hurricane, Florida was facing its largest crop in history. Suddenly the hurricane changed prospects dramatically for the Florida citrus grower. Because the United States was in the midst of World War II, approximately 14 million cases of processed Florida citrus products were purchased by the government. On January 18, 1945, a "freeze order" was issued to confine grapefruit juice shipments only to the armed forces as an emergency measure resulting from increased military requirements. On January 30, 1945, the "freeze order" was expanded to include orange juice. Effective February 17, 1945, the War Food Administration (WFA) increased the amount of canned citrus juices to be set aside for the military to the point where production would not meet military requirements and thus no product was available to the consuming public. Due to the wartime situation, the Office of Price Administration (OPA) placed price ceilings on fruit for processing as well as processed products. However, through the Pepper-Stewart Amendment, Senator Claude

Pepper succeeded in getting the Price Control Act amended from the floor of the Senate allowing an upward revision of the price ceilings to compensate for the hurricane damage (Citrus Magazine, 1944A).

Another complicating factor during the war was the inability to export fruit to Europe leaving Canada as the only major export market. This served to depress prices during these tumultuous and difficult times.

Because of the need to replace trees that were destroyed by the storm, nursery sales of all citrus varieties increased in 1945-16 (Table 7-10). While hurricane damage accounted for a small part of this increase, most came from the explosive expansion of the Florida citrus industry after the end of World War II.

Personal Observations

G. W. "Buck" Mann remembers that, "during the '44 hurricane, it blew over 100 big oak trees in Bartow and blocked most of the streets, but it didn't damage many of the houses. The military was at Bartow Airfield then and they came and cleared out the streets. No grapefruit was left on the trees around Winter Haven.

TABLE 7-10. SALES OF GRAPEFRUIT NURSERY TREES TO FLORIDA GROWERS IN THE YEARS PRECEDING AND THE YEAR AFTER THE OCTOBER 18-19, 1944 HURRICANE.

Season	No. trees sold		
	Grapefruit	Orange	Tangerine
1939-40	80,588	403,775	3,441
1940-41	85,954	592,208	6,198
1941-42	64,069	579,809	8,812
1942-43	104,754	533,802	5,876
1943-44	136,637	701,977	7,782
1944-45[1]	125,135	611,854	5,902
1945-46	218,642	794,373	12,244

[1]Hurricane year.
Source: Wilson, 1941-46.

The next year or so, there was another hurricane through Bartow and five trees hit five houses. There weren't so many oak trees left then. It was compulsory to get rid of fever ticks in those days. We had to drive 4,000 head of cattle from a pasture on the south side of Lake Marion in Osceola County west to the Micklejohn pen to be inspected. The cattle were spread out over ten to fifteen thousand acres and normally it would have taken 16 men a week to round them up. However, the hurricane blew them north toward the lake and the crevice to the pen and we had them into the pen in half a day."—G. W. "Buck" Mann, Jr., 1998.

Thinking back to 1944, Dudley Putnam recalls: "I had my first eyeball to eyeball meeting with Mark Edwards up in Pasco County. Mark told me that we would have three freezes and three tropical storms (or hurricanes) every ten years. He was pretty close in those days, except we had more hurricanes than freezes during the 1940s"—Dudley Putnam, July 1998.

1945

The North Atlantic Basin hurricane season of 1945 produced ten tropical storms and hurricanes, three of which affected Florida. These were a June hurricane, originating in the western Caribbean, which diminished to tropical storm force as it crossed Florida, a September tropical storm which brushed the Florida Gulf Coast without reaching hurricane velocity, and a major Atlantic hurricane which affected the entire state from September 15-17.

The Hurricane of June 24-25, 1945

Meteorological Considerations

The first storm of the 1945 season formed in the Western Caribbean Sea between Swan Island and the coast of Honduras during the night of June 19 (Sumner, 1946A). A definite circulation was noted on the 20th about 100 miles west northwest of Swan Island, as the storm moved toward and through the Yucatan Channel into the Gulf of Mexico. At a latitude of approximately 27.5°N, longitude 86.5°W, it made a sharp turn to the northeast and developed winds of hurricane force as it approached the gulf coast of Florida. About 120 miles south of Apalachicola, a hurricane hunter plane estimated winds exceeding 100 mph near the center on June 23. The hurricane then lost force as it reached the coast and slammed inland between Brooksville and Dunnellon at about 4:00 P.M., June 24, and crossed the Florida peninsula exiting into the Atlantic between Daytona Beach and St. Augustine. The circulation remained intact as it crossed Florida with heavy rains and gale force winds. Over the Atlantic, it regained hurricane intensity.

The Damage

The heavy rains which accompanied the storm were considered beneficial as they broke a 12 month drought. However, there was crop damage in some areas due to the very excessive rains. The 24-hour rainfall at Tampa was 10.42 inches which broke the existing record at that station. Lakeland recorded 10.43 inches, Lake Alfred 12.60 inches and Orlando 8.60 inches. Wind damage was minor.

The Hurricane of September 15-16, 1945

"Highlands County was probably the hardest hit section in the state as far as fruit damage is concerned."—Eaves Allison, 1945.

Meteorological Considerations

The hurricane was detected in its formative stages east of the Leeward Islands near 19°N, 56°W. It proceeded on a westerly course and crossed 60°W on September 12, and made a gradual shift to a west northwest course passing north of Puerto Rico during the day on the thirteenth and north of the Dominican Republic on the evening of that same date with 150 mph wind gusts being reported at Turks Island. Beginning a slow curve to the northwest, it passed through the Bahamas during the night of September 14-15 and howled ashore along a 100 mile wide front on the southeast Florida coast with the center passing over the north end of Key Largo at about 3:30 P.M. on the fifteenth (Figure 7-2). The storm's next tar-

FIGURE 7-2. PATH OF THE SEPTEMBER 15-16, 1945 HURRICANE FROM KEY LARGO THROUGH HOMESTEAD AND THE LAKE OKEECHOBEE AREA AND NORTH THROUGH CENTRAL FLORIDA TO ST. AUGUSTINE AND THE FLORIDA-GEORGIA BORDER. SOURCE: CLIMATOLOGICAL DATA, SEPTEMBER 1945.

get was the airbase at Homestead which experienced a 50 minute period of calm and a barometer reading of 28.09 inches at 4:55 P.M. on the fifteenth. Two hours earlier the lighthouse at Carysfort Reef had experienced the highest sustained wind velocity reported during the hurricane's passage through Florida at 138 mph from the southwest. Carysfort Reef reported a 20 minute lull and a barometer reading of 28.15 inches. The highest wind velocity at Carysfort Reef was unusual in that it was a few miles west of the storm center's point of landfall, the highest wind velocities being normally reported east of the center. At the Homestead Airbase, winds were estimated at 125 to 130 mph with gusts to 150 mph, but this was after the wind instruments had failed (Sumner, 1946A).

Leaving the coastal area, the hurricane crossed the Everglades to Lake Okeechobee where Belle Glade felt its full ferocity. Shifting to a northerly course, the storm hammered central and north Florida before reaching the east coast again on September 16 between St. Augustine and the Florida-Georgia state line. Fortunately, winds had diminished to 75 mph or less by the time the center reached the heart of the citrus belt in Polk, Lake, Orange and adjacent counties. Thus, major damage to the state's citrus crop was narrowly averted. However, the citrus industry of the 1990s, much of which is located further south, would suffer significant damage if such a hurricane followed that course today. At it was, losses were extensive in the Highlands County communities of Lake Placid, Sebring and Avon Park and Frostproof in southern Polk County (Citrus Magazine, 1945). The path of the storm when it struck the Florida east coast was about 100 miles wide, but as it turned north it had narrowed to 40 to 50 miles through Highlands, DeSoto and Hardee counties. The storm's width narrowed further to only 30 to 40 miles as it approached Lakeland and Orlando. Meteorological data are shown in Tables 7-11 and 7-12.

The Damage

Agricultural losses were estimated to be $10,000,000 for crops, $2,000,000 for ornamentals, shrubs and timber and $15,000 for livestock. Animal losses included 400 cattle and 7,000 chickens (Sumner, 1946A). Approximate citrus losses were 3 million boxes of grapefruit and one million boxes of oranges. Most of the citrus losses were in the corridor from Lake Placid north to Frostproof and were the heaviest ever suffered in Highlands County up to that time (Citrus Magazine, 1945).

Warren O. Johnson (1945), who was stationed at the U.S. Weather Bureau in Miami when the hurricane passed, reported a 100% loss of the on-tree lime crop in Dade County. On a trip to the

TABLE 7-11. METEOROLOGICAL DATA FOR THE FLORIDA HURRICANE OF SEPTEMBER 15-16, 1945 WITH STATIONS LISTED FROM SOUTH TO NORTH.

Station	Minimum pressure (inches)	Maximum sustained winds (mph) direction	Wind gusts (mph)
Carysfort Reef Light	28.15	138 SW	—
Homestead	28.09	—	150[1]
Miami	29.17	107 SE	—
Belle Glade	29.51	75 ENE	79
Clewiston	29.42	75 E	83
Naples	29.46	60 SSW	—
Fort Myers	29.09	—	79
Egmont Key	29.62	52 N	—
Lakeland	29.33	66 NNE	—
Tampa	29.55	40 NNE	50
Orlando	29.29	66 SE	78
Melbourne	29.54	69 SE	72
Sanford	29.29	55 NE	77
DeLand	29.24	44 NNE	55
Cape Canaveral Light	29.53	75 S	—
Daytona Beach	29.27	68 ESE	—
Jacksonville	29.37	67 N	78

[1]Estimated.
Source: Sumner, 1946A.

north end of Key Largo, he observed that half of the lime trees were "blown out of the ground" and the avocados and mangos severely damaged with the remainder of the season's crop a total loss. Many trees were uprooted or broken off. Truck crop damage was estimated to be 75 to 100% due to the heavy rains.

The Lyons Fertilizer Company compiled the following reports from their fieldmen at key locations in the citrus belt (Allison, 1945):

Southwest Florida and Highlands County: "Highlands County was probably the hardest hit section in the state by the recent storm as far as fruit damage is concerned. Marsh seedless grapefruit suffered the heaviest loss with many growers estimating that their loss

TABLE 7-12. RAINFALL TOTALS ASSOCIATED WITH THE HURRICANE WHICH PASSED THROUGH FLORIDA SEPTEMBER 15-16, 1945. LISTINGS ARE FROM SOUTH TO NORTH ALONG THE PATH OF THE HURRICANE.

	Rainfall (inches)	
Station	September 15	September 16
Homestead	1.25	4.65
Kendall	0.00	6.15
Miami	3.81	0.11
Canal Point	2.37	2.48
Belle Glade	0.22	4.86
Everglades City	0.28	1.38
Clewiston	1.08	3.42
Moore Haven	0.64	3.30
Okeechobee City	0.45	3.00
LaBelle	0.38	3.70
Lake Placid	0.51	5.74
Fort Pierce	0.30	3.75
Avon Park	1.70	2.30
Mountain Lake	0.20	3.52
Bartow	0.34	2.52
Vero Beach	0.66	2.78
Melbourne	1.80	7.04
Winter Haven	Trace	2.01
Lake Alfred	0.46	2.02
Orlando	1.64	8.43
New Smyrna Beach	0.00	6.55
Daytona Beach	0.09	3.75
Jacksonville	0.40	6.61

Source: Climatological Data, September 1945.

would be as high as 70%. Oranges and tangerines suffered less damage, but the overall picture is pretty bad. The hurricane did very little damage in other sections of this territory. However, heavy and continued rains in the Ruskin area have destroyed tomato and pepper seed beds and have delayed field operations to a considerable extent. Further south conditions have not been so bad, with less precipitation in the Palmetto-Bradenton-section. Celery plantings are proceeding as usual in the Sarasota area with favorable soil moisture conditions."—Eaves Allison.

Polk County: "The recent hurricane did some damage in this county, and most of the fruit loss occurred in the ridge section where considerable grapefruit was lost and some orange loss. However, from close observation it appears that the damage might not be as heavy as was at first thought. Marsh seedless grapefruit was hit hardest, and growers were unable to do very much of a salvage job because of fruit being too immature. We are now hoping that we can get through the remainder of the hurricane season without further damage."—J. M. (Jim) Sample.

Hillsborough County and Pinellas County: "Groves in many places were showing water damage so you can see that we were worried when we received news that the hurricane was approaching the citrus section of Florida. However, we were very fortunate as the wind blew very little and we did not get more than a couple of inches of rain."—C. S. (Charlie) Little.

West Central Florida: "The hurricane that blew over part of Florida several weeks ago did some damage in parts of the state but we were very fortunate in this territory and I am very happy to report that we had absolutely no damage."—E. A. (Mac) McCartney.

North Central Florida: "The recent hurricane did some damage in this territory—knocking off some fruit and of course more will drop as a result of thorn-pricks and the general hammering around that had to be withstood."—V. E. (Val) Bourland.

Damage to Tropical Fruits

The September 1945 hurricane blasted a 40 mile wide path of destruction from South Miami to slightly south of Florida City, an area including 75 to 90% of the production of avocados, limes and mangos in the state. Winds of hurricane force in this area lasted for about four hours, with the lull lasting 45 to 50 minutes.

The damage to avocados varied with the size and height of the trees, and to a lesser extent on variety and condition. Australian pine windbreaks provided some protection in winds of 90 to 100 mph, but at higher velocities the windbreak trees blew over on the avocado trees causing more damage than the wind. Tall trees suffered more damage than smaller trees. Damage was classified in three types: root damage, bark damage and breakage of limbs and loss of foliage. Root damage was severe in trees that were not blown over as well as those which were blown over. Bark damage was largely due to sunburn before the trees could be pulled back up, but also the result of damage from twisting of the limbs and trunk. Breakage of limbs and loss of foliage resulted from the winds which broke limbs as large as 10 to 12 inches in diameter. On the average, the cost of rehabilitating large trees varied from $40.00 to $140.00 per acre, with the largest single cost for hauling out the brush and debris.

Damage to limes varied with rootstock. Limes on lemon rootstocks were blown over easily whereas trees on grapefruit rootstock were more resistant to the wind. Damage to tree canopies was severe due to twisting and splitting of limbs. Approximately 50% of the foliage was lost.

The damage to mangos was probably more severe than to any other type of tropical fruit. Mango trees with limbs of 14 to 16 inch diameter were either broken off three to five feet above the ground, blown over, or in some cases blown totally out of the ground. Small trees fared much better than large trees.

Personal Observations

Bill Krome (1997) recalls the 1945 hurricane in the Homestead area:

"From 1935 to 1945, we suffered no great damage from storms, although the 1940 freeze did more damage than most hurricanes. I graduated from the University of Florida and by 1945 was in charge of our grove operations. We had about 200 acres, mostly avocados. The hurricane of September 15, 1945 was comparable in strength to Hurricane Andrew in 1992. It is the only time I felt any anxiety about the safety of our house. During the height of the storm, the house shuddered continually, which it had never done before. The eye passed mostly north of us but we had a short lull, during which I was struck by a noxious smell. We found out later it was from the planes and equipment burning at the Richmond Base, seven miles away.

"As in the '26 hurricane, all the trees in the groves were overturned. I had two crews, 20 German war prisoners and 20 Bahamians under federal contract. We started resetting the trees immediately after the storm and continued until January 15, 1946. By that time, the tree roots that had remained unbroken underneath the overturned trees had taken such a set that they broke when we put the trees back up.

"I should like to mention that while both Germans and Bahamians worked well, the German boys were the best. I had the first chain saw in the area, a big, two-man machine made by Mercury outboard motors. The Germans regarded it as an honor to be assigned to operate that big saw. On one occasion, the German crew encountered a rattlesnake. For some reason, the snake was quite active and behaved in what we considered to be an aggressive manner. I picked up a rock about the size of my two fists and cautiously approached the snake, hoping to make it stop and coil. When I got as close as I dared, I tossed the rock at the snake. What happened then was almost unbelievable. The snake struck the rock as it came through the air; the impact must have broken the snake's skull or neck. It writhed uncontrollably and died before our very eyes."

Buck Mann recalls the backlash from the hurricane in Osceola County:

"We had been suffering from a drought and the Kissimmee River was no more than ankle deep. My sisters and I rode our horses from the east side of the river over to where River Ranch Acres is today, then the storm came. When we were finally able to go back, we had to swim our horses to get to the east side. I have seen the river run backward. It was two miles wide during one hurricane in the 40s."—G. W. "Buck" Mann, Jr., 1998.

With the succession of hurricanes in the 1940s, Dudley Putnam remembers, "the fruit trees being on the ground in a circle. With oak trees down all over Bartow, German war prisoners from the airbase came with derricks to lift the trees"—Dudley Putnam, 1998.

1946

The Hurricane of October 7-8, 1946

"As usual, south Polk and Highlands counties were a pet area for the wind to blow strongest during a hurricane."—R. L. (Bob) Padgett, Citrus Industry Magazine, 1946.

Meteorological Considerations

The October hurricane began as a tropical storm with 50 mph winds just off the northeast coast of Belize, near 18°N, 87°W, during the night of October 5-6. It moved on a northeast course and during the late afternoon of the 6th, winds of 85 mph were reported by both ships and aircraft. The central pressure at this time had dropped from 29.68 to 29.32 inches (Sumner, 1946B).

During the night of October 6-7, the storm crossed the tip of western Cuba, with reports of wind gusts to 112 mph, and emerged into the Gulf of Mexico west of the Dry Tortugas where an extreme wind velocity of 84 mph from the south was reported at 12:30 P.M. on October 7. Ships west of the Tortugas reported winds exceeding 100 mph and a reconnaissance plane flying into the center estimated winds of 132 mph as the storm moved inexorably on a north northeast course toward the Florida west coast. At this point, the potential damage to the Florida citrus crop appeared enormous, but fortunately for the Florida grower the storm began to lose its intensity as evidenced by reports from the reconnaissance plane. By 4:18 P.M. winds had diminished to less than 100 mph with the center located 100 miles west southwest of Fort Myers.

With the storm continuing to advance toward Tampa Bay (Figure 7-3), the ring of hurricane force winds was destroyed at the surface leaving a poorly defined and relatively flat central area surrounded by a broad circulation of gale force winds which gradually decreased from 40 to 50 mph to 15 to 20 mph, then gradually increased again without any relatively calm period intervening (Sumner, 1946B). Some stations did report a calm period lasting for up to one hour, but without the rapid wind shifts characteristic of the eye of a hurri-

FIGURE 7-3. PATH OF THE OCTOBER 7-8, 1946 HURRICANE FROM THE FORT MYERS-PUNTA GORDA AREA NORTH THROUGH TAMPA BAY AND ACROSS THE STATE TO NEAR JACKSONVILLE. SOURCE: CLIMATOLOGICAL DATA, OCTOBER 1946.

cane. However, while wind speeds remained less than hurricane force at the surface, winds a short distance above the surface continued to approach hurricane force and were reported to twist and shear the taller pine trees. The maximum winds along the Florida west coast, reported from stations in the Fort Myers-Punta Gorda section to the right of the storm center, were 75 mph, but when the actual center passed over Tampa at midnight, the maximum wind was only 47 mph (Table 7-13). The forward movement of the hurricane accelerated from 14 mph to over 30 mph as it passed over Florida into southeastern Georgia. Rainfall amounts as the hurricane traversed the state varied from 1.06 inches at Punta Gorda to 2.21 inches in Tampa to 6.20 inches in Ocala (Table 7-14). The duration of gale force winds was 9 hours at Miami, 5 hours at Fort Myers, 3.5 hours at Okeechobee City and 4 hours at St. Petersburg.

The Damage

When the October 1946 hurricane crossed western Cuba with wind gusts to 112 mph and pointed its destructive force toward the Florida west coast, growers and farmers feared major damage, but

TABLE 7-13. METEOROLOGICAL DATA FOR THE FLORIDA HURRICANE OF OCTOBER 7-8, 1946 WITH STATIONS LISTED FROM SOUTH TO NORTH.

Station	Minimum pressure (inches)	Maximum sustained winds (mph) direction	Wind gusts (mph) direction
Dry Tortugas	—	84 S	—
Key West	29.50	38 SW	—
Miami	29.61	52 SE	—
Fort Myers	29.31	58 ESE	80 SSE
Okeechobee City	29.53	64 SSE	—
Punta Gorda	29.16	75 SSE	—
Tampa	29.14	47 W	—
St. Petersburg	29.16	62 NE	—
Ocala	29.12	—	—
Daytona Beach	29.33	65 SSE	—
Jacksonville	29.21	33 ESE	60

Source: Monthly Weather Review, December 1946.

TABLE 7-14. RAINFALL AT SELECTED STATIONS DURING THE HURRICANE OF OCTOBER 7-8, 1946, WITH STATIONS LISTED SOUTH TO NORTH.

Station	Rainfall (inches)
Key West	2.01
Homestead	2.44
Ft. Myers	1.04
Belle Glade	0.87
LaBelle	0.77
Fellsmere	2.32
Punta Gorda	1.06
Bradenton	2.61
Tampa	2.21
St. Petersburg	2.73
Lakeland	1.91
Orlando	2.73
Ocala	6.20
DeLand	1.61
Jacksonville	2.55

Source: Climatological Data, October 1946.

when the hurricane began to dissipate before making landfall, "many in Florida attributed it to the miraculous intervention of Divine Providence in answer to statewide invocations" (Norton, 1948). The greatest damage was to the citrus crop in Lee, Charlotte and Sarasota counties where the estimated loss was 10% of the oranges and 15 to 20% of the grapefruit. However, as these three counties were not major citrus producers in 1946, these losses were small when compared to the total crop. Losses to the total crop were estimated at only 2%, or 2,000,000 boxes at $2.50 per box for a total of $5,000,000 (Sumner, 1946B). The Lyons Fertilizer Company compiled the following reports from their fieldmen in various citrus producing parts of the state (Padgett, 1946).

Polk County: "The recent hurricane had everyone much worried, but I am glad to report that we were damaged only very lightly. There was some grapefruit blown from the trees and some oranges were thorn pricked and bruised to the extent that we are having some droppage."—J. M. (Jim) Sample.

North Central Florida: "There was some damage to our fruit crop by the hurricane, but we are so thankful that the loss was light that we do not even discuss the few fruit on the ground."—V. E. (Val) Bourland.

Southwest Florida: "The hurricane missed us, but some fruit was blown off in the Wauchula-Arcadia area and in the Fort Myers section. However, the damage is light. Rains accompanying the disturbance were heavy in some areas and light to medium in others."—Eaves Allison.

South Polk County and Highlands County: "As usual, it seems that this section is the pet area for the wind to blow strongest during a hurricane. It seems that we had as much damage as any other part of the fruit belt. While the loss of grapefruit was not heavy, we do estimate that we lost 5 to 25% of the crop which was already light. The orange loss will come from fruit that was thorn pricked and will drop from this cause. This is especially true in the case of Hamlin oranges with their very delicate skin."—R. L. (Bob) Padgett.

Hillsborough County and Pinellas County: "The hurricane did only slight damage in this section."—C. S. (Charlie) Little.

West Central Florida: "The recent hurricane did practically no damage in this territory."—E. A. (Mac) McCartney.

Cucumbers: The cucumber acreage in Lee County was the same as in the 1945-46 season, but yields were expected to be smaller due to damage from the October hurricane.

The Tropical Storm of November 1-2, 1946

Meteorological Considerations

This storm developed from a tropical wave near 21°N, 71°W, or just off the northern coast of the Dominican Republic, on October 31. A fall in pressure was noted over the Bahamas on November 1, and a reconnaissance plane located signs of a weak but poorly defined center (Sumner, 1946B). Winds of 39 to 46 mph were reported at West End, Grand Bahama Island, from where the storm moved on a northwesterly course and crossed the Florida coastline near Palm Beach at 4:30 P.M. on November 1. Though wind velocities of 30 to 40 mph were reported along the Florida east coast from

Miami to Daytona Beach, winds over Florida never exceeded minimal tropical storm force. The rainfall at representative stations is shown in Table 7-15.

After crossing the Florida coastline, the center of the storm continued northwest until it reached a point between Lakeland and Orlando at about 7:30 A.M. on November 2, after which it deteriorated rapidly with the remnants drifting back into the Atlantic Ocean near Jacksonville.

The Damage

Florida suffered no wind damage from the disturbance, but floods caused by the heavy rains resulted in the loss of 50 to 60% of the 10,000 acres of potatoes, celery and cabbage in the Lake Okeechobee area. No damage was reported to vegetables farther north in the state (Florida Grower, 1946).

1947

The 1947 season featured hurricanes which struck Florida on September 17-18 and October 11-12. In addition, there were tropi-

TABLE 7-15. RAINFALL AT SELECTED STATIONS DURING THE TROPICAL STORM OF NOVEMBER 1-2, 1946.

Station	Rainfall (inches)
Belle Glade	6.95
Canal Point	4.71
St. Lucie Lock	5.19
South Bay	3.88
Felda	2.65
Ft. Myers	2.63
West Palm Beach	2.53
Moore Haven	2.12
Miami	1.16
Ft. Lauderdale	1.08
Orlando	0.50
Lakeland	0.07

Source: Climatological Data, November 1946.

cal storms which affected the state on September 23 and October 6. The September hurricane was a very important event, the October hurricane less important. The tropical storms were not significant.

The Major Florida Hurricane of September 17-18, 1947

"The overall damage to the citrus crop was greater than first estimated. Some growers with lakefront groves are reporting 25 to 40% of their grapefruit on the ground."—R. L. (Bob) Padgett, Citrus Industry Magazine, October 1947.

Meteorological Considerations

This was a classic "Cape Verde" hurricane. It was first reported by the Pan American Airways station at Dakar, French West Africa as a low pressure area, over the country known today as Senegal, on September 2. As the low moved west into the Atlantic Ocean, it became a tropical depression and drenched Dakar with 3.36 inches of rain (Sumner, 1947). The circulation was followed until it reached the Cape Verde Islands on September 5, but was lost at this point due to the lack of ship reports. It was later reported by the S. S. Arakala as a well-developed tropical storm near 15°N, 49°W.

At this time, both Army and Navy aircraft began to track the storm closely as it followed a west northwest course to a position just east of Abaco Island in the Bahamas on September 15 where it came to an almost complete standstill for 24 hours, with the observatory at Hopetown, Abaco reporting winds of 160 mph. Prior to this pause, the hurricane's west northwest movement appeared to put it on a course for northeast Florida. However, when it slowly resumed forward movement, the hurricane had shifted to a west southwesterly direction and aimed its ferocious category 4-5 winds directly toward Fort Lauderdale. The hurricane center shrieked ashore between Fort Lauderdale and Lake Worth shortly after noon on September 17, with a peak gust of 155 mph at Hillsboro Light near Pompano at 12:56 P.M. (Table 7-16). The lowest reliable pressure reading, 27.97 inches, was recorded at this same time (Sumner, 1947).

Winds of 100 mph or more from this gigantic hurricane were felt along the Florida east coast from the northern suburbs of Miami 70

TABLE 7-16. METEOROLOGICAL DATA FOR THE MAJOR FLORIDA HURRICANE OF SEPTEMBER 17-18, 1947 WITH STATIONS LISTED FROM EAST TO WEST.

Station	Minimum pressure (inches)	Maximum sustained winds (mph) direction	Wind gusts (mph) direction
Carysfort Reef Light	29.29	76 SW	—
Fort Lauderdale	28.22	—	127 S
Hillsboro Light	27.97	—	155 ENE
Miami	28.71	90 SSW	—[1]
West Palm Beach	29.02	100 NNE	110
Daytona Beach	29.86	43 ENE	60
Everglades City	28.81	—	60-65
Moore Haven	29.09	92 NE	—
Naples	28.80	105 NW	—
Fort Myers	28.82	90 NNW	110 NNW
Sanibel Light	28.67	—	120 S
Lakeland	29.64	44 ENE	75 ENE
Tampa	29.53	38 NE	—
Apalachicola	29.69	67 SE	—
Pensacola	29.54	91 SE	—

[1]Aneometer cups blew away.
Source: Summer, 1947.

miles to well north of Palm Beach, while winds of hurricane force stretched from Cape Canaveral 240 miles south to Carysfort Reef Lighthouse. Such a wide expanse of destructive winds from a storm moving at right angles to the coastline classifies the hurricane of September 17-18 as one of the most severe storms to strike the United States mainland. Fortunately, by passing between Miami and West Palm Beach, it spared those cities from the level of destruction they might have felt otherwise. However, Pompano, Deerfield Beach, Boca Raton and Delray Beach were not so fortunate.

After crossing the coastline, the hurricane roared directly west at 10 mph until the center emerged into the Gulf of Mexico at approximately 10:00 P.M. on September 17, 1947 after sweeping the area from Lake Okeechobee to Naples to Fort Myers with the full force of its winds and heavy rains (Figure 7-4). Sanibel Light recorded

FIGURE 7-4. PATH OF THE SEPTEMBER 17-18, 1947 HURRICANE FROM FORT LAUDERDALE ACROSS THE EVERGLADES TO NAPLES. SOURCE: CLIMATOLOGICAL DATA, SEPTEMBER 1947.

gusts of 120 mph while Fort Myers endured sustained winds of 90 mph with gusts to 110 mph (Table 7-16), illustrating the intensity of the hurricane even after passing across the entire south Florida peninsula. The farms and pastures, already saturated from previous rains, collected up to 8.72 inches of rain at Ft. Myers from the hurricane (Table 7-17). At Naples, a one hour lull was felt between 9:00 and 10:00 P.M. on the seventeenth, with the wind dropping to 12 mph at 9:45 P.M., as the center moved offshore and sped toward a rendezvous with the Mississippi and Louisiana coasts on September 19. Florida growers and farmers began to evaluate the damage.

TABLE 7-17 RAINFALL AT SELECTED STATIONS DURING THE HURRICANE OF SEPTEMBER 17-18, 1947 WITH STATIONS LISTED FROM EAST TO WEST.

Station	Rainfall (inches)
Ft. Lauderdale	2.90
Pompano	6.00
Miami	3.82
West Palm Beach	4.75
Belle Glade	7.30
Moore Haven	5.67
Okeechobee	4.10
Lake Placid	6.00
Bartow	6.71
Lakeland	5.05
Everglades City	3.84
Naples	3.17
Ft. Myers	8.72
Tampa	4.08

Source: Climatological Data, September 1947.

The Damage

Crop damage from the hurricane of September 17-18 was estimated at $10,000,000 including 6,000,000 boxes of citrus (Climatological Data, September 1947). Damage to trees, shrubbery and tropical ornamentals was estimated at $2,000,000. The citrus losses were heaviest to grapefruit.

Immediately after the hurricane, Commissioner of Agriculture Nathan Mayo visited the storm damaged areas and estimated the loss of 95% of the citrus crop in the Fort Myers area, and 50 to 75% loss in adjacent counties (Florida Grower, 1947). Further north to DeSoto County, he estimated that 10% of the oranges and 35% of the grapefruit crop was lost. In the vegetable growing area around Lake Okeechobee, he found flood waters up to 5 feet deep and predicted the complete loss of the first plantings of celery, beans and potatoes, and a serious setback to 35,000 acres of sugarcane. With the exception of the Fort Myers area, the Indian River section had the heaviest citrus losses. In St. Lucie County, a combination of wind and high water killed more than 35,000 trees, and gale force winds destroyed an estimated 30 to 50% of the grapefruit crop and 5 to 20% of the orange crop. The rain total from the hurricane was 5.95 inches. Damage from bruises and wind scar was widespread, and lower fresh fruit grades were predicted in the Indian River area. However, industry spokesmen indicated that with a pre-season estimate of 49.5 million boxes of oranges and an expected 26 million boxes of grapefruit, the losses from the hurricane would not severely impact the industry. William G. (Bill) Strickland, manager for the Indian River District of the Florida Citrus Exchange, estimated that the Indian River District had lost a third of its total citrus crop in the September 17-18 hurricane (Citrus Magazine, 1947).

Groves in inland counties, particularly along the ridge, were much more fortunate (Citrus Industry, 1947). It was predicted that no more than 1% of the fruit was lost or damaged in Lake and Orange counties, 1% of the oranges and 10% of the grapefruit in northern Polk County, 3 to 5% of the oranges and 20 to 25% of the grapefruit in southern Polk County. In DeSoto and Highlands counties, the orange loss was expected to be 5% on budded groves to as much as 40% in taller seedling groves, while 20% of the grapefruit in protected locations would be lost, but up to 60% in locations exposed to the winds. Lyons Fertilizer Company fieldmen made the following reports:

Southwest Florida: "As far as our fruit losses are concerned, some localities were hit hard while others came through in good shape. DeSoto County was suffering from excessive rains before the hurricane so many groves are showing serious water damage. Fruit losses were quite heavy with many old seedling trees blown over."—Eaves Allison.

South Polk, Highlands and Hardee counties: "The overall damage to the citrus crop is greater than first estimated. Some growers with lakefront groves are reporting 25 to 40% of their grapefruit on the ground. Keen Fruit Corporation in Frostproof reports some trees blown over and up to 20% grapefruit loss and that rains had been excessive and damaging. Mr. Mark Smith, a prominent Hardee County grower reports heavy damage to large seedling trees."—R. L. (Bob) Padgett.

Polk County: "The hurricane passed Polk County to the south doing no great damage. In the Winter Haven area, wind gusts reached 65 mph, but a check of groves showed only 3 to 5% of grapefruit and less than 1% of oranges and tangerines on the ground. However, further south losses as high as 10 to 15% were indicated. No estimate could be made on fruit which was bruised and thorn-pricked and may drop later."—J. M. (Jim) Sample.

Hillsborough County and Pinellas County: "We have completed a thorough check to determine just how much fruit and tree damage we actually had, and have concluded that our losses were not too severe. About 10% would cover the loss in even the hardest hit groves. There is more grapefruit on the ground in Pinellas County than in Hillsborough County."—C. S. (Charlie) Little.

West Central Florida: "The recent hurricane gave everyone a scare, but now that it has passed over and we have had the opportunity to check the damage, we are glad to report that the loss to citrus is negligible."—E. A. (Mac) McCartney.

North Central Florida: "We suffered only a very slight loss of fruit from the hurricane."—V. E. (Val) Bourland.

The final pickout for the 1947-48 season (Florida Agricultural Statistics Service, 1948) was 57.530 million boxes of oranges (not including Temples), 29.3 million boxes of grapefruit and 4.44 million boxes of specialty fruit for a record total of 91.27 million boxes. Even such an intense hurricane as struck the state September 17-18 did not slow the explosive growth of the Florida citrus industry.

The Florida Panhandle: "Hitting the gulf coast from Santa Rosa County, Florida to New Orleans with an eye 25 miles in width, the hurricane did extensive damage to crops in the west Florida area" (King, 1972).

With the heaviest summer rains in south Florida since 1930, and memories of the horrible disasters in 1926 and 1928, the advisories from the hurricane center prompted some 15,000 residents to evacuate the Lake Okeechobee area. Some went as far as Lake City and Jacksonville. However, the levee held, not a cup of water came over the top or leaked through the 85 mile barrier (Hanna and Hanna, 1948), although serious flooding resulted from the heavy rains during and after the hurricane.

Flood Damage

Officials of the Everglades Drainage District surveyed the 11 counties of that district and estimated agricultural damage for each county (Table 7-18). Their report, published on December 10, 1947 (Everglades Drainage District, 1947) indicated that these estimates were for flood damage only and did not include wind damage. At that time, the 11 county area annually produced $25 million in value of vegetables shipped by rail alone, nearly $9 million in citrus fruits, $15 million in sugarcane and $12 million in cattle, poultry and dairy products.

Personal Recollections

Pete Spyke (1998) writes that:

"My grandfather bought his first grove in Davie in 1947. The trees were under a year old when the '47 storm hit. There was a barn at the grove built from corrugated aluminum. The barn was on a couple of feet of fill. That storm caused so much flooding that the water kept rising in Davie for days after the storm as the glades drained through the South New River drainage. The water was sufficiently acid that it etched the aluminum of the barn, so we had a permanent record of the maximum elevation. Also, there were lines where the water level would drop ¼ inch or so and then pause long enough to etch a watermark in the aluminum.

"It always made an impression on me to stop and study those lines. As I looked out across the grove, I can safely say that the only dry ground must have been on the sand or rock ridges. EVERYTHING else was under water, for a long time. The entire southern peninsula must have been covered by slowly-moving water.

"Newer residents cannot understand what happened in those days. It's true that we've modified the glades irretrievably, but we have also eliminated the possibility of another flood of that magnitude. High and dry feet allow people to have a higher social conscience. If they ever had to deal with water levels such as those indicated on the old barn, I suspect they would have an entirely different attitude."

TABLE 7-18. AGRICULTURAL LOSSES IN 11-COUNTY EVER-GLADES DRAINAGE DISTRICT FROM FLOODS ASSOCIATED WITH SEPTEMBER 1947 HURRICANE (FIGURES ARE IN 1947 DOLLARS).

Type of loss	Amount
Broward County	
Citrus	11,066,000
Vegetables	925,000
Beef cattle and related losses	534,500
Dairy cattle and related losses	958,200
Miscellaneous (damage to 2,000 farm buildings, nurseries, sod grass farms, etc.)	2,872,000
TOTAL	$16,477,350
Collier County	
Crop loss	$ 2,000
TOTAL	$ 2,000
Dade County	
Citrus and tropical fruits	35,000
Vegetables and seed beds	125,000
Dairy cattle and related losses	5,049,200
Beef cattle and related losses	150,000
Poultry and related losses	475,000
TOTAL	$ 5,834,200
Glades County	
Vegetables and seed beds	6,000
Livestock, dairy and poultry	725,000
Improved pasture	400,000
Sugarcane	200,000
Ramie	15,000
TOTAL	$ 1,346,000
Hendry County	
Citrus	100,000
Vegetable	30,000

Source: Everglades Drainage District, 1947.

TABLE 7-18. AGRICULTURAL LOSSES IN 11-COUNTY EVER-GLADES DRAINAGE DISTRICT FROM FLOODS ASSOCIATED WITH SEPTEMBER 1947 HURRICANE (FIGURES ARE IN 1947 DOLLARS).

Type of loss	Amount
Sugarcane	374,400
Livestock and related losses	175,000
TOTAL	$ 679,400

Highlands County

Citrus	775,000
Vegetables, seed beds, etc.	500,000
Livestock	5,000
Improved pasture	570,000
Dikes, etc.	110,000
TOTAL	$ 1,960,000

Martin County

Citrus	170,000
Vegetables	1,198,400
Livestock and poultry	555,000
Dikes, etc.	300,000
TOTAL	$ 2,223,400

Monroe County

(No estimate available)

Okeechobee County

Cattle, pastures and related losses	4,000,000

Palm Beach County

Citrus and tropical fruits	130,000
Vegetables and seed beds	2,820,000
Sugarcane	1,250,000
Ramie	390,500
Dairy cattle and related losses	60,000

Source: Everglades Drainage District, 1947.

TABLE 7-18. AGRICULTURAL LOSSES IN 11-COUNTY EVER-
GLADES DRAINAGE DISTRICT FROM FLOODS ASSOCIATED WITH
SEPTEMBER 1947 HURRICANE (FIGURES ARE IN 1947 DOLLARS).

Type of loss	Amount
Beef cattle and related losses	267,000
TOTAL	$ 4,917,500
St. Lucie County	
Citrus	382,000
Vegetables	3,960,000
Livestock and related losses	2,610,000
TOTAL	$ 6,925,000
GRAND TOTAL AGRICULTURAL LOSSES IN ELEVEN COUNTIES	$ 44,391,850

Source: Everglades Drainage District, 1947.

At Fort Pierce, Reuben Carlton (1998), reminiscing about the 1947 hurricane, remembers that, "the family all went to mother's two story house at sixth street and Delaware. At midnight, I decided to go home and see about the house. As I pulled up with the car facing into the east wind, the debris was flying by and the wind was whistling with an awful noise."

West of Fort Pierce, Alto "Bud" Adams (1998) remembers that, "state highway 70 was under a foot of water with cows up and down the road. Cows had to be able to lie down, they couldn't stand for a long period of time like horses, so the ranch had to lease land inside the drainage district so the cows could lie down. Some also got into groves."

The Hurricane of October 11-12, 1947

Meteorological Considerations

This hurricane developed on October 9 as a tropical storm in the southwestern Caribbean Sea, just north of Panama, near 15°N, 82°W. Its initial movement took it on a northward course passing just east of Cape Gracias, Nicaragua (Climatological Data, October

1947) after which it crossed western Cuba. The storm intensified to hurricane force on October 11, recurved to the northeast and smashed across southern Florida from Cape Sable to Pompano on the night of October 11-12. As this part of the state had already felt the wrath of the September hurricane, very little additional wind damage occurred.

Before reaching the Florida coast, the maximum wind velocity of the hurricane was estimated at 150 mph by an observer in the Dry Tortugas. As the eye of the hurricane moved off the east coast of Florida, the calm lasted from 3:30 to 4:30 A.M. at Hillsboro Lighthouse with a low pressure reading of 29.27 inches recorded at 2:45 A.M.

The Damage

While there was no significant wind damage from this hurricane, the heavy rainfall hurricane contributed to flooding conditions initiated by the September hurricane (Table 7-19). The resulting flood was said to be the most severe ever experienced in southeast Florida (Climatological Data, October 1947) and covered a 12 county area

TABLE 7-19. METEOROLOGICAL DATA FOR THE HURRICANE WHICH CROSSED SOUTH FLORIDA ON OCTOBER 11-12, 1947.

Station	Minimum pressure (inches)	Maximum sustained winds (mph) direction	Wind gusts (mph)	Rainfall (inches)
Dry Tortugas	29.31	84[1]	150[2]	—
Key West	29.49	65 SSE	65	3.90
Homestead	29.52	60[2]	—	9.71
Miami	29.47	62 S	—	4.96
Hillsboro Light	29.27	92	—	—
West Palm Beach	29.46	62 NNE	68	8.84
Belle Glade	—	48 E	—	4.35
Canal Point	—	60 NE	—	3.57
Ponce de Leon Light	—	44 N	—	—

[1]Anemometer froze at this value.
[2]Estimated.
Source: Climatological Data, October 1947.

from Osceola County to the southern tip of the state. Standing water around Dania and Davie killed up to 70% of the citrus trees in the area. Overall agricultural damage from the September and October floods was estimated to exceed $20 million.

1948

The 1948 season again produced two hurricanes and two tropical storms. However, while neither hurricane produced damaging winds comparable to the 1947 hurricane, the rainfall resulted in severe flooding. Hurricane dates were September 21-22 and October 4-5.

The Hurricane of September 21-22, 1948

"We didn't think so much about hurricanes back in the 40s. They were a natural thing, but of course, there wasn't so much to be destroyed."—G. W. "Buck" Mann, Jr., 1998.

Meteorological Considerations

The September 1948 hurricane formed in the Western Caribbean on September 18 from a tropical wave, near 13°N, 80°W. It moved to the north northwest on the nineteenth, passed just east of the Isle of Pines on the morning of the twentieth, crossed Cuba between Havana and Matanzas and emerged into the Florida Straits. The storm had obtained hurricane force prior to hitting Cuba, with 100 mph winds in the easterly quadrant at Matanzas and 90 mph left of the center at Havana. Considerable damage was incurred at Matanzas.

By noon on September 21, the hurricane center was slightly east of Key West, where 122 mph winds and a low pressure of 28.45 inches were recorded at Boca Chica Airport, eight miles east northeast of the Key West Weather Bureau Office (Sumner, 1948). Boca Chica was in the edge of the eye for about 15 minutes at 11:00 A.M. on the twenty-first. From this point, the hurricane moved northeast toward the southwest Florida coast and moved inland just east of

Everglades City, wobbled across Lake Okeechobee between Clewiston and Belle Glade with gusts to 96 mph, and emerged into the Atlantic near Stuart in Martin County with peak gusts still exceeding hurricane force (Figure 7-5). Meteorological data for key Florida cities are shown in Tables 7-20 and 7-21.

FIGURE 7-5. PATH OF THE SEPTEMBER 21-22, 1948 HURRICANE AS IT PASSED ACROSS SOUTH FLORIDA FROM EVERGLADES CITY TO BELLE GLADE AND INTO THE ATLANTIC OCEAN NEAR STUART. SOURCE: CLIMATOLOGICAL DATA, SEPTEMBER 1948.

TABLE 7-20. METEOROLOGICAL DATA FOR THE FLORIDA HURRICANE OF SEPTEMBER 21-22, 1948 WITH STATIONS LISTED FROM SOUTH TO NORTH.

Station	Minimum pressure (inches)	Maximum sustained winds (mph) direction	Wind gusts (mph)
Key West	28.45	122 NNW	150[1]
Carysfort Reef	29.24	80 NNE	—
Everglades City	—	120 NNW	—
Naples	28.99	86 W	87
Fort Myers	29.05	50 NNW	58
Miami	29.09	75 SSE	90
Clewiston	28.47	96 ENE	—
Canal Point	28.50	—	—
Hillsboro Light	29.05	87 NE	—
Pahokee	—	90-100[1]	—
West Palm Beach	28.35	58 W	84
Stuart	28.51	—	—
Fort Pierce	28.82	58 ESE	—
Vero Beach	28.83	80[1]	70-80
Melbourne	29.02	52 NE	74
Lake Placid	29.15	65 NW[1]	81
Lakeland	29.33	29 NW	—
Daytona Beach	29.34	—	52

[1]Estimated.
Source: Climatological Data, September 1948.

Obviously, this was a significant hurricane for citrus and tropical fruit growers, vegetable growers and cattlemen. Citrus and tropical fruit took a thrashing from the winds, and fall truck crops and pastures in the Okeechobee mucklands suffered heavy losses from floods brought on by an 8 to 11 inch rainfall. Flood waters in Clewiston were 2 to 3 feet in depth and LaBelle was under water for several days. Flooded pastures made it necessary to evacuate herds to higher ground and some cattle were lost.

This was a very unusual hurricane in that there were two separate and distinct eyes with calm winds and blue skies, each spaced several hours apart. Well-known hurricane forecaster Grady Norton (Sum-

TABLE 7-21. RAINFALL FROM THE FLORIDA HURRICANE OF SEPTEMBER 21-22, 1948.

Station	Rainfall (inches)
Key West	6.33
Tavernier	6.02
Naples	4.90
Fort Myers	8.69
Miami	11.00
Clewiston	10.07
Belle Glade	8.47
Okeechobee City	4.89
West Palm Beach	9.04
Stuart	6.36
Lake Placid	8.43
Melbourne	8.53
Vero Beach	5.25
Orlando	5.71
New Smyrna Beach	2.25
Daytona Beach	2.09

Source: Climatological Data, September 1948.

ner, 1948) reported that, "there were so many eyes reported at so many widely separated places, and the movement at 8 to 10 mph was so slow, we were reminded of an oxcart - moving leisurely through the Florida Everglades ogling every community in the southeastern part of the state."

The Damage

Crop damage of all kinds was estimated at $6.5 million (Norton, 1948). Vegetable growers in Dade, Palm Beach and Lee counties and the entire vegetable growing areas around Lake Okeechobee were especially hard hit with early beans, potatoes, tomatoes, peppers and other vegetables left under water after the torrential rains totaling as much as 10 to 11 inches (Table 7-21). Damage to the citrus crops in the Indian River area was estimated at 50%, but only 10% of the state's crop was produced in that section so the losses on a statewide basis were not severe. Winds and rain caused heavy damage to sugarcane, and flooding of pastures caused some loss of cattle. Areas other than Indian River suffered some punctured fruit and loss of grapefruit, but these losses were small on a percentage basis (Florida Grower, 1948). In fact, the first crop estimate placed the total crop at 99,000,000 boxes, up from 95,100,000 boxes including Temples produced in 1947-48. However, it should be noted that the grapefruit estimate was for 33 million boxes, down 2 million boxes from the previous year. This was probably due to damage to the grapefruit crop in the Indian River area.

Personal Observations

Bill Krome describes problems in south Dade County from the 1947 and 1948 hurricanes.

"A major problem was flooding. Hurricanes in 1947 and 1948, while they did little wind damage in south Dade, there were very wet areas which had standing water for weeks; the residential areas in Hialeah and Davie were under water for over a month. This upset many more people than damage to the groves would have, with the result that a system of canals, from Lake Okeechobee south, was started with a vowed goal of preventing such floods in the future. It achieved its purpose but also lowered the normal water table as much as four feet. This made much more land available for residential development and also made it necessary to provide irrigation to most of the agricultural land in Dade County. This was done with the "big gun" pumps, which could water ten acres from four wells, and with permanent systems using a single pumping unit on ten or more acres and distributing the water through buried pipes and impact type sprinklers.

"We used the latter method and soon found that the relatively warm water (67 to 69°F) was effective for frost protection. In all but the most severe freezes, applying water at a rate of 0.15 to 0.25 inches an hour gave excellent protection. Even in the 1977 freeze, when a properly exposed thermometer registered 19°, the trees protected by irrigation came through with minor damage, while mature avocados and mangos without such protection were killed back to 3 and 4 inch wood.

"During the '40s, some equipment and the techniques to use it to mitigate hurricane damage were being developed. The first was a device attached to the front of a heavy tractor such as a Caterpillar D-8 which was used to cut trenches in the soft rock to plant the trees. The trenches were usually about 14 inches wide and 16 inches deep. The land to be planted was first trenched with trenches spaced apart the same distance as the tree rows, and then cross-trenched with trenches spaced the distance apart the trees would be in the row. The trees would then be planted at the intersections of the two sets of trenches. This gave the trees much better anchorage and reduced their tendency to be blown over in a hurricane.

"Shortly after this, the giant toppers were developed, and most groves in south Dade were topped either every year or every two years. Groves planted in trenches and topped to 18 or 20 feet could withstand a considerable amount of hurricane winds. These practices didn't eliminate all damage from a hurricane but they greatly reduced the damage, and trenching and topping are standard practices in south Dade groves today.

"In addition to trenching, topping and irrigating, I have done one other thing to reduce hurricane damage. Whatever I plant must mature its crop before hurricane season. Limes and mangos mature during the summer as well as some varieties of avocados. It doesn't bother me to know that avocados will bring higher prices from December on; I'm willing to let others go for the higher prices. I want my fruit off the trees by mid-September."

And Buck Mann reminisces about the problems with high water in Belle Glade:

"We had some cattle down at Belle Glade when the hurricane gave us 13 inches of rain and broke our dike. We had 400 head of cattle in water up to their bellies. We couldn't get the horses in to herd them out for us when along came a man in a frog boat and offered to herd them out for us. Unfortunately, he didn't know anything about herding cows and he couldn't move them. Finally, I borrowed his frog boat, got the hang of it, got behind those cows and air blasted them right out of the water and into the pen in 15 minutes."—G. W. "Buck" Mann, Jr., 1998.

Reuben Carlton (1998) vividly remembers the problems with flooding which resulted in St. Lucie County from the hurricane:

"The rains were heavy in September 1948 and the dike broke in two places outside the North St. Lucie Water Drainage District. Water poured through so high that I could dive off the front porch of our house 10 miles west of the Indian River. I put my mother into a rowboat and pulled her up McCarty Road. We got to Okeechobee Road before the boat touched the pavement.

"With the whole south end of the state under water and people putting cattle up on the highways so they would have a place to lie down, my brother Thad and I set out to get something done politically. We picked up Senator Claude Pepper in a boat at the northwest corner of Lake Okeechobee and took him to an Indian mound so everybody could have a chance to talk to him. We then took him by car to Fort Pierce so he could catch a train back to Washington. I rode on the fender of the car to guide it through the flood.

"The next week, I picked up Senator Spessard Holland and flew him over the flooded areas. It was two to three years before everybody got their cattle straightened out. We started paying taxes in 1948 to create the flood control district, but we got no help until 1962. They spent all the money in Dade and Broward counties."

Ann Wilder (1998) recalls that, "everybody went out to work on sandbags along Adam's personal dike. The permanent dike was put in later."

The Hurricane of October 4-5, 1948

Meteorological Considerations

Less than two weeks after the September 21-22 hurricane, another hurricane had formed in the western Caribbean and south Florida was again under the gun from hurricane force winds. This hurricane was first detected as a tropical storm on October 3 near 16°N, 82°W or just off the coast of Honduras. The storm moved on a northwesterly course across the 85th parallel, after which it turned to the northeast and reached hurricane force as it approached western Cuba late on October 4. The hurricane continued its northeasterly movement and the center passed just west of Havana at 6:00 A.M. on October 5, with a peak wind gust of 132 mph (Sumner, 1948). Crop losses in Cuba were heavy.

The hurricane continued on its steady northeast course after emerging into the Florida Straits and was pounding the Florida Keys with 100+ mph winds by noon on the fifth, Figure 7-6. A 45 minute lull was felt at Marathon, after which the storm continued its march to the northeast at 18 to 20 mph, reaching Miami at approximately 7:30 P.M. Miami Airport reported a low pressure of 28.92 inches with a lull lasting from 7:00 to 7:30 P.M. At the Weather Bureau office in Miami, the lull lasted from 7:00 to 7:45 P.M. with a low pressure reading of 28.96 inches. The center of the storm had an estimated diameter of 15 miles. The highest winds at the Miami Weather Bureau office were 86 mph with gusts exceeding 90 mph (Table 7-22). The hurricane finally emerged into the Atlantic Ocean near Fort Lauderdale at about 9:30 P.M. after battering the Dade County agricultural areas with gale and hurricane force winds and heavy rains for over four hours. The Experiment Station at Homestead recorded sustained winds of 90 mph (Table 7-22) and was deluged by almost nine inches of rain (Table 7-23).

The Damage

Crop damage in southeast Florida, mostly in Dade County, was estimated at $1.5 million (Sumner, 1948). The loss of citrus was estimated at two million boxes, mostly in Lee, Sarasota and Charlotte counties. This was a major blow to growers in those areas, but as the crop was the greatest in history at the time, the overall loss to the industry was not significant (Climatological Data, October 1948).

1949

Florida only experienced one hurricane in 1949, but it was one of the most destructive hurricanes to ever hit the Indian River area.

FIGURE 7-6. PATH OF THE OCTOBER 4-5, 1948 HURRICANE AS IT CROSSED SOUTHEAST FLORIDA STRIKING THE AGRICULTURAL AREAS OF DADE COUNTY. SOURCE: CLIMATOLOGICAL DATA, OCTOBER 1948.

The Hurricane of August 26-27, 1949

"There wasn't a leaf left on a single tree at the Indian River Field Laboratory."—Herman J. Reitz, July 1997.

The August 1949 hurricane was one of the most destructive of the modern hurricanes to strike peninsular Florida, particularly for the havoc it wreaked on the Florida citrus industry. Striking first at

TABLE 7-22. METEOROLOGICAL DATA FOR THE FLORIDA HURRI-CANE OF OCTOBER 4-5, 1948 WITH STATIONS LISTED FROM SOUTH TO NORTH.

Station	Minimum pressure (inches)	Maximum sustained winds (mph) direction	Wind gusts (mph) direction
Key West	29.25	56	—
Boca Chica	29.16	85+ N[1]	120 N[1]
Sombrero Key	28.80	100 S[1]	—
Tavernier	29.38	60-70 SSW	—
Homestead Expt. Station	28.95	90 NW	—
Miami	28.96	86 NW	90 NW
Everglades City	29.56	—	35 N
West Palm Beach	29.45	55 N	62
Belle Glade Expt. Station	29.69	30 E	—

[1]Estimated.
Source: Sumner, 1948

the bountiful Indian River grapefruit producing counties with the full force of its 150 mph winds, the storm then moved across Florida's major citrus producing district, the ridge, with wind gusts still approaching 100 mph as it ripped through Polk and Highlands counties. Later it passed through Lake and Orange counties with wind gusts still at minimal hurricane force (Figure 7-7).

Meteorological Considerations

The hurricane of August 26-27 formed from an easterly wave which was first noted on August 23 near 19°N, 61.5°W, or about 125 miles northeast of St. Martin in the Leeward Islands (Norton, 1949). The wave moved on a west northwest course and reached the Bahamas on August 25, at which point a well-developed center had developed. The center passed north of Nassau at about 5:00 A.M. on August 26, after which it pointed its fury toward the southeast Florida coast with increasing intensity. The calm center was felt first at Delray Beach, at about 6:00 P.M. on the twenty-sixth, and then moved over the West Palm Beach airport where a flat calm was experienced for

TABLE 7-23. RAINFALL FROM THE FLORIDA HURRICANE OF OCTOBER 4-5, 1948.

Station	Rainfall (inches)
Key West	4.55
Boca Chica	3.73
Tavernier	6.40
Homestead Expt. Station	8.81
Miami WBAS	9.95
Miami WBO	6.02
Everglades City	2.50
West Palm Beach	7.48
Belle Glade Expt. Station	1.16

Source: Sumner, 1948.

22 minutes from 7:20 to 7:42 P.M. The lowest pressure recorded was at West Palm Beach, 28.17 inches at sea level (Table 7-24). The anemometer was blown down when sustained winds reached 110 mph with gusts to 125 mph. The official in charge of the West Palm Beach weather bureau estimated the highest sustained winds from the storm to be 120 mph with gusts to 130 mph or higher. A private observer on Palm Beach recorded gusts to 155 mph.

As expected, the strongest winds were in the northeast quadrant of the storm near Jupiter and Stuart. The anemometer failed at Jupiter Lighthouse after reaching a peak of 153 mph with stronger winds afterward. Stuart recorded 125 mph sustained and Cape Canaveral 83 mph.

After devastating the southern most counties of the Indian River district, leaving thousands of boxes of scratched and battered grapefruit on the ground, the hurricane took direct aim at vegetable growing areas around Lake Okeechobee and the citrus producing areas of Florida's central ridge (Figure 7-7). In the vegetable growing area, peak gusts reported were Canal Point 120 mph, Belle Glade 113+ mph, Port Mayaka 124 mph and Okeechobee City 100 mph (Table 7-24). Moving further inland away from the warm waters needed to fuel its engine, the hurricane diminished in intensity as it moved over the ridge with 90 mph sustained winds at Lake Placid, 100 mph gusts at Bartow and 75 mph at Clermont, destroying poultry and livestock and battering orange groves throughout

FIGURE 7-7. PATH OF THE AUGUST 26-27, 1949 HURRICANE AS IT SMASHED ITS WAY ACROSS THE INDIAN RIVER, EVERGLADES AND RIDGE AGRICULTURAL AREAS. SOURCE: CLIMATOLOGICAL DATA, AUGUST 1949.

mid-Florida before crossing the state line into Georgia, just north of Jasper where winds of 55 mph were felt. Lowest pressures and rainfall throughout the state are shown in Table 7-24.

The Damage

Early Reports: On August 31, five days after passage of the hurricane, J. C. Townsend, Jr. (1949), Statistician at the USDA Bureau of Agricultural Economics, released a preliminary estimate indicating

TABLE 7-24. METEOROLOGICAL DATA FOR THE FLORIDA HURRI-CANE OF AUGUST 26-27, 1949 SHOWING STATIONS FROM LAND-FALL TO POINT OF EXIT FROM FLORIDA.

Station	Minimum pressure (inches)	Maximum sustained winds (mph) direction	Wind gusts (mph) direction	Rainfall (inches)
Miami	29.36	60 SW	—	1.95
West Palm Beach	28.17	120 SW[1]	130[1]	5.93
Jupiter Light	28.34	153 NE	—	—
Stuart	28.42	125 NE	150	—
Melbourne	29.44	78 SE	78	7.63
Daytona Beach	29.65	50 ESE	65	3.85
St. Augustine Beach	29.61	80 SE	85	3.85
Belle Glade Expt. Station	28.72	104 NNW	113+	8.18
Moore Haven	28.95	65 NNW	81	4.98
Canal Point	28.24	—	120 N	4.53
Point Mayaka	28.36	—	124 SW	6.85
Okeechobee City	28.35	—	100 S	7.10
Lake Placid	28.35	90 N	100 S	—
Bartow	28.64	—	100 N	4.00
Lakeland WBO	28.64	61 NNE	75	6.37
Orlando	29.40	55 SSE	84	5.63
Ocala	29.25	52 ENE	—	2.75
Tampa	29.11	54 SW	67	4.80
Cedar Key	29.02	45 NE	60 NE	2.50
Jasper	29.30	55 NE	—	—

[1]Estimated.
Source: Climatological Data, August, 1949.

the loss of 10 to 11 million boxes (30 to 35%) of the grapefruit crop, 3 to 4 million boxes (5 to 8%) of the orange crop and 200 to 300 thousand boxes of tangerines out of a total citrus crop believed to be 95 to 100 million boxes (Florida Grower, 1949). As this estimate was made immediately after the storm, it did not include damaged fruit which might fall from the trees before harvest. Townsend indicated that Polk, Indian River, St. Lucie and Highlands counties suffered the greatest losses. Looking at these counties individually, Townsend forecast grapefruit losses at 90% in St. Lucie County, 75% in Indian River County and 55 to 65% in Polk County; and orange losses at 40% in St. Lucie County and 30% in Indian River County, noting, however, that these two counties combined produced only 5% of the early and midseason oranges and 8% of the Valencia oranges. Valencias appeared to be less affected in most areas than were early and midseason oranges. As damage on tangerines usually develop with time, Townsend indicated that it would probably be the end of September before the tangerine damage could be fully evaluated.

Townsend reported that tree losses would be confined mostly to trees which had been blown over by earlier hurricanes as well as old seedling trees with foot rot. However, much fruit would be reduced in grade due to windscar.

Quoting growers who were considered to be reliable sources, the Citrus Industry (1949A) placed injury to grapefruit in the lower Indian River section where the hurricane made landfall (St. Lucie County and Palm Beach County) at 90% of the grapefruit and 65 to 70% of the oranges. However, these sources indicated that further north into Indian River County the damage lessened to grapefruit 60% and orange 10 to 25%.

For groves in the direct path of the storm, along the ridge in Polk County and Highlands County, the sources predicted the loss of 65 to 70% of Marsh grapefruit, 40 to 50% of Duncan grapefruit, 25% of early oranges and 10 to 15% of Valencia oranges, with a 15 to 17 million box loss of all varieties statewide, most of which would be grapefruit. Robert C. Evans, general manager of the Florida Citrus Commission estimated the loss for the state at large at 13 to 17 million boxes, with 10 to 12 million boxes of grapefruit accounting for the bulk of the loss.

An initial estimate published by Citrus Magazine (September 1949A) placed statewide losses at 35% of the grapefruit crop, and 6% of the orange crop. However, losses in the lower Indian River section, where the storm came ashore, as well as in Polk County and Highlands County were much heavier. In the hardest hit areas, the writer estimated that 50 to 80% of the grapefruit and 5 to 15% of the oranges were on the ground, and as the hurricane had struck so early in the season, only a small amount of this fruit could be salvaged. Orange County and Lake County also suffered some losses but Pinellas County and Hillsborough County, on the opposite coast from the hurricane landfall, came through surprisingly well.

Final Reports: The loss of grapefruit and oranges in the most affected counties are well-illustrated in Tables 7-25 and 7-26, where it can be seen that grapefruit production in St. Lucie County dropped from 1,178,000 boxes for the 1948-49 season to only 359,000 boxes for 1949-50, the hurricane year, and then recovered to 1,239,000 boxes the year after the hurricane. Precipitous losses in grapefruit production are also indicated for Indian River, Polk and Highlands counties with strong recoveries in the following season. Orange losses were not as severe except in St. Lucie County where the drop in production was almost as severe as for grapefruit.

Lyons Fertilizer Company fieldmen made the following reports (Citrus Industry, 1949B) on damage in the interior and west coast counties:

South Polk, Highlands and Hardee counties: "The hurricane of August 26 that passed in full force over this territory has left quite a bit of speculation as to the damage to our present fruit crop. I too, would like to join this predicting group and give out with the damage as I see it. In Highlands County and south Polk County, we estimate a loss of from 65 to 75% of seedless grapefruit with the loss on seeded grapefruit not to exceed 50%, and probably much lower. The loss on oranges will not exceed 10% even with fruit that will continue to drop. Hardee County suffered heavy losses on seedling

TABLE 7-25. EFFECT OF THE AUGUST 1949 HURRICANE ON GRAPEFRUIT PRODUCTION IN THE MOST AFFECTED COUNTIES.

County	1948-49	1949-50[1]	1950-51
	---------- thousand boxes ----------		
St. Lucie	1,178	359	1,239
Indian River	1,375	886	1,464
Polk	10,136	7,704	12,802
Highlands	2,238	1,056	2,251

[1]Hurricane year.
Source: Wilson, 1950.

TABLE 7-26. EFFECT OF THE AUGUST 1949 HURRICANE ON ORANGE PRODUCTION IN THE MOST AFFECTED COUNTIES.

County	1948-49	1949-50[1]	1950-51
		thousand boxes	
St. Lucie	1,201	581	1,421
Indian River	596	539	617
Polk	14,021	15,102	17,166
Highlands	2,455	2,680	3,398

[1]Hurricane year.
Source: Wilson, 1950.

oranges and the overall tree loss was heavy. There was heavy foliage loss over this entire area, but we have had a very fine flush of new growth to replace this loss. While the new growth has not been very vigorous it has certainly shown that we do not have as much actual wood damage as was at first anticipated. The fall application of fertilizer will be applied somewhat earlier than usual because of the hurricane damage with the idea that it will build up tree strength and vitality during the coming months"—R. L. (Bob) Padgett.

Polk County: The hurricane damage to this area has been reported by various official agencies and leaves little to add by this report. However, new growth has now covered most trees where defoliation occurred. This growth is not strong or vigorous in appearance, and many growers are now applying their fall application of fertilizer to strengthen the trees and aid the new growth to maturity. In some cases, increased amounts of nitrogen are being applied, but generally, the regular fall application is calculated and put on the grove. About six inches of rain fell during the storm and scattered rains have occurred since, but generally this section is short of rainfall. The lakes are well below their record high of last year and some concern is felt for an early fall drought. This condition seems to be another factor for applying an early fall fertilization while some moisture is present"—J. M. (Jim) Sample.

Southwest Florida: "This area escaped any damage from the late August hurricane except for very minor loss of fruit in the Arcadia section and some water damage to citrus from the abnormal amount of rainfall in August. It was only on poorly drained soil that this latter occurred. Heavy rains necessitated some replanting of early field tomatoes, but on the whole no disastrous effects of wind or water are evident down this way. Fall crops are progressing nicely. Early glads are well along and look good. First plantings of celery in the Sarasota section are going ahead with prospects good for profitable returns. The tomato plantings seem to be in usual volume in the Ruskin section and in Manatee County, and so far growing conditions point to a good fall season"—Eaves Allison.

Hillsborough County and Pinellas County: "The recent hurricane did very little damage to either groves or buildings in this section of the state, and now growers are feeling very optimistic about their prospects for the approaching season. While most growers are not too happy to reap a nice harvest at the expense of their fellow growers, they do feel that this particular section will have a very profitable year if we can avoid further hazards of climate"—C. S. (Charlie) Little.

North Central Florida: "The hurricane damage to this territory cannot be considered as too serious. In some localities, we had quite a bit of damage but the territory as a whole came through in very good shape"—V. E. (Val) Bourland.

West Central Florida: "The hurricane in August caused considerable damage to grapefruit throughout this territory, but only slight losses to oranges and tangerines. We were lucky"—E. A. (Mac) McCartney.

Table 7-27 shows the effect of the hurricane as well as other factors, on carlot shipments. Statewide, Florida's total production for 1949-50 (Florida Agricultural Statistics Service, 1950) was 57.79 mil-

TABLE 7-27. SHIPMENTS FOR FLORIDA GRAPEFRUIT AND ORANGES FOR THE YEARS BEFORE AND AFTER THE AUGUST 1949 HURRICANE.

County	1948-49	1949-50[1]	1950-51
	--------------- thousand carlots ---------------		
Grapefruit	30.20	24.20	33.20
Oranges	57.38	57.79	66.20

[1]Hurricane year.
Source: Wilson, 1950.

lion boxes of oranges, 24.2 million boxes of grapefruit and 5.97 million boxes of specialty fruit. Orange production increased 410,000 boxes over the previous season, specialty fruit production increased 450,000 boxes, but grapefruit production decreased by 6 million boxes, demonstrating the vulnerability of the grapefruit crop to hurricane damage.

Market Considerations

H. F. Wilson (1950) summarized the citrus season as follows:

"The Florida hurricane in late August blew off much of the early bloom citrus, especially grapefruit, and this reason combined with higher quality standards resulted in consistently light supplies during the early part of the season.

"There were a few shipments of grapefruit immediately after the storm of August 26, then followed a brief period without any movement whatsoever. Through Saturday, October 15, straight rail shipments from the state totaled only 217 cars compared to 1,805 cars the year before or approximately one-eighth. Much of the early bloom fruit was blown from the trees by the hurricane which considerably delayed the movement of grapefruit in volume. It was not until the week ending October 22nd that the combined shipments by rail, boat and truck exceeded 400 cars weekly."

In the October issue of Florida Grower (1948), it was reported that the USDA estimate of a 13 million box loss in the August 1949 hurricane would mean a dollar loss of $32 million at prices which were then current.

In the same issue, it was noted that Homestead area lime growers were predicting higher prices for their fruit as a result of hurricane damage done to lime groves in the Lake Placid and Avon Park areas of Highlands County.

The Citrus Industry (1949C) magazine quoted R. L. (Bob) Padgett of Lyons Fertilizer Company who indicated that, as a result of storm damage, some pretty bad quality fruit from south Polk, Highlands and Hardee counties had been diverted from the packinghouses to processing plants.

Government Actions

Florida Senator Claude Pepper encouraged the Federal Crop Insurance Corporation to insure Florida citrus agents against any damage from future hurricanes (Florida Grower, 1949). In a letter to Secretary of Agriculture Brannan, Pepper noted that, "in view of the fact that the corporation does have the authority to insure Florida citrus products, and in view of the great losses which Florida citrus growers have incurred, I believe that it is essential that the corporation undertake at once a crop insurance program for Florida citrus." Author's Note: Federal crop insurance has been helpful to many growers in replanting their groves after destructive freezes particularly the freezes of the 1980s (Attaway, 1997).

As a result of Senator Pepper's action, Secretary of Agriculture Brannan designated certain Florida counties where Florida farmers would be eligible for disaster loans from the Farmer's Home Administration. To be eligible for a loan, it was necessary for the applicant to demonstrate damage as the result of the disaster, and inability to obtain sufficient credit from commercial sources. Counties designated in alphabetical order were: Alachua, Baker, Bradford, Brevard, Broward, Charlotte, Citrus, Clay, Columbia, DeSoto, Duval, Flagler, Gilchrist, Glades, Hardee, Hendry, Hernando, Highlands, Hillsborough, Indian River, Lake, Lee, Levy, Manatee, Marion, Martin, Nassau, Okeechobee, Orange, Osceola, Palm Beach, Pasco, Pinellas, Polk, Putnam, Sarasota, Seminole, St. Johns, St. Lucie, Sumter, Suwannee, Union and Volusia.

At the state government level, the Florida Citrus Commission made a preliminary reduction in its advertising budget of $365,000 due to the crop losses from the hurricane (Citrus Magazine, 1949B).

Affect on Citrus Prices

After the hurricane, the price for a carload of Pope Summer oranges from the Indian River area peaked at $10.90/box in New York. This was thought at that time to be the highest prices ever recorded for Florida oranges in the New York market (Citrus Magazine, 1949C). Price averages recorded by the Vero Indian River Producers Association were Florigold Golden, $9.08; Florigold

Bronze, $8.23 and Flo, $7.92. Florida grapefruit blown off the trees by the August 27 hurricane were said to bring as high as $6.00/box on the New York auction market.

Tropical Fruits

Avocados: There was an 80% loss of avocados in Highlands County with considerable tree damage (Townsend, 1949), but avocado damage in Dade County was negligible. During the previous season (1948-49), Highlands County and Polk County produced about 32,000 bushels of avocados and a good crop had been set for 1949-50. As a result of severe freezes, these two counties are no longer important avocado producers.

Damage to Vegetables

Planting of fall crops in the Pompano and Everglades vegetable producing areas was well advanced. Losses to snap beans, lima beans, cucumbers, eggplant, pepper and squash in the Pompano area was heavy, requiring extensive replanting. In the Everglades, the loss to the important snap bean area was so heavy that only light crops remained for harvest. Again, reseeding was required.

Tomatoes: Fall tomato plantings in both the Fort Pierce and the Manatee-Ruskin sections suffered heavy losses. Most of the 4,000 to 5,000 acres of tomatoes in the Fort Pierce section were just coming up and most of the acreage required replanting.

Celery: Major damage was confined to young celery seed beds in the Everglades which received heavy wind and rains. Older plantings were in fair condition and celery in other parts of the state came through in relatively good condition.

Cucumbers, Squash, Peppers and Eggplant: A few early plantings were damaged.

Personal Observations

Reuben Carlton (1998) recalls a 20 acre grove on Newell Road near Fort Pierce. "We had figured to pick 6,000 boxes of oranges but we picked only 307 boxes. Some people from Orlando came to look at another grove and found that the wind had stripped the bark off the limbs. They figured that it took a 165 mph wind to do that. In another interesting incident, a train was creeping up the tracks from Belle Glade at 20 mph. When it reached the corner of Shin Road and Glades Cut-Off Road, the wind blew five cars and the caboose off the tracks. Fortunately no one was in the cars."

When John King (1998) arrived in Fort Pierce in 1955 he saw., "many trees which had been set up after the 1949 hurricane with half the root system still in the air and half in the ground, but they were producing a good crop."

Herman Reitz (1997) remembers that, "the Australian Pine windbreaks in the Indian River area fell over like the pages of a book. There were miles of them. There was more damage from the windbreaks than from the wind."

Reitz continued, "I didn't know what to recommend to the growers whose trees were bare of leaves. I finally suggested a heavy application of fertilizer in September and it worked. There was a big flush of growth the next spring followed by a big crop in the 1950-51 season.

Louis Forget (1998) considers the "1949 hurricane did more economic damage to St. Lucie County than any other hurricane or any freeze. We spent six months straightening the trees with two by fours and croker sacks. Some growers never bothered to straighten the trees, but 10 years later they were making big crops with half the root systems exposed."

Alto "Bud" Adams (1998) recalls that, "in 1949, it was fortunate that we knew it was coming. The ranch had 15,000 acres with a dike around 3,000 of those acres. We turned the cows into those 3,000 acres and started the pump to keep the water out. After a while, the pumphouse blew away but the pump kept running. Finally it began to blow the truck away so we shut down the pump and went home. The next morning we saw citrus groves with all the fruit and leaves blown off and many trees blown away. We had never seen that before. Usually there were only 20 to 30% of the fruit on the ground after a hurricane, not the total destruction we saw in 1949. We went over to the east side of Lake Okeechobee to check on a friend. The highway was covered with cattails and hyacinths and every telephone pole between the lake and Palm Beach was flat on the ground."

Prominent Fort Pierce grower and member of the Florida Citrus Commission, O. C. Minton (Citrus Magazine, October 1949D) said after the 1949 hurricane that, "out of 50,000 boxes of citrus estimated before the hurricane, I doubt that I will have enough left to feed my family."

Buck Mann recalls the tremendous problems in herding cattle in Osceola County during the storm.

"We had 1,200 head of cattle on the east side of Kissimmee Lake in Osceola County. With the wind behind them, they drifted up against the lake and finally crowded into the lake. Some of them made it all the way across the lake, some of them made it to Brahma Island, and some of them drowned.

"In that same year, Joe Jackson, the Civil Engineer, had built a dike for us around the muck on Lake Jackson, east of Kissimmee Lake in Osceola County. I told him he better raise it another four feet and he did, but the hurricane still washed it away.

"There was a seedling grapefruit grove along Highway 542 east of Winter Haven. The hurricane stripped off all the fruit"—G. W. "Buck" Mann, Jr., 1998

Chapter 8

From 1950 through 1959

"I was a senior at Florida Southern College in Lakeland, FL when Hurricane King moved up the peninsula. Expecting classes to be canceled, I stayed in Haines City to monitor the anemometer at Haines City Citrus Growers Association headquarters. Unfortunately for me, the college administration did not see fit to cancel classes and I was charged for cuts in all my courses. Colleges were stricter about such things in those days."—John A. Attaway.

During the 1950s, the hurricanes which had battered Florida with great regularity during the 1940s turned away from the sunshine state to make landfall elsewhere. Only in 1950, the first year of the new decade, did the destructive trend which had plagued Florida annually since 1944 continue. The years from 1951 through 1959 were relatively quiet, and Floridians were not heard to complain.

Naming Hurricanes

To avoid confusion, especially when two or more hurricanes were in progress at the same time, a system of hurricane identifiers was developed during the second World War. These identifiers were short words, easily pronounced and remembered, and were much more convenient than cumbersome latitude-longitude descriptions. Initial identifiers were based on the phonetic alphabet, Able, Baker, Charlie, etc., used by the military services. However, in 1953, the system was changed to use girl's names, Hurricane Alice in 1953 being the first use of a girl's name to identify a hurricane (Dunn and Miller, 1964).

The 1950 Atlantic Basin season produced eleven hurricanes and two tropical storms in the Atlantic. Three of the hurricanes, Easy, King and Love struck peninsular Florida, and one hurricane, Baker, made landfall in the Pensacola area. Both Florida coasts felt a hurricane's wrath, Hurricane Easy on the gulf coast and Hurricane King on the east coast. Hurricane Easy was considered the worst hurricane to strike Cedar Key, pelting the area with hurricane force winds for almost ten hours with peak gusts up to 125 mph. The three-day rainfall at Cedar Key totaled 24.50 inches (Table 8-1).

Hurricane King was a small but violent hurricane when it came ashore over Miami on October 17 with sustained winds of 122 mph and gusts to 150 mph along a narrow band. Leaving Miami, King crossed Lake Okeechobee and raked east and central Florida with gales and hurricane force winds as it sped northward to Georgia.

Considering these hurricanes in chronological order:

TABLE 8-1. METEOROLOGICAL DATA FOR HURRICANE EASY DURING ITS PASSAGE THROUGH WEST AND CENTRAL FLORIDA SEPTEMBER 4-7, 1950.

Station	Minimum pressure (inches)	Maximum sustained winds (mph) direction	Wind gust (mph) direction	Rainfall (inches)
Key West[1]	29.51	60 E	71	—
Miami[1]	29.62	63 SE	—	2.51
Sarasota[1]	29.53	60 SW	—	—
Cedar Key	28.30	102 N	125 N	24.50
Clearwater[2]	29.32	—	—	12.00
Tampa[2]	29.19	45 S	51 S	7.67
Anclote Light[2]	29.08	76 SW	—	8.42
Brooksville[2]	29.25	—	—	19.50
Cross City[2]	29.29	65 NE	—	7.40
Lakeland[2]	29.24	32 S	—	7.40
Jacksonville[3]	29.47	52 NE	—	14.08

[1]Data for September 3, 1950.
[2]Data for September 5, 1950.
[3]Data for September 6, 1950.
Source: Climatological Data, September 1950.

Hurricane Baker, August 30-31, 1950

Meteorological Considerations

Hurricane Baker developed on August 20 from a tropical depression east of the Leeward Islands near 16°N, 55°W. It intensified rapidly and moved westward across Antigua with 90 to 120 mph winds, then lost force and was only a tropical disturbance with 35 to 40 mph winds as it passed over Cabo Rojo, Puerto Rico on the twenty-third. Moving west northwest, the disturbance deteriorated to a tropical wave as it followed the north coast of the Dominican Republic and Haiti before crossing Cuba into the Caribbean Sea on the twenty-fifth where it began to reintensify to a tropical storm. Turning north on August 29, Baker again reached hurricane force with winds of 115 mph and moved rapidly north northeast to strike the coast between Pensacola and Mobile with 75 to 85 mph winds. Moving north into Alabama, wind gusts to 50 mph were felt as far inland as Birmingham. Rainfall totals for Hurricane Baker are shown in Table 8-2.

The Damage

As Hurricane Baker came ashore, the strongest winds were felt to the east of the center heavily damaging crops as far east as St. Marks, Florida. Damage to crops was estimated at $100,000.

TABLE 8-2. RAINFALL IN INCHES AT SELECT WEST FLORIDA STATIONS DURING PASSAGE OF HURRICANE BAKER, AUGUST 30-31, 1950.

Station	Rainfall (inches)
Apalachicola	6.14
Blountstown	7.23
Chipola	7.18
DeFuniak Springs	10.67
Marianna	6.75
Panama City	8.10
Pensacola	7.38
Pensacola Airport	6.77
Tallahassee	7.34

Source: Climatological Data, August 1950.

Hurricane Easy, September 4-7, 1950

Meteorological Considerations

Hurricane Easy formed on September 1 from a tropical disturbance in the western Caribbean Sea south of the Isle of Pines near 20°N, 83°W (Norton, 1951). It remained almost stationary for two days then moved northward over Cuba and shifted to the north northwest passing between Key West and the Dry Tortugas on September 3 with winds of minimal hurricane force. Easy then paralleled the Florida gulf coast, at a speed of 10 to 12 mph to the north northwest, remaining 30 to 50 miles offshore until it reached 28°N, 83.5°W, about 70 miles northwest of Tampa on September 4. Reconnaissance aircraft estimated the maximum wind speed at 125 mph. At this point, the hurricane made the first of two loops and began a slow northeasterly movement to just south of Cedar Key. On the morning of September 5, Easy made its second loop moving the eye over Cedar Key from the southeast and then toward the south (Figure 8-1). As a result, Cedar Key was battered by the same side of the hurricane twice with 2.5 hours with the calm center in between (Norton, 1951). During the first loop, hurricane force winds with gusts to 125 mph continued (Table 8-1), and as a result of the looping motion of Easy, Cedar Key had winds of hurricane force continually from 6:00 A.M. to 6:00 P.M. on September 5, 1950, excepting the 2.5 hours of calm when the eye was over the town from 11:00 A.M. to 1:30 P.M. Cedar Key residents described Easy as the worst storm experienced in that area for 70 years (Norton, 1951).

After exiting the Cedar Key area, Easy moved south to a point 30 miles north of Tampa, losing hurricane strength as the eye remained over land. It then turned east and back to the north on the sixth, finally dissipating as it exited into south Georgia on the seventh.

Having spent the better part of three days over the Florida peninsula, Easy deposited enormous amounts of rain (Table 8-1). Cedar Key recorded 24.5 inches of rain during the three day period and several other stations recorded 10 to 20 inches.

FIGURE 8-1. PATH OF HURRICANE EASY AS IT MADE ITS INFA-MOUS LOOP AROUND CEDAR KEY, FLORIDA, SEPTEMBER 4-7, 1950. SOURCE: CLIMATOLOGICAL DATA, SEPTEMBER 1950.

The Damage

Crop damage was estimated at $200,000 in 1950 dollars. Lyons Fertilizer Company fieldmen assessed the damages as follows (Citrus Industry, 1950A).

Southwest Florida: "Very little damage was sustained by citrus in this area from the Labor Day hurricane. In fact, the large amount of rainfall accompanying the winds over this whole area was most welcome. We now have sufficient moisture in the ground to properly size up the fruit and supply all needs of the tree. Vegetable growers in the Ruskin-Palmetto area had flood conditions and severe loss of plants, but as the season was just getting started the loss was not a catastrophe. In the Immokalee and Ft. Myers area, the rainfall amounted to a life saver—the summer rains having been delayed so long as to approximate a drought. Farms in that section are now in condition to begin the fall plantings with an ample water supply. The acreage being planted seems to be an increase over last year."—Eaves Allison.

West Central Florida: "We had some damage to citrus groves due to heavy rains in Pasco and Hernando counties, but it was spotty and the damage was confined to hills where groves were planted. Some fruit in the same section was blown from the trees but the losses could not be considered as very serious. The rains accompanying the hurricane were very beneficial as most groves were beginning to suffer from dry weather."—E. A. (Mac) McCartney

North Central Florida: "I think we were very lucky as far as hurricane damage was concerned, with not too much damage and plenty of water which was very much needed."—V. E. (Val) Bourland.

Polk County: "The Labor Day hurricane could have been a catastrophe in this section but fortunately it was beneficial. We lost very little fruit from wind and we had the best rain that we have had in some time."—J. M. (Jim) Sample.

Hillsborough County and Pinellas County: "The recent hurricane did some damage in areas throughout this territory but we also had some benefits too. The water stood in some low groves due to the heavy rainfall but it certainly didn't do very much damage, and the winds blew off some fruit but it was not a severe loss. The benefits derived from the rain far outweighed any damage done by both winds and water."—T. D. (Tillman) Watson.

South Polk County, Highlands County and Hardee County: "We did not experience much damage from the hurricane and if we can get by without a bad storm and get some additional rains we should have an excellent crop."—R. L. (Bob) Padgett.

Hurricane King, October 17-18, 1950

Meteorological Considerations

Hurricane King developed from a low pressure area along the north coast of Nicaragua near 16°N, 84°W on October 11 (Norton, 1951). Initial movement was very slow to the northeast passing the western tip of Jamaica as a tropical storm, after which King intensified, turned north and crossed Cuba just west of Camaguey as a very small storm on the night of October 16-17. The hurricane then intensified rapidly and moved north northwest across the Florida straits and passed directly over the city of Miami near midnight on October 17-18 as a small but very violent hurricane. The calm center was only five miles in diameter. The highest sustained wind velocity was 122 mph for a 1-minute period at the Weather Bureau office in downtown Miami, 97 mph sustained for a 5-minute period. The lowest pressure readings recorded were 28.25 inches at the downtown Weather Bureau office and 28.34 inches at the Miami Airport.

Leaving the lower east coast district, Hurricane King passed directly over Lake Okeechobee and moved north northwest over the agricultural areas of Okeechobee, Polk, Osceola, Orange, Lake, Seminole, Marion, and counties to the north (Figure 8-2). Lying to the east of the hurricane center, the agricultural areas of Broward, Palm Beach, Martin, St. Lucie, Indian River and Broward counties felt the full fury of the storm. Meteorological data for Hurricane King is shown in Table 8-3.

The Damage

With the path of Hurricane King, bringing gusts to hurricane force northward through the Indian River district, the greatest agricultural damage may have been to grapefruit (Norton, 1951), but damage to vegetable crops in the Pompano and Everglades areas was also severe. It was estimated that the hurricane destroyed 90% of the anticipated yield of tomatoes in St. Lucie, Glades, Okeechobee and Indian River counties (Florida Grower, 1950). Shortly after the hurricane, the USDA estimated the total citrus loss

FIGURE 8-2. PATH OF HURRICANE KING AS IT SWEPT FROM SOUTH TO NORTH UP THE EAST COAST OF PENINSULAR FLORIDA, OCTOBER 17-18, 1950. SOURCE: CLIMATOLOGICAL DATA, OCTOBER 1950.

to be between 2.5 and 3 million boxes, with the grapefruit loss believed to be as much as 2 million boxes and oranges somewhat less. Reports from fieldmen from Lyons Fertilizer Company were as follows. Unfortunately, Lyons did not furnish a report from either the Indian River or the Everglades districts.

South Polk County, Highlands County and Hardee County: "In summarizing the October 18th hurricane damage, it looks about like this to us. There is a loss of 5% grapefruit blown from the trees

TABLE 8-3. METEOROLOGICAL DATA FOR HURRICANE KING AS IT PASSED FROM SOUTH TO NORTH UP THE FLORIDA PENINSULA OCTOBER 17-18, 1950.

Station	Minimum pressure (inches)	Maximum sustained winds (mph) direction	Wind gust (mph) direction	Rainfall (inches)
Miami	28.25	122 S	150[1]	3.50[1]
Miami Airport	28.34	80 NNE	125	2.00
Pompano	—	85 E	105[1]	4.45
West Palm Beach	29.30	45 E	65	4.94
Vero Beach	29.32	72 ENE	85-95[1]	—
Melbourne	29.30	55 ENE	71	—
Cape Canaveral	29.36	75 NE	—	—
Belle Glade Expt. Sta.	29.01	85 E	—	4.37
Clewiston	28.76	—	93 NE	5.50
Moorehaven	29.03	—	80 ENE	3.13
Lake Placid	29.26	55 NNE	60	3.94
Lakeland	29.27	32 N	45	0.91
Haines City[2]	—	40 NNE	57[2]	—
Orlando	29.15	48 NE	82	14.19
Sanford	29.22	—	75-80[1]	9.83
Gainesville	29.28	35 NE	—	4.68
St. Augustine	29.38	60 NE	—	5.00
Jacksonville	29.43	72 E	81[1]	5.75
Fernandina	29.42	60 NE	85[1]	5.61

[1]Estimated.
[2]Value recorded by John A. Attaway at Haines City Citrus Growers Association headquarters.
Source: Climatological Data, October 1950.

and it is difficult to tell just how much more will drop as a result of the hurricane, but we expect to have some heavy droppage. The loss was very light as far as fruit blown from the tree is concerned, but here too it is difficult to tell what additional loss we will have from droppage. The loss to avocados will run from 30 to 50% which is a heavy loss. However, the avocado growers will be able to save a large percentage of this fruit. Vegetable crops in the Lake Istokopoga section were hard hit by both winds and water. Hardee County vegetables also suffered heavily from water damage. The rainfall during the two day storm period totaled about five inches all over the territory."—R. L. (Bob) Padgett.

West Central Florida: "We had slight damage from the recent hurricane but we cannot estimate how much fruit we will lose from droppage during the next few weeks as a result of being buffeted around by high winds. The rains that came along with the hurricane was of much value to the citrus growers but did some damage to vegetable crops where they had an excess of moisture. However, we were lucky and came through in pretty good shape."—E. A. (Mac) McCartney.

Pasco County: "In scattered places some fruit is falling as a result of the hurricane injury. Yet, in an overall sense this is negligible."—George C. James.

Reports from other areas indicated no significant damage to either citrus or truck crops. These reports included Polk County by J. M. (Jim) Sample, Hillsborough and Pinellas counties by T. D. (Tillman) Watson, North Central Florida by V. E. (Val) Bourland and Southwest Florida by Eaves Allison.

Citrus Industry (1950B) magazine for commented that while Hurricane King did cause "serious damage in both the Indian River and lower east coast districts, the overall loss for the state as a whole was not extensive and in some interior localities the benefit derived from the heavy rainfall largely, if not entirely, offset the loss."

Hurricane Love, October 21, 1950

Meteorological Considerations

Love was a minimal hurricane briefly over the open gulf, but had dissipated to a tropical storm before coming ashore. Love began forming on October 17 in the Gulf of Mexico south of Louisiana (Norton, 1951), near 27.5°N, 89°W at the same time that Hurricane King was sweeping up the middle of the Florida peninsula. In some respects, it could be considered an "offshoot" (Norton, 1951) of Hurricane King as it developed from an elongated trough of low pressure ahead of King. This low finally formed a center of circulation of its own on October 18, looped down into the central gulf on the nineteenth and intensified to hurricane force. Reconnaissance aircraft reported winds of 85 to 98 mph on the twentieth as Love completed its curve and moved to the northeast toward the upper west Florida coast. However, during the night, dry air encircled the storm center and it lost force quickly and when it reached the coast early on October 21 its winds had diminished to only gale force. Rainfall at selected stations for Hurricane Love is shown in Table 8-4.

The Damage

No significant damage was reported from Hurricane Love.

Summary

As far as the citrus industry was concerned, the three hurricanes did not slow the rapid growth which began at the end of World War II. Orange production jumped to 66.2 million boxes, grapefruit to 33.2 million boxes and specialty fruit to 5.98 million boxes for a total of 105.38 million boxes, up 17.43 million boxes from the 1949-50 season. It was the first season to top the 100 million box level for all varieties (Florida Agricultural Statistics Service, 1951).

TABLE 8-4. RAINFALL IN INCHES AT SELECT NORTH FLORIDA STATIONS DURING THE PASSAGE OF DISSIPATED HURRICANE LOVE, OCTOBER 21, 1950.

Station	Rainfall (inches)
Cross City	1.79
High Springs	4.75
Hilliard	2.68
Jasper	1.04
Lake City	2.19
Mayo	2.55
Madison	1.43
Raiford	2.60

Source: Climatological Data, October 1950.

1951

An interesting feature of the 1951 hurricane season was a pre-season storm, Hurricane Able, which developed in the Atlantic on May 17, 1951 about 100 miles east of the Fort Pierce-Stuart area (Anderson, 1951). Although it came no closer to the mainland and its only effects were fresh winds and light showers along the east coast, it is worthy of mention because the development of an Atlantic hurricane during the month of May is so very, very unusual. The only hurricane to strike the mainland of North America in 1951 was Hurricane Charlie which formed in the eastern Atlantic near latitude 12°N, longitude 45°W, and moved on a course due west, crossing Jamaica, the Yucatan and finally into the east coast of Mexico near Tampico. Other Atlantic hurricanes turned northward very early and affected only shipping as they finally moved off to the northeast. There were a total of ten hurricanes and tropical storms during the season.

A tropical storm formed in the central Gulf of Mexico on October 1, 1951 and crossed Florida on October 2 along a line from Ft. Myers to Vero Beach. Very heavy rains accompanied this storm with amounts from 8 to 13 inches being recorded, but no strong winds were felt over the peninsula and no citrus damage occurred. However, the storm did produce extreme flooding of farms and pasture lands over a broad belt extending from Naples, Ft. Myers and Punta Gorda on the west coast to Stuart, Ft. Pierce and Vero Beach on the east coast causing extreme losses to early fall crops in Lee, Hendry, Okeechobee, Martin, Indian River, St. Lucie, Palm Beach and adjacent counties (Anderson, 1951). Much replanting had to be done, and cattle had to be moved from the flooded areas to avoid death from drowning or starvation. Monetary damage was estimated at $2 million. Rainfall totals in selected cities are shown in Table 8-5.

1952

No hurricanes or tropical storms affecting peninsular Florida or the Florida panhandle between June and December 1952. However, an unusual tropical storm formed in February and crossed the southern tip of the state, producing gusts of 68-84 mph in the

**TABLE 8-5. RAINFALL TOTALS IN KEY SOUTH FLORIDA AGRICUL-
TURAL AREAS FOR THE THREE-DAY PERIOD OCTOBER 1-3, 1951.**

Station	Rainfall (inches)
Belle Glade	7.02
Bonita Springs	15.72
Clewiston	8.42
Fort Myers	10.44
Fort Pierce	6.99
LaBelle	10.08
Naples	13.95
Punta Gorda	9.84
Stuart	6.36
Titusville	4.49
Vero Beach	5.09
West Palm Beach	4.59

Source: Climatological Data, October 1951.

Miami area before becoming extratropical in the Atlantic (Williams and Duedall, 1997). The only hurricane to strike the United States in 1952 was Hurricane Able which formed as a tropical depression near the Cape Verde Islands on August 18, reached tropical storm intensity at latitude 16°N, longitude 52°W, was classified as a hurricane at 22°N, 65°W, approached the Florida coast just east of Jacksonville on August 30, at which point it turned due north and made landfall at Beaufort, South Carolina. The season total for Atlantic and Caribbean tropical storms and hurricanes was only seven.

1953

The 1953 season was very active with a total of 14 hurricanes and tropical storms. The first tropical storm of the year formed in the southeastern Gulf of Mexico on June 1, at the very beginning of the formal hurricane season. The storm paralleled the west Florida coast before moving inland near Panama City on June 6 with 40-45 mph winds. As the month of May had been unusually dry, the rains associated with this storm were beneficial (Climatological Data, July 1953). This storm was given the name Alice, and was the first hurricane or tropical storm to receive a woman's name.

The first hurricane to affect the state was minimal Hurricane Florence which formed in the Caribbean, moved through the Yucatan Channel on September 24 and came ashore between Ft. Walton Beach and Panama City on September 26 (Figure 8-3). Florence was one of only two hurricanes to strike the U.S. mainland in 1953. This storm did not affect the Florida citrus industry, but rainfall amounts between three and 11 inches caused agricultural damage in the panhandle. Pecan trees lost some immature nuts to the wind and a

FIGURE 8-3. THE TRACK OF HURRICANE FLORENCE WHICH HIT THE FLORIDA PANHANDLE SEPTEMBER 26, 1953. SOURCE: CLIMATOLOGICAL DATA, SEPTEMBER 1953.

small amount of cotton was damaged by the rains. Corn stalks were blown down making harvesting difficult. Meteorological data is shown in Table 8-6.

The only storm in 1953 with any possible impact on the citrus industry was Tropical Storm Hazel which formed in the Yucatan Channel on October 8 and crossed the Florida coastline between Ft. Myers and Punta Gorda about 11:30 A.M. on October 9. The winds were slightly below hurricane force except for an 80 mph gust at Okeechobee City which was thought to be in association with a tornado-like squall. Okeechobee City recorded the lowest pressure at 29.15 inches. The storm moved northeastward across the state, exiting into the Atlantic near Vero Beach, on a course which would cause great destruction to the southwest Florida and Indian River citrus areas if followed by a major hurricane in the 1990s or 2000s (Figure 8-4). Some typical wind gust velocities are shown in Table 8-7. Rainfall near and to the north of the center ranged from 3 to 5 inches which contributed to an already flooded condition. Crop damage ranged from total loss of some plantings to the added expense of pumping water from the fields.

1954

As far as Florida interests were concerned, 1954 was a kind year as Atlantic storms preferred the Carolina and New England coasts,

TABLE 8-6. METEOROLOGICAL DATA FOR HURRICANE FLORENCE, SEPTEMBER 26, 1953.

Station	Maximum sustained winds (mph) direction	Rainfall (inches)
Pensacola	75	9.08
Niceville	80	11.85
Panama City	87	10.70
DeFuniak Springs	59	6.51
Marianna	55	6.88
Apalachicola	52	8.95

Source: Climatological Data, September 1953.

FIGURES 8-4. TROPICAL STORM HAZEL ENTERED FLORIDA BETWEEN FT. MYERS AND PUNTA GORDA ON OCTOBER 9, 1953, CROSSED THE STATE AND EXITED INTO THE ATLANTIC NEAR VERO BEACH. THIS STORM SHOULD NOT BE CONFUSED WITH MAJOR HURRICANE HAZEL WHICH DEVASTATED EASTERN NORTH CAROLINA A YEAR LATER ON OCTOBER 15, 1954. SOURCE: CLIMATOLOGICAL DATA, OCTOBER 1953.

and the only gulf storm, Hurricane Alice, struck northeast Mexico and the Rio Grande valley. There were a total of 11 tropical storms and hurricanes this season, the most unique being a rare winter hurricane, again using the name Alice, which formed in the mid-Atlantic near 22°N, 54°W on December 30. Alice moved southwest

TABLE 8-7. WIND GUSTS DURING THE PASSAGE OF TROPICAL STORM HAZEL ACROSS SOUTH FLORIDA, OCTOBER 9, 1953.

Reporting station	Wind gusts (mph)
Ft. Myers	62
Captiva	70
Okeechobee City	80
Patrick AFB, Cocoa	64

Source: Climatological Data, October 1953.

crossed the Leeward Islands as a minimal hurricane, and finally dissipated in the eastern Caribbean on January 5, 1955. The feature storm of the Atlantic season was Hurricane Hazel which formed east of the Antilles on October 5, moved west to longitude 75° at which it turned sharply, crossed Haiti and passed just to the east of the Bahamas before slamming into Wilmington, North Carolina on October 15, with 135 mph winds and maintained hurricane force winds through the Raleigh-Durham area to the Virginia border. Two other storms skirted the outer banks of North Carolina, Hurricane Carol on August 30 and Hurricane Edna on September 10-11. Carol ultimately moved rapidly north and crossed inland over Long Island and Connecticut on August 31. Edna followed a north northeast path making landfall along the Maine coast on September 11-12. Florida growers and ranchers were delighted to see these ladies take a heading further to the north.

1955

The hurricanes again bypassed peninsular Florida in 1955, preferring to continue their assault on the coastline of North Carolina. Hurricane Connie smashed into the Outer Banks on August 12, Hurricane Diane struck Wilmington, North Carolina on August 17, and Hurricane Iona hit the Outer Banks on September 19.

The nearest system affecting Florida was Tropical Storm Brenda which moved inland near the Louisiana-Mississippi border on August 1 causing heavy rain as far east as Niceville which received 14.44 inches, and wind gusts to 34 mph at Pensacola (Climatological Data, 1955).

1956

Peninsular Florida was again spared the ravages of a hurricane in 1956. However, the only hurricane to cross into the United States, Hurricane Flossy, affected west Florida. Flossy formed as a depression in the southwest Caribbean on September 21, crossed the Yucatan Peninsula and became a tropical storm in the Gulf of Mexico on September 22. She then moved due north intensifying into a hurricane late on September 23, after which she moved over Grand Isle, Louisiana during the night of September 23-24, turned east northeast, and passed between Pensacola and Panama City about 7:00 P.M. on September 25. The strongest winds measured in Florida were 64 mph with gusts to 88 mph at Forrest Sherman Field near Pensacola (Climatological Data, September 1956). Three tornadoes were spawned in Florida on September 24, one at Wewahitchka in Gulf County, one at Eridu in Taylor County and the third in the southern part of Suwanee County. Rains soaked open cotton, mature corn and peanut stacks and flooded pastures, while the wind broke some limbs on pecan trees. However, it was felt that the beneficial effect of the rains across northern Florida exceeded the storm damage in the northwest counties. Panama City received 6.65 inches of rain, Niceville 8.20 inches and Pensacola 10.69 inches.

1957

Florida's luck continued in 1957—no hurricanes! The major tropical feature of the year was Hurricane Audrey which formed in the Bay of Campeche on June 25 and crossed the Louisiana-Texas coastline on June 27. North and west Florida felt two tropical storms, an unnamed storm on June 8 and Tropical Storm Debbie on September 8. These storms resulted in very heavy rains which reduced the yield of field crops, but there was no wind damage. Some representative rainfall amounts were Panama City 10.5 inches, Niceville 5.52 inches, Pensacola 6.84 inches, Perry 9.95 inches and Tallahassee 8.98 inches.

1958

Although the season produced seven hurricanes and three tropical storms, no part of Florida was affected by either a tropical storm or a hurricane in 1958. The major feature of the season was Hurricane Helene which passed just east of the Outer Banks of North Carolina on September 27.

1959

Two tropical systems affected peninsular Florida in 1959 and one system affected the panhandle. Tropical Storm Irene moved northeastward out of the Gulf and moved inland over Pensacola on October 8. Wind gusts were only 40 to 50 mph and rainfall three to five inches. Flooding was minor, but unharvested cotton, soybeans and stacked peanuts were damaged. A tropical depression crossed from Tampa Bay to Cape Canaveral on June 18 and briefly became an unnamed hurricane in the north Atlantic. Judith was briefly a hurricane in the southwest Gulf of Mexico before moving out of the gulf between Fort Myers and Punta Gorda as a tropical storm early on October 18 and proceeding east northeast to enter the Atlantic on the afternoon of the same day. Highest winds were 50 to 55 mph near the point of landfall. Storm rainfall totals were mostly four to seven inches in a band 100 miles wide, 50 miles north and south of a line from Fort Myers to Fort Pierce (Butson, 1959). Judith later reached hurricane status again in the mid-Atlantic.

The major storm of the 1959 season was Hurricane Gracie which formed in the southeast Bahamas on September 22 and crossed the South Carolina coastline between Beaufort and Charleston on September 29.

Chapter 9

The 1960s, Including Hurricanes Donna and Betsy

"Hurricanes Cleo, Dora and Isbell, all in 1964, caused the greatest vegetable damage for any one year."—G. Norman Rose, Florida Agricultural Statistics Service (1973).

1960

The ten-year vacation from hurricanes which Florida had enjoyed after Hurricane King roared up the east coast in October 1950, ended abruptly and forcefully in September 1960 when the state was visited by two hurricanes, Donna on September 9-11 and Ethel on September 14-15, and one tropical storm, Florence on September 21-25. The damage to agriculture from these storms was accentuated by the fact that above average rainfall had fallen throughout the state prior to the arrival of the storms. Record or near record totals, up to 150 to 250 percent of average, hit some sections of the state, particularly the southeast coastal areas and the western counties of the Panhandle (Butson, 1960).

Early September rainfall totaled two to five inches in central and southern Florida which, coupled with late August rains, left the peninsula unusually wet prior to the arrival of Hurricane Donna which deposited seven to ten additional inches of rainfall across extreme south Florida and an additional five to seven inches on southwest, central and northeast Florida. Rainfall along the east coast was moderate, generally less than three inches from Palm Beach County to Brevard County. The highest total for the month was 29.50 inches at Perrine, just south of Miami.

Hurricane Ethel's effects were felt only in the western counties of the panhandle where rainfall totals ranged from eight to ten inches in Escambia and Santa Rosa counties to near two inches in the Tallahassee area. The greatest one-day precipitation was 10.84 inches on September 16 at the Milton Experiment Station.

Hurricane Donna—September 9-11, 1960

"When I came home from the service after the hurricane, I couldn't recognize the place. The trees were all stripped of leaves. My father only picked one trailer load of fruit that year." Hugh English, 1997.

Hurricane Donna stands alone as the most destructive hurricane ever to strike the Florida citrus industry, and was the first major hurricane to impact the industry since 1950. Its awesome power was demonstrated by the fact that it was the first storm to maintain hurricane force winds from the time it slammed into Florida until it dissipated in New England. The intervening 38 years, to 1998, have not produced a storm comparable in its effects on the citrus industry. In fact, the past 38 years have not produced any hurricanes sufficiently severe to cause widespread damage to Florida citrus. If a major hurricane followed Donna's course today, the damage to groves in Hendry and Collier counties, which did not exist in 1960, would be extreme as the strong right quadrant of the storm would have swept through those counties with great ferocity, little diminished due to the flatness of the landscape.

Meteorological Aspects

Donna was a classic Cape Verde hurricane. If today's satellite technology had been available in 1960, it would probably have been seen first as a low pressure area moving off the coast of Africa. It is known (Dunn and Miller, 1964) that a strong easterly wave with heavy rain passed over Dakar, Senegal on August 29, and that heavy rain occurred in the Cape Verde Islands on August 31. The storm probably reached hurricane intensity on September 1 near 12° north latitude and 40° west longitude. It was located by a Navy hurricane hunter plane on September 2 at approximately 14° north latitude and 49° west longitude. At that point, it was already a major hurricane with 138 mile per hour winds and a central pressure of 28.73 inches. Donna moved west-northwest at approximately 19 miles per hour, passing through the Leeward Islands the evening of September 4, causing severe damage in Barbuda and St. Maarten. Fortunately, Puerto Rico, the Virgin Islands, and the Turks Islands were south of the center and suffered little damage. However, Mayaguana, Acklins Island, Fortune Island and Ragged Island in the Bahamas suffered major damage to houses and buildings from 12 or more hours of hurricane force winds with gusts in excess of 150 miles per hour.

When the eye crossed the Florida Keys after midnight on September 10, the central pressure had dropped to 27.46 inches and sustained winds were 140 miles per hour with gusts to 180 to 200 miles per hour. It was a category 4 hurricane by the Saffir-Simpson scale (Table 15-4), and as the hurricane left the Keys, it began to recurve northward at a forward speed of 8 miles per hour as it aimed its ferocious wind gusts toward Naples and Fort Myers.

Track Across Florida

It would be difficult to plot a storm track with greater potential for damage to Florida citrus (Figure 9-1) than that followed by Hurricane Donna as the eye passed over Naples and Fort Myers, then followed U.S. Highway 17 north over Arcadia, Wauchula, Bartow, Winter Haven, just west of Orlando and DeLand and reentered the Atlantic Ocean at Flagler Beach, with the powerful right quadrant passing through Collier, Lee, Charlotte, Hendry, Glades, DeSoto, Hardee, Highlands, Polk, Lake, Orange and Volusia counties—a blueprint for disaster. At Fort Myers, the sustained wind was 92 miles per hour with gusts to 121 miles per hour (Butson, 1960). At Everglades City and Naples, sustained winds were estimated at 100 miles per hour with gusts from 140 to150 miles per hour. Heavy rainfall ranged from 5 to 10 inches across a band 80 to 100 miles in diameter with rain concentrated 50 to 70 miles to the right of the storm's path and 30 to 40 miles to the left. Tables 9-1 and 9-2 show meteorological data at key stations mainly in agricultural areas.

FIGURE 9-1. THE PATH OF HURRICANE DONNA AS IT PASSED OVER NAPLES AND FORT MYERS AND FOLLOWED U.S. HIGHWAY 17 NORTH THROUGH ARCADIA, WAUCHULA, BARTOW, WINTER HAVEN TO JUST WEST OF ORLANDO AND DELAND AND MOVED OFFSHORE NEAR FLAGLER BEACH, SEPTEMBER 9-11, 1960. SOURCE: CLIMATOLOGICAL DATA, SEPTEMBER 1960.

The Damage

Prior to the hurricane, there were unofficial predictions that Florida would see its first 100 million box orange crop and 40 mil-

TABLE 9-1. METEOROLOGICAL DATA AT REPRESENTATIVE FLORIDA STATIONS AS HURRICANE DONNA MOVED FROM SOUTH TO NORTH THROUGH THE FLORIDA PENINSULA, SEPTEMBER 9-11, 1960.

Station	Minimum pressure (inches)	Maximum sustained winds (mph) directions	Time	Wind gusts (mph)
Tavernier[1]	28.80	120	—	160[3]
Everglades City	28.15	150 NE	9:15 A.M.	150[3]
Ft. Myers	28.08	92 NE	1:31 P.M.	121
Naples[2]	28.04	100[3]	12:20 P.M.	150[3]
Homestead	29.17	60 ESE	4:36 A.M.	89
Tampa	29.11	62 N	9:40 P.M.	75
Lakeland	28.60	68 NE	9:35 P.M.	90[3]
Leesburg	28.88	62 NE	1:30 A.M.	96
Orlando	28.66	46 ENE	2:36 A.M.	69
Daytona Beach	28.73	58 ENE	3:42 A.M.	86

[1]Sustained wind above 120 mph for 45 minutes, above 100 mph for 7 hours. Gusts above 100 mph for 13 hours.
[2]At Naples, the station was abandoned before time of highest winds. Eye over city 12-1 P.M.; 3 to 4 feet of water over most of city.
[3]Estimated.
Source: Climatological Data, September 1960.

lion box grapefruit crop for the 1960-61 season (Florida Grower and Rancher, 1960). However, Donna's strong, gusty winds prevented this from occurring. Gale force winds with gusts to hurricane force lashed south and southwest Florida the afternoon of September 10, south central Florida into the afternoon and evening and finally north central Florida during the early morning hours of September 11. Many trees in unprotected locations and along the central Florida ridges were stripped of leaves and fruit. Damage to seedling trees was especially severe.

The September 19, 1960 Triangle provided a complete, county-by-county estimate of hurricane damage similar to that provided after major freezes. The results were as follows, beginning with the most heavily damaged southwest Florida counties and ending with the least damaged north central counties:

TABLE 9-2. RAINFALL AMOUNTS IN PENINSULAR FLORIDA AS HURRICANE DONNA MOVED NORTH FROM FORT MYERS THROUGH CENTRAL FLORIDA TO DAYTONA BEACH, SEPTEMBER 9-11, 1960.

Location	Rainfall (inches)			
	Sept. 9	Sept. 10	Sept. 11	Monthly total
South Florida				
Fort Myers	0.21	4.63	0.17	11.93
Homestead Exp. Sta.	2.93	4.41	0.45	19.04
LaBelle	T	1.85	2.05	9.36
Miami WB	2.36	6.07	0.05	24.40
Central Florida				
Avon Park	0.63	2.75	3.75	14.82
Clermont	0.14	1.55	4.00	11.38
Lake Alfred Exp. Sta.	0.17	8.10	1.40	19.44
Lakeland	T	6.63	0.28	10.25
Orlando WB	0.01	2.85	1.62	11.21
Tampa WB	T	3.42	0.18	8.17
North Florida				
Daytona Beach	T	1.97	2.88	18.15
DeLand	0.30	2.00	4.26	13.83
Jacksonville	0.01	0.62	4.12	8.57

Source: Climatological Data, September 1960.

Lee County and Hendry County: Very severe damage to citrus in this area. Ninety to 100% of the grapefruit was lost, 60 to 65% of the early and midseason oranges and 50% of the Valencias. There was a 15 to 20% tree loss and a very heavy foliage loss. No water damage.

Collier County: No report due to scarcity of plantings in that county in 1960.

Charlotte County: This was one of the more severely damaged counties in the state. Loss of grapefruit was set at 75% and oranges 40 to 50%. Some groves suffered a greater loss, but a few got by surprisingly well. Tree and foliage damage in older groves was severe.

Glades County: No report due to few plantings in that county in 1960.

DeSoto County: Extensive damage to citrus in this county. There was quite a variation in the loss between individual groves. However, most old groves were damaged severely, especially the old seedling orange groves. Damage to the orange crop in this county was estimated at from 40 to 50% with the loss on some individual groves even higher. The grapefruit loss was from 80 to 90% and the tangerine loss at 25 to 30%. An additional loss of fruit was anticipated from bruising and thorning in the high winds. Tree damage was heavy especially to seedling oranges. One grower reported a 30 acre seedling orange grove to be a total loss. The younger, low bush trees were not damaged so badly even though there was a heavy loss of leaves and numerous broken limbs. All groves were completely waterlogged with standing water in many places. This may have resulted in the loss of additional fruit from weakened stems and trees.

Hardee County: Damage was very severe. Estimates of fruit loss were 75% of grapefruit, 40 to 50% of oranges and 20 to 25% of the tangerine crop. There was severe damage to trees and foliage with seedling orange trees the hardest hit. Younger trees, under 10 to 12 years of age, got by pretty well with an occasional tree split or uprooted. Water was not the problem expected. Rainfall was from four to five inches and except in the lowest spots no damage was anticipated from excess water.

Highlands County: Fruit loss was very heavy with grapefruit estimates ranging from 75 to 90%. Marsh grapefruit suffered a heavier loss than Duncans. Estimates of orange losses ranged from 15 to 35% and tangerines at 20%. Tree damage ranged from some to total loss in all old groves, especially grapefruit. Foliage loss was heavy, particularly on tall trees. There was considerable tree splitting in tangerines which resulted in later fruit loss. There was also a heavy loss of foliage, especially on the taller trees. The younger, low bush trees got by pretty well. No water damage was reported in this county although lakes were almost at record high levels.

Sarasota County: Loss of grapefruit was estimated at 40 to 50% with some groves losing as much as 75% of the fruit. Loss of early and midseason oranges was 25% and Valencias 15%. There was some tree loss in older groves but not severe.

Polk County: Grapefruit losses ranged from as little as 10% to as high as 95% with most between 20 and 60%. A calculated average loss for the county was between 35 and 40%. Seedless grapefruit suffered a greater loss than seedy varieties. Early and midseason orange varieties suffered losses from 5% to as high as 70% with an average for the county of 20%. Valencias had much less damage with a range of 1 to 50%, averaging about 10%. Tangerine losses ranged from 5 to 50%, averaging 10 to 15% with continuing heavy drop due to fruit damage. Loss of trees due to splitting and uprooting ranged from heavy to serious in exposed locations but was nil in most groves. Large, high, old trees with heavy crops of fruit and a wet soil were the most severely damaged.

Hillsborough County: The county lost from 15 to 20% of its orange crop with early and midseason oranges generally taking a greater loss than Valencias. Approximately 35 to 40% of the grapefruit were on the ground and in the hardest hit areas 15 to 20% of the tangerines were on the ground. The eastern part of the county (the Plant City area) lost 20 to 25% of its orange crop and 50% of its grapefruit. The southwest section (the Riverview area) lost 6 to 10% of its oranges and 35 to 40% of its grapefruit while the northwest section (the Lutz area) suffered only slight damage. Much of the fruit left on the trees was bruised, thorned or scarred and a portion of it was certain to drop. Tree and foliage damage ranged from very severe in the eastern section, to severe to moderate in the southwest section, to slight in the northwest section. The older seedling trees were, of course, hurt the worst; however, budded trees, specially on rough lemon rootstock were also hurt badly. In some cases where the soil had excessive moisture, three to four year old trees were blown over. Approximately 3% of the trees in the Plant City area were blown over or broken up and the foliage on the trees left standing was sparse and badly damaged.

Broward County (Davie): Estimates varied with the consensus at 20% loss of grapefruit, 15 to 20% of early and midseason oranges and 10% of Valencias. Tree and foliage damage was very light.

St. Lucie County and Indian River County: There was about a 5% grapefruit loss, but no orange loss and no tree or foliage damage.

Brevard County: This county suffered a grapefruit loss of about 20% and an orange loss of 3 to 5%. There was some tree and foliage damage as well as water damage.

Pasco County: About 10 to 15% of the grapefruit and 2 to 3% of the oranges were on the ground. Early and midseason oranges, especially seedlings, and Marsh grapefruit were hardest hit. Individual groves lost as much as 50% of their fruit and as little as 1%. Trees and foliage damage was generally slight but was very severe in some of the older seedling blocks and the old tall budded trees. Less than 1% of the trees in the county were severely damaged.

Lake County: Damage was light. Grapefruit loss was put at about 10%, oranges 1 to 2% and losses to tangerines, tangelos and Temples was negligible. Some further loss of bruised and scarred fruit was expected, but it was too early to determine the extent of the loss. Tree damage was very light.

Orange, Osceola and Seminole counties: Orange and Seminole counties suffered near equal loss from the hurricane. These percentage estimates are overall averages and there are variations both ways. Grapefruit loss was estimated to be 15 to 20%, orange loss 3 to 5%, and tangerine loss 5 to 8%. Osceola County experienced somewhat heavier losses. Grapefruit was set at 50%, oranges at 10 to 15%, and tangerines 8 to 10%. Over the entire area, seedling trees as well as fruit suffered the most damage. Loss of fruit from navel trees was also above average. Damage to trees of other varieties was mostly light. Valencias appear to have come through with the least amount of damage. Some further loss from punctured and bruised fruit was expected, but there was no way at the time to estimate this loss. Some water damage was certain. In many cases, there was no way to remove the excess water.

Hernando County and Pinellas County: Orange loss in these counties was estimated at 1 to 2% and grapefruit loss at 5 to 7%. Tree and foliage damage was negligible.

Citrus County and Sumter County: These counties experienced only gale winds and hardly any damage to either fruit or trees.

Marion County: There was only a small amount of fruit and tree damage. Only a few trees which had been damaged in the 1957 freeze were lost. Some water damage was expected in low pockets.

Volusia County: Damage to citrus in the western part of the county was not severe. Less than 3% of the fruit was lost and tree damage was negligible. However, in east Volusia County, grapefruit loss was estimated at 20% and oranges 5 to 10%. Some groves had as much as 3 feet of standing water immediately after the hurricane.

Tree damage was confined to seedling oranges and trees damaged severely in the 1957 freeze.

In commenting on the storm, Bob Rutledge, General Manager of Florida Citrus Mutual (Triangle, Sept. 14, 1960) indicated that growers in Polk, Highlands, Hardee, DeSoto and Lee counties might have experienced losses of 50 to 90% of grapefruit and 35 to 50% of oranges. However, he emphasized that Florida was still very much in the citrus business with, "a merchantable grapefruit crop of 23 to 25 million boxes, compared with the previous season's crop of 31,500,000 boxes, and an orange crop which would not exceed the previous season's 91,500,000 boxes." Rutledge also cautioned growers that maturity standards should be maintained to protect the reputation and integrity of the Florida citrus industry.

A reassuring message was also issued by Homer E. Hooks, General Manager of the Florida Citrus Commission (Florida Grower & Rancher, 1960), who reported to 10,000 members of the trade-buyers, brokers, distributors and retailers all over the U.S. and Canada, one week after the storm, that "despite substantial losses, Florida is still very much in the citrus juice business, and there has been no relaxation of Florida's well-known high quality standards for fresh and processed citrus."

A preliminary estimate issued by Charlie Townsend of the Crop Reporting Service of the USDA in Orlando on September 14, only four days after the storm, indicated that 25 to 35% of the state's grapefruit crop and 5 to 10% of the orange and tangerine crops would be lost (Citrus & Vegetable Magazine, 1960).

When the official USDA crop estimate was released on October 10, 1960, the prediction was 90.5 million boxes of oranges and 30 million boxes of grapefruit, down only slightly from the 91.5 million boxes of oranges and 30.5 million boxes of grapefruit harvested during the 1959-60 season (Citrus & Vegetable Magazine, 1960). The biggest surprise in the estimate was the grapefruit figure which confirmed the suspicions of many that Hurricane Donna had bailed out the state's grapefruit growers by knocking off a large amount of fruit which would have gone unsold. Homer Hooks, General Manager of the Florida Citrus Commission, said the loss estimates of between 12 and 14 million boxes of grapefruit, "indicated there may have been 42 to 44 million boxes in the groves, nearly a third more than was successfully utilized last year." Hooks added that, "the October estimate indicated a 10 million box loss of oranges due to the hurricane, or total hurricane damage between 22 and 24 million boxes. J. Dan Wright, Chairman of the Florida Citrus Commission added that Donna presented growers with an "economic reprieve" (Citrus Industry, 1960).

Governor LeRoy Collins, after touring the storm-damaged industry, noted that, "I'm glad the damage is not worse than it is. It seems to me that the citrus industry always appears able to adjust itself to these disasters (Citrus & Vegetable Magazine, 1960).

Follow-up Damage Reports

Four weeks after the initial damage reports appeared in the Triangle, Florida Citrus Mutual followed with a second county by county assessment (Triangle, Oct. 14, 1960) again beginning with the most severely damaged southwest counties.

Lee County and Hendry County: Groves continued to look weather beaten although in Hendry County a vigorous new growth was coming in defoliated areas. The job of cleaning up after the hurricane continued with the removing of trees which could not be propped up to save. Some fertilization and cultivation had started but this was not general. There was water damage in low areas but this was not serious. Fruit damage continued to appear in all varieties but was slowing down. No confirmed sales reported.

Collier County: Again no report due to few groves present.

Charlotte County: No report was given, but as damage was comparable to that in Lee County and Hendry County, the situation was similar.

DeSoto County: Grove condition, especially old groves, was bad. Some groves had been cleaned and the trees straightened, but some seedling groves had been neglected. No fertilization or cultivation was taking place due to water and continued rains, but water damage was limited to young trees on low, sandy soils. Droppage continued high in all varieties. Sales were limited as growers who had not sold before the storm were expecting higher prices.

Hardee County: Excess water continued to be a major problem in this area, but growers were generally engaged in cleaning up debris left by the hurricane. Many were leaving trees that were down but still green with the hope of maturing the fruit on the trees. Buyers had been very active during the week.

Highlands County: The general appearance of groves was good. Most broken trees had been removed and the up-rooted trees had been either straightened or removed. Trees that lost foliage in the hurricane were putting out healthy new foliage. Dropping of fruit continued in all varieties of oranges and grapefruit. Growers were generally cultivating and fertilizing. Buying activity had been slow.

Sarasota County and Manatee County: Tree condition was generally good. Damaged trees were being removed or straightened. Some water damage was apparent to trees on light soils. Dropping of damaged fruit continued, but the loss was not excessive. Groves on higher ground were being cultivated and fertilized. No confirmed sales were reported.

Polk County: In general, storm damaged trees had been erected or removed, although some growers were leaving downed, green trees to salvage the fruit before removing. The flush of growth replacing storm lost foliage had been extremely gratifying. Fruit drop since the storm had been significant but had almost ceased. The larger sized grapefruit were mostly blown off by the hurricane which had added delay to early movement. Even though there was considerable scarring of fruit from the storm, ample quantities of clean, good quality fruit remained on the lower branches.

Offers and sales of grapefruit and oranges were taking place, but tangerine loss and fruit scarring had been heavy and sales slow. However, now that severe dropping was over and more fruit were beginning to size more activity was expected. The Temple crop was much short of the previous season and post-hurricane fruit drop had been as great or greater than immediate hurricane loss. This, along with wind scars on fruit, was causing a wait and see attitude on the Temple crop.

Hillsborough County: Fruit continued to drop from injuries received from hurricane winds, but splitting was the chief cause of drop which varied from grove to grove.

Broward County: No report, but tree conditions should have been good.

Brevard, St. Lucie and Indian River counties: Tree condition very good and growers were following normal production practices.

Pasco County and Hernando County: Fruit drop continued and was more severe on Hamlins and seedlings than other varieties.

Lake County: Trees in the higher land areas were in good condition with healthy foliage. The loss of leaves due to the hurricane was comparatively light. Droppage of fruit was about normal although it was rather severe in all varieties immediately after the hurricane. There was some scarring of tangerines and Temples. Fruit buyers were active.

Orange, Osceola and Seminole counties: Trees were recovering rapidly from the effects of the hurricane. New growth was exceptional on trees that lost foliage due to the high winds. Buying activity normal.

Pinellas County: Fruit drop due to splitting and hurricane winds was severe in some groves and negligible in others. Fruit buyers were fairly active.

Citrus County and Sumter County: No report as damage from the hurricane was minimal.

Marion County: Some water damage around land-locked lakes, but no problems resulting from hurricane damage.

Volusia County and Putnam County: Trees exposed to hurricane winds were showing abundant new growth on defoliated limbs. There were aphid problems on the new growth. Tangerines were sizing and showing no wind damage.

Tropical Fruits

The avocado crop was hit hard. Most of the fruit was blown from the trees and the trees themselves sustained heavy damage (Climatological Data, October 1960).

Vegetable Damage

Early pepper plantings in the Fort Myers-Immokalee area were severely damaged which required heavy replanting. The entire pepper producing area was adversely affected by the rainfall, not only from Donna, but also from the unusually heavy fall rains which followed the hurricane (Rose, 1974).

Excessive rainfall from Hurricane Donna saturated most celery producing regions of Florida which interrupted and delayed planting operations. Snap bean plantings were also interrupted and in many cases replanted. However, 800 acres were lost and not replanted (Rose, 1975A, B).

Tomato growers suffered very heavy losses as Hurricane Donna passed over the state's fall tomato acreage (Rose, 1973). Major vegetable crops in Florida affected by Hurricane Donna are shown in Table 9-3.

Damage to Vegetation in Southern Florida

The slow forward movement of Hurricane Donna subjected the area to extremely damaging winds for almost 36 hours. In addition, high storm tides, 12 feet above normal at Flamingo, added to the damage. In the mangrove belt from Madeira Bay west to Shark River, most trees over two inches in diameter were sheared off six to ten feet above the ground, and from the Shark River to Lostman's River 50 to 75% of the mature mangroves were dead leaving open stands with no canopies at many locations. North from Lostman's River to Everglades City only 10 to 25% of the mangroves were lost.

In the Florida Bay Keys, coconut palms and cabbage palms were severely damaged (Craighead and Gilbert, 1962), but thatch palms suffered less damage. Severe damage also occurred among the 10,000 islands and Cape Sable where much of the vegetative growth including shrubs, yuccas, agaves, cacti, grasses and small hardwoods was stripped. Pines on Long Pine Key escaped major damage, but nearer the path of the storm center, in the Ft. Myers, Naples-Corkscrew areas west of Immokalee in Collier County, the pinelands suffered severe damage.

Vegetation in the sawgrass areas of the open Everglades suffered much less, although there was wind breakage and defoliation of the larger trees in the bayheads. The taller cypress trees in Everglades National Park were toppled. The excellent study by Craighead and Gilbert (1962) provides great detail regarding the hurricane's effect on the many species of small plants in the park. The authors concluded that, "it is obvious that the hurricane will have profound effects on plant succession in the area," as many of the more vulnerable species would be replaced by more storm tolerant vegetation.

TABLE 9-3. ACREAGE OF MAJOR COMMERCIAL VEGETABLE CROPS IN COUNTIES AFFECTED BY HURRICANE DONNA, SEPTEMBER 9-11, 1960 (COUNTIES LISTED ALPHABETICALLY).

County	Acreage
Brevard	2,060
Broward	8,050
Charlotte	1,160
Collier	10,355
Dade	37,785
DeSoto	800
Flagler	5,350
Glades	1,530
Hardee	4,120
Hendry	12,975
Hillsborough	10,195
Indian River	600
Lake	1,525
Lee	6,945
Manatee	4,370
Marion	9,820
Martin	2,460
Okeechobee	1,980
Orange	11,810
Palm Beach	109,640
Polk	3,525
Putnam	7,855
St. Johns	16,900
Sarasota	1,180
Seminole	4,420
Sumter	7,370

Source: Florida Agricultural Statistics Service, 1960.

Economic Considerations, Citrus

As Hurricane Donna roared up the Florida peninsula on September 10, one month to the day before the release of the 1960-61 crop estimate on October 10, 1960, it is difficult to calculate the number of boxes lost and the resultant economic consequences with any degree of certainty. According to the Florida Agricultural

Statistics Service, the actual production for the 1959-60 Florida crop was 91.5 million 90 pound boxes of oranges, including Temples, 30.5 million 85 pound boxes of grapefruit, and 2.8 million 95 pound boxes of tangerines. However, before the hurricane, industry guesstimates had ranged to as high as 100 million boxes of oranges and 40 million boxes of grapefruit. How close those guesstimates were will never be known, but production was certainly in an upswing, as evidenced by the jump to 113.4 million boxes of oranges in the 1961-62 season, Florida's first 100 million box crop (Table 9-4) and the increase in the number of bearing trees (Table 9-5).

Government Response

At its October 1960 meeting, the Florida Citrus Commission responded to the havoc wreaked by Hurricane Donna by cutting its operational budget by 12% (Citrus & Vegetable Magazine, 1960).

Personal Observations

William H. "Bill" Krome (1997) of Homestead in south Dade County wrote, "Donna, in September 1960, was the first bad hurricane in over ten years. The center of Donna missed us, passing over

TABLE 9-4. FLORIDA CITRUS PRODUCTION IN MILLIONS OF BOXES FOR THE YEARS BEFORE AND AFTER HURRICANE DONNA.

Crop year	Oranges and Temples	Grapefruit	Tangerines
	---------------- million boxes ----------------		
1958-59	86.0	35.2	4.5
1959-60	91.5	30.5	2.8
1960-61[1]	86.7	31.6	4.9
1961-62	113.4	35.0	4.0

[1]Hurricane year.
Source: Florida Agricultural Statistics Service, 1960.

TABLE 9-5. INCREASE IN NUMBERS OF BEARING TREES FROM THE 1958-59 SEASON THROUGH THE 1961-62 SEASON.

Season	Early and midseason oranges	Valencia oranges	Grapefruit
	----------------- thousands of trees ----------------		
1958-59	12,946	12,005	6,563
1959-60	13,577	12,890	6,654
1960-61[1]	12,825	13,034	6,000
1961-62	13,258	14,684	6,105

[1]Hurricane year.
Source: Sugg, 1966 and Buston, 1965.

the Cape Sable area about 50 miles southwest of Homestead. But the storm was so violent and covered such a broad area that it did major damage to the groves in Dade County. We had not yet developed the practice of topping all grove trees every one or two years, so we had a lot of trees to put back up, and it reinforced our determination to keep the tops low and to plant summer varieties. I was struck by the damage it did to the mangrove forest at Cape Sable, as well as to all the mangroves in the middle Keys. The mangroves at the Cape will probably never come back, while the smaller mangroves that fringe most of the Keys seem to have recovered pretty well after 30 years."

Dr. William "Bill" Grierson (1998), packinghouse researcher extraordinaire remembers, "It was pretty lively until the eye of the hurricane reached us with its expected period of delusive calm. That must have been somewhere near midnight. I stepped outside just as my neighbor, Marion Gill, did also. Marion had one of those huge flashlights that looked a bit like a kid's lunch pail. He shone it on the row of tall palm trees that, at that time, ran along the back of both of our properties. *Every frond on those big palms was straining upwards.* It was quite eerie to see, looking like arms raised in supplication! Though startled, I realized that this was quite logical. The "eye" is a small area of intense low pressure. The "calm at the eye of the storm" was actually an area of *upward* wind."

Mr. Dudley Putnam (1998), prominent Bartow area grower, recalls that it seemed an oak tree was down between every two power poles. They had no electricity for two weeks. Everything was out including the ice plant and they scrounged around the county looking for a block of ice to buy. Mr. Putnam also remembers efforts to straighten citrus trees blown down by Hurricane Donna. "We had big trees blown over, 25 box trees. First we cut the tap root, then we used a ¾ ton truck, a former munitions carrier, with a winch on the front. We wrapped a chain in burlap bags so as not to scrape the trees anymore than necessary. The chain went around a limb and we pulled it up. We could do four trees per hour."

Hurricane Ethel—September 14-15, 1960

According to Keith Butson (1960), state climatologist at the U.S. Weather Bureau, Gainesville, Florida, Hurricane Ethel developed rapidly in the central Gulf of Mexico on September 14 with winds near the center estimated at 125 to 150 mph. However, the storm lost intensity almost as fast as it had developed and when the center reached the Gulf Coast on September 15 near the Alabama/Mississippi border, no point on the Florida Gulf Coast felt hurricane force winds. The lowest pressure at Pensacola airport was 29.77 inches at 7:45 A.M. on September 15 and the fastest wind was east 32 mph at 4:48 A.M. on the 15th with gusts to 51 mph at 7:48 A.M. Apalachicola recorded gusts to 50 mph at 10:52 A.M. on the fifteenth. The right forward quadrant of Hurricane Ethel spawned several tornadoes, waterspouts and funnel clouds. Two tornadoes were confirmed in the Panama City area and a third near Port St. Joe. No deaths or serious injuries were reported in Florida.

Rainfall occurring with the passage of Ethel is shown in Table 9-6.

The Damage

No major agricultural damage resulted from either the wind gusts or the rainfall from Ethel which was only a tropical storm when it made landfall. However, the Milton Experiment Station suffered a major deluge (Table 9-6).

1961

Florida was blessed with a year to recover from the ravishes of Hurricane Donna in 1960. Although there were a total of eleven

TABLE 9-6. RAINFALL AT SELECTED WEST FLORIDA STATIONS DURING THE PASSAGE OF TROPICAL STORM ETHEL, SEPTEMBER 14-16, 1960.

Station	Rainfall (inches)
Apalachicola	1.85
Carrabelle	2.05
Crestview	2.64
DeFuniak Springs	2.98
Milton Experiment Station	12.94[1]
Panama City	2.56
Pensacola Airport	7.36
Quincy	4.30

[1]10.84 inches fell on September 16, 1960.
Source: Climatological Data, September 1960.

tropical storms and hurricanes in the Atlantic Basin, using names from Anna to Inga with one storm unnamed, neither tropical storm nor hurricane passed near the Florida coastline. In fact, only one system struck the U.S. mainland. This was Hurricane Carla, which hit the coast of Texas on September 11.

1962

There were only five tropical systems in the Atlantic Basin in 1962, three hurricanes and two tropical storms. Of these, only tropical storm Alma which passed along the east coast from Miami to Titusville on August 26-27 affected the Sunshine State, but winds were not damaging and precipitation was moderate.

However, although Florida was spared hurricane damage in 1962, it was a disastrous year weatherwise for Florida agriculture which was heavily damaged by one of the most severe freezes of all time on December 12-13, 1962 (Attaway, 1997).

1963

Fortunately, the debilitating freeze of December 1962 was not followed by a hurricane or tropical storm in 1963. Despite the generation of seven hurricanes and two tropical storms during the year in the Atlantic Basin, none of these systems affected either peninsular Florida or the Florida panhandle.

1964

Florida's vacation from destructive tropical weather systems ended decisively in 1964. Of twelve systems in the Atlantic Basin, three hurricanes and two tropical storms affected the state. The hurricanes were Cleo, Dora and Isbell. These three hurricanes caused the greatest damage to Florida vegetable production for any one year. Taking these hurricanes in the order in which they occurred:

Hurricane Cleo—August 27-28, 1964

Meteorological Considerations

Hurricane Cleo was of Cape Verde origin. According to Dunn and Staff. (1965), it developed from a disturbance which came off the African coast south of Dakar on August 15 as a 1010 mb (29.83 inches) low accompanied by continuous rain and thunderstorms. It was first recorded and photographed by a TIROS VIII satellite on August 18, 1964 as a broad area roughly centered at 10°N, 34°W, and on that same date a German ship recorded a light east wind, continuous rain and a pressure of 29.79 inches at 12°N, 33.5°W. As the system moved west northwest at about 16 mph, it was first located by a hurricane hunter plane on August 20 near 13.2°N, 44.5°W. The plane recorded it as a tropical depression with a minimum pressure of 29.71 inches with 35 mph winds in squalls. However, on the twenty-first, Cleo rapidly intensified to hurricane force with a central pressure of 29.32 inches. Moving west northwest at 20 to 25 mph, an unusually rapid speed for a hurricane at this low latitude, Cleo reached Guadeloupe early in the afternoon of August 22 with gusts to 81 mph and a minimum pressure of 29.62 inches being measured at a point 15 miles north of the eye. Agricultural damage in the French West Indies was severe with the banana crop said to be destroyed.

The hurricane continued to the west northwest passing south of the Virgin Islands and Puerto Rico on August 23 and continued to intensify with a central pressure of 28.05 inches and peak winds of 140 mph. However, it was a very concentrated storm with hurricane force winds limited to a small area near the center. A Navy hurricane hunter plane was badly damaged on the twenty-third with seven crewmen injured as it left the eye of the storm and entered the wall cloud encountering winds which increased by 90 mph with a distance of one mile. As a result, no attempt was made to penetrate the eye wall on August 24 and its winds and pressure on that date are not known. On the twenty-fourth, Cleo passed south of the Dominican Republic and over the southwest peninsular of Haiti inflicting considerable damage. However, after passing over land, the hurricane lost strength and did not reintensify in the Caribbean.

Shifting to the northwest, Cleo entered Cuba as a minimal hurricane near longitude 79°W (Dunn and Staff, 1965). The hurricane crossed Cuba on a north northwest course with winds of less than hurricane force and emerged into the Atlantic at 7:00 A.M. EST on August 26. Damage in Cuba was not serious. Again over warm water, Cleo continued to the north northwest, being carefully monitored by Key West and Miami radar, and rapidly regained hurricane force with winds at 85 mph and a central pressure near 29.06 inches. At 2:00 A.M. EST on August 27, the eye reached Key Biscayne, the first hurricane to strike Miami head on since Hurricane King on October 17, 1950. The minimum pressure reading recorded in the Miami area was 28.57 inches with maximum sustained winds of 100 to 110 mph with gusts to 135 mph (Butson, 1964A).

As Cleo moved north up the east coast of Florida (Figure 9-2), the eye varied from 8 miles to 16 miles in diameter. Fortunately for Dade County growers, damage in the Perrine and Homestead areas and west of Bird Road was described as minor. Meteorological data for Hurricane Cleo is shown in Table 9-7.

The Damage

Citrus: Damage estimates indicated a loss of 10% of the oranges and 20 to 30% of the grapefruit in the Indian River area, with

FIGURE 9-2. THE COURSE OF HURRICANE CLEO AS IT STRUCK MIAMI ON AUGUST 27, 1964 AND MOVED UP THE EAST COAST OF PENINSULAR FLORIDA DAMAGING CITRUS GROVES IN THE INDIAN RIVER AREA AND VEGETABLE FIELDS IN THE POMPANO BEACH AND FT. PIERCE AREAS. SOURCE: CLIMATOLOGICAL DATA, AUGUST 1964.

heavier damage in Palm Beach and Martin counties than in Indian River and St. Lucie counties. A tornado touched down in Broward County, along a path one-quarter mile wide and one-half mile long, uprooting 2,400 of 6,000 trees in a 100-acre grove. In this severely

TABLE 9-7. MINIMUM PRESSURES, SUSTAINED WINDS, PEAK GUSTS AND RAINFALL AS HURRICANE CLEO MOVED NORTH ALONG THE FLORIDA EAST COAST AUGUST 27-28, 1964.

Location	Minimum pressure (inches)	Maximum sustained winds (mph) direction	Wind gusts (mph) direction	Rainfall (inches)
Key West	29.72	22 NE	28 NE	0.15
Miami (NHC)[1]	28.74	110 N[2]	135 N[2]	6.80
West Palm Beach	29.14	86 ESE	104 ESE	3.94
Orlando	29.43	32 N	46 NNE	3.20
Daytona Beach	29.43	40 E	60 E	3.82
Jacksonville	29.58	43 NE	44 ENE	2.71

[1]NHC = National Hurricane Center.
[2]Estimated.
Source: Butson, 1964A.

damaged area, the tree loss by variety was 5 to 40% of the grapefruit trees, 65% of the Navel oranges and 35% of the Valencia oranges.

As Hurricane Cleo approached the Florida east coast, the Florida Citrus Mutual Triangle for August 27, 1964 commented as follows:

> As this issue of the Triangle goes to press, hurricane Cleo has shifted back and forth in the Caribbean and apparently has settled on a course slightly inland and parallel with the east coast of Florida.
>
> Naturally we hope that those who are in the path of this storm, or who have interests in that projected path, will escape with a minimum of damage. In any event, a complete and factual survey of damages to citrus areas will be made by the Mutual staff and the results of that survey issued as rapidly as possible.
>
> Be sure and watch the next edition of your Triangle for a hurricane report.

And, the Crop Condition Report and Outlook Summary in the September 3, 1964 issue of the Triangle noted the following:

> The East Coast of the state was the only area affected by the hurricane. The most severe damage occurred in the extreme

southern part where there was some tree loss and limb breakage while almost no damage was reported from the northern part. Although fruit losses on some individual groves range much higher, the overall loss of grapefruit in the affected area is reported to be 25% to 30% or 2½ to 3 million boxes. The loss of oranges was negligible. There have been no guesses made as to additional losses as a result of bruising or thorning or reduction in grade.

In the September 18, 1964 issue of the Triangle, the following Hurricane Cleo footnote was added:

A survey made by Mutual's fieldman, Sebert Hall, in the Davie area of Broward County this week, pointed up some severe damage to a small area in Davie which appears to have been caused by a tornado accompanying Hurricane Cleo.

There is an area approximately one-fourth mile wide and one-half mile long that lies northwest of Davie in the area south of State Road 84 and west of Flamingo Road. In this area, tree loss ranges from 5 to 40% and fruit loss was substantial. Eighty-five percent of the grapefruit was lost; 65% of Navel oranges and 35% of Valencias and Pineapples.

More than one-third of the 6,000 trees in a 100-acre grove were lost to the tornado.

Except for this relatively small area, the balance of Broward County was as reported in an earlier Triangle, although fruit of all varieties continues to drop due to bruising, thorn damage, stem damage, etc.

Broward County: Tree condition, color, fruit crop, size and texture was very good prior to Hurricane Cleo. However, the storm changed the picture somewhat. This area had hurricane winds for about seven hours with 3 to 5 inches of rain. There was some tree loss, 2 to 3% in old groves, and considerable limb breakage.

Estimates indicate fruit losses will be severe and as follows: 70% loss of grapefruit, 50% loss of Navel oranges and tangelos, 30% loss of early and mid-season oranges, and 15% loss of Valencias.

The 1961-62 orange production was 810,000 boxes. Anticipated 1964-65 production is 75% or 608,550 boxes.

Martin County: Hurricane Cleo passed through or near Port Mayaca and the Indiantown citrus area causing considerable loss of fruit in the older groves. Actual loss has not been determined but early estimates indicate a loss of 35% to 50% of Navel oranges and grapefruit with only slight losses to other varieties. Damage to trees was light with no expected water damage to young plantings as all were well drained. There was no water standing in groves a few hours after the storm which dropped 5 to 6 inches of rain.

Production figures from this county for 1961-62 are not available. An increase of 150% is anticipated due to young plantings.

St. Lucie County: Tree condition and foliage color, new growth, etc. were very good before the hurricane. Now the picture has changed. Although there was a surprising lack of tree damage, heavy defoliation did occur in some areas. No water damage is expected since canal levels were low and heavy pumping started immediately after the hurricane had passed.

Prior to the storm, grapefruit had good external quality, size and shape, and oranges were of good quality and size. Some shipment of grapefruit was expected in September. This latter will not now happen since Cleo got the larger size fruit.

The older grapefruit groves suffered the heaviest losses to the storm while the younger groves got by much better. Some individual losses were above 75% but the average for the county is probably less than 50%. Orange losses were much less, probably around 5%. There will be an undetermined secondary loss from droppage of damaged fruit and loss in grade for fresh shipment.

The orange production for 1961-62 was 4,009,000 boxes. Anticipated 1964-65 production is 95% or 3,810,000 boxes. Indian River County: Tree condition, color, new growth, etc. is generally very good. Moisture conditions, which were not good before the storm, are fine now. Fruit sizing on all varieties is very good with excellent shape and texture on grapefruit.

There was little or no tree damage from the hurricane and the five to six inches of rain will help more than hurt. Estimates of grapefruit loss range from 25 to 50% while loss to oranges is not expected to exceed 5%. There will be a further monetary loss from down grading due to wind scars.

The 1961-62 orange production was 1,811,000 boxes. Anticipated orange production for 1964-65 is 95% or 1,720,000 boxes.

Okeechobee County: Tree condition and color is very good. New growth is at a minimum as it has been dry during the summer. Fruit has sized reasonably well considering dry weather and is of good quality and texture. Young tree plantings have been delayed because of dry weather. Hurricane Cleo relieved the drought condition but did only minor damage to fruit. This damage was very minor since the bulk of plantings in this county are oranges.

Brevard County: This county was more fortunate than her neighbors to the south from the effects of Cleo, but in past years has been less fortunate from freezes, so it all averages out.

Tree condition, color, new growth, etc. is very good throughout the county. Moisture conditions have been good enough to insure good fruit sizing. It is reported that sizes, quality, shape, and texture of fruit are well above normal. Crops are heavy with the possible exception of Valencias which are a little lighter than last year. There is little or no damage reported from the hurricane other than what will appear later in the form of wind scars, bruising, etc. 1961-62 orange production was 2,880,000 boxes. Anticipated 1964-65 production is 100%, or 2,880,000 boxes.

Highlands County: General tree condition is excellent for old and new trees. Foliage has a good healthy color and normal new growth. Moisture conditions are good with very little irrigation being done at this time. Cultivation is normal for this time of year and spraying activities are at a low level.

All fruit is sizing well with good outer appearance and quality. Grapefruit is well advanced with good shape and texture. Maturity is a little late but some movement is expected by September 15. There was no damage from Hurricane Cleo except possibly some wind scarring. Rainfall was about 1½ inches over the county. 1961-62 orange production was 5,919,000 boxes. Anticipated 1964-65 production is 90% or 5,327,000 boxes.

Looking ahead one year to Mutual's Crop Condition Report published in the Triangle for the week ending August 28, 1965, further references were made to the effects of Hurricane Cleo in two of the hardest hit counties as follows:

Broward County: Groves that were damaged by last year's Hurricane Cleo are showing a good recovery.

St. Lucie County: Tree condition is very good except for some groves that were damaged by last year's hurricane and freeze. These latter groves are light of foliage and fruit.

Vegetable Damage

Hurricane Cleo caused heavy damage to vegetable crops along the lower east coast, particularly vineripe tomatoes in the Pompano area, mature green tomatoes in the Ft. Pierce area and early pepper plantings in the Pompano-Martin County area (Rose, 1973).

Vegetable damage from Hurricane Cleo was summarized in Florida Weather and Crop News for September 1, 1964 as follows:

A total of 3,050 acres of mature green tomatoes planted in the Ft. Pierce area were damaged by the hurricane winds and rain and replanting was necessary in some fields. In the Pompano area, 850 acres of vine ripe tomato plants were stripped of their leaves and blown out of the ground. The acreage was replanted.

The fringe of the storm hit the Everglades damaging about 1,600 acres of the sweet corn crop which was only a few inches high. Planting was continued. The celery crop in the Everglades and at Zellwood escaped damage.

Only a small acreage of pepper had been planted in the Pompano area, but most of it was lost. Eggplant fields in the Pompano area were damaged and subsequently replanted.

Field Crop Damage

Damage to sugarcane was significant in the area east of Belle Glade where stalks were blown over, but fortunately only a small percentage of the cane was uprooted or received terminal bud or leaf damage. The Fellsmere area was near the center of the hurricane, and cane there suffered significant terminal bud and leaf damage.

Hurricane Dora—September 9-10, 1964

Meteorological Considerations

Hurricane Dora, a Cape Verde storm, had the distinction of being the first hurricane recorded which came off the Atlantic, directly from the east, into the St. Augustine/Jacksonville area and proceeded to the west across northern Florida (Figure 9-3). When the eye of Dora passed over St. Augustine, Jacksonville felt sustained winds of hurricane force (82 mph) for the first time in the nearly 80 years during which records had been kept (Butson, 1964B).

Dora developed from a low pressure system which moved through the Cape Verde Islands on August 28. On August 31, the TIROS satellite transmitted a picture showing a cloud mass near 11°N, 41°W indicating that the system had moved on a west southwest course after leaving the vicinity of the Cape Verde Islands, and on September 1, a hurricane hunter plane found that the central pressure in the cloud mass was 29.47 inches and the system was christened tropical storm Dora. The storm moved on a west northwest course and reached hurricane intensity on September 2, and turning more to the northwest, it reached a point 300 miles south of Bermuda as a large and severe hurricane with a central pressure of 27.82 inches. Dora then returned to a west northwest path at a slower speed, allowing the beach front communities of northeast Florida to prepare for the onslaught. As she hammered ashore over St. Augustine, with 125 mph winds from the southwest and a central pressure of 28.52 inches, hurricane force winds were felt from southeast Georgia to Flagler County, Florida. St. Augustine was in the eye of the hurricane for an hour and 15 minutes from 12:15 to 1:30 A.M. on September 10. The winds diminished as Dora moved inland but extremely heavy rains persisted across northern Florida and southern Georgia with over six inches recorded from Brunswick and Waycross, Georgia to Tallahassee and Orlando, Florida. The heaviest rains were in Lafayette and Suwannee counties in Florida where Mayo measured 23.73 inches from September 10-13 with 14.62 inches during the 24-hour period ending at 6 P.M. on September 12. Live Oak recorded 18.62 inches during a four-day period. Minimum pressures, sustained winds, peak gusts and rainfall for representative stations are shown in Table 9-8.

FIGURE 9-3. IN MOVING INLAND OVER NORTHEAST FLORIDA ON SEPTEMBER 9, 1964, HURRICANE DORA FOLLOWED A VERY UNUSUAL COURSE. HURRICANES DO NOT USUALLY CROSS THE STATE AT THIS LATITUDE. SOURCE: CLIMATOLOGICAL DATA, SEPTEMBER 1964.

The Damage

Agricultural damage to unharvested corn, cotton and peanuts in north Florida and south Georgia from the heavy rains was extensive, particularly in low lying fields. Total losses from the hurricane were estimated at $220 to 230 million in Florida and $9 million in Georgia (Dunn and Staff, 1965).

TABLE 9-8. METEOROLOGICAL DATA AS HURRICANE DORA MOVED FROM EAST TO WEST ACROSS NORTHERN FLORIDA AND SOUTHERN GEORGIA.

Location	Minimum pressure (inches)	Maximum sustained winds (mph) direction	Wind gusts (mph) direction	Rainfall (inches)
Coastal Cities				
St. Augustine	28.52	125 SW	—	7.10
Marineland	28.80	60 NW	—	5.85
Jacksonville	29.05	82 N	85	8.67
Daytona Beach	29.04	46 W	71	9.32
Mayport	29.03	74 NNE	101	—
Brunswick, GA	29.40	—	90 NE	6.23
Inland Cities				
Mayo	—	—	—	23.73
Live Oak	—	—	—	18.62
Gainesville	29.08	—	—	10.74
Lakeland	28.80	60 NW	—	5.85
Orlando	29.41	40 WSW	—	5.99
Tallahassee	29.29	35 N	44 N	6.11

Source: Butson, 1964B.

The Florida Citrus Mutual Triangle for September 11, 1964 made these comments regarding Hurricane Dora:

> Citrus growers in the northeastern section of the citrus belt were expressing relief and "thanks to the Good Lord" today as Hurricane Dora moved out of citrus land. Although damage to citrus was lighter than indicated from hurricane warnings, some losses will be sustained.

> Losses in grade due to wind scarring will be noticed in Lake, Orange, Putnam, Volusia and Brevard counties. Loss of grapefruit in Brevard County is estimated at 10%, and excessive water conditions in Brevard and Volusia counties will also be a problem.

> Mutual's field staff also reported some minor limb breakage on cold damaged groves across the northern fringe of the cit-

rus belt. In general, the Florida citrus industry has escaped Hurricane Dora with only minor losses. Detailed four county report was as follows:

Brevard County: Damage in the Cocoa, Merritt Island area was negligible—certainly not as severe as further up the coast. Continued rains through September 10 could become a problem.

In the Titusville, Mims area it is estimated that not more than 10% of grapefruit and 2 to 3% of the orange crop has been lost, although some individual losses are higher. Winds of 50 to 60 mph were reported with only minor tree damage. Heavy water and loss of grade are the big problems as well as fruit splitting. Splitting was already occurring as a result of Hurricane Cleo. Six to seven inches of rain fell up until early Thursday morning with rains continuing to fall the morning of the tenth.

In the October 17, 1964 issue, The Florida Citrus Mutual Triangle indicated that the rains from the hurricane were more beneficial than the winds were destructive.

Volusia County: In east Volusia County (Oak Hill area) six to seven inches of rain were reported by early Thursday morning with continued heavy rains throughout the day. Some palm trees are down in groves causing some minor tree damage. There is some loss of fruit—grapefruit loss is about 20%, several individual groves as high as 50% loss. Loss in grade will be the big factor, plus the potential water damage.

A follow up report dated October 17, 1964, commented that, trees in east Volusia County are showing yellow leaves as a result of too much water from Hurricane Dora and droppage of fruit has been more than expected.

Putnam County: Seventy-plus miles per hour winds reported with some limb breakage on cold-recovering trees. Some loss of fruit, but the big worry is loss of grade in tangerines.

On October 17, 1964, Mutual reported that oranges and tangerines continued to drop as a result of Hurricane Dora and quality and grade of fruit left on the trees would be affected.

Lake County: Some grapefruit and a few oranges are on the ground but loss will not be severe. Winds of 50 miles caused some minor limb breakage. Through Thursday morning, two to three inches of rain had fallen by 8:00 A.M., but this was a welcome rainfall and growers could have used even more.

Hurricane Isbell—October 14-15, 1964

Meteorological Considerations

As is characteristic of many late season storms, Hurricane Isbell formed from a weak tropical disturbance over the western Caribbean near 13°N, 80°W. Although it was noted as early as October 7, it was so poorly organized that the first bulletin was not issued until October 10, and it did not become a named tropical storm until late on October 12 (Dunn and Staff, 1965), at which time it had drifted to the northwest to near 18°N, 86°W, then turned to the north toward the western tip of Cuba. Slow intensification is typical of mid-October storms in the western Caribbean.

Isbell's winds reached hurricane force on October 13 as it neared and crossed Cuba with a minimum pressure of 28.91 inches. Winds to 70 mph were recorded at Havana well to the east of the storm center. Moving to the northeast after crossing western Cuba, the hurricane's calm eye crossed the Florida coastline near Everglades City at 4:00 P.M. on October 14, with a calm period felt from 4:15 to 4:40 EST. Isbell proceeded across the Florida peninsula at a forward speed of 20 mph and exited into the Atlantic Ocean near Jupiter at 9:30 in the evening of the fourteenth (Figure 9-4). Highest winds of near 90 mph were felt along both the Gulf and Atlantic coasts. A peak gust of 63 mph was recorded in Miami, 47 miles to the southeast of the center. The lowest pressure reported was 28.47 inches when the storm center was southwest of Key West. The lowest pressure in Miami was 29.54 inches. Meteorological data for Hurricane Isbell are shown in Table 9-9.

The Damage

The heaviest damage to Florida agriculture from Hurricane Isbell was to vegetable crops in the Everglades, the Devil's Garden section of the Immokalee area, Dade County and in the Pompano area (Rose, 1973).

Leaf crops and sweet corn received the most damage in the mucklands of the Everglades with many cabbage plants washed out of the ground by heavy rains and strong winds, and two acres of

FIGURE 9-4. ISBELL WAS A MINIMAL HURRICANE AS IT CROSSED SOUTH FLORIDA FROM EVERGLADES CITY TO JUPITER ON OCTOBER 14, 1964, THE THIRD HURRICANE TO HIT THE PENINSULA DURING 1964. SOURCE: CLIMATOLOGICAL DATA, OCTOBER 1964.

recently transplanted celery blown out of the ground. In the Pompano area, the north end was impacted more heavily than the south end, with blowing sand burning plant foliage and causing heavy bloom drop in snap beans. Older plantings of peppers were also heavily damaged with leaves burned black by wind and sand where there were no windbreaks. Some replanting was required. Vine ripe tomatoes in the Pompano area also suffered wind burn and leaves stripped by the wind. Fruit on the older plants was blown off or

TABLE 9-9. METEOROLOGICAL DATA FROM HURRICANE ISBELL AS IT CROSSED SOUTH FLORIDA FROM EVERGLADES CITY TO JUPITER ON OCTOBER 14, 1964.

Location	Minimum pressure (inches)	Maximum sustained winds (mph) direction	Wind gusts (mph)	Rainfall (inches)
Key West	29.47	73 S	76	1.43
Everglades City	28.75	90	—	—
Miami	29.54	46	63	2.28
Fort Lauderdale	29.45	52	81	—
Belle Glade	29.07	65 N	80	4.25
Everglades Expt. Station	28.98	67	—	5.09
West Palm Beach	29.05	74 SSE	87	2.38

Source: Dunn and Staff, 1965.

badly scarred. Green tomatoes in Martin and Palm Beach counties and in the southeast side of the Immokalee area also sustained severe damage, with fruit blown off or scarred.

Citrus: The Florida Citrus Mutual Crop Condition Report published in the Triangle for the week ended October 17, 1964 made the following comments regarding citrus damage in the counties affected by Hurricane Isbell, and further follow up comments on earlier Hurricane Cleo:

Martin, Palm Beach and Okeechobee counties: Tree condition is excellent in these counties. Moisture is fine and not excessive. Fruit sizes are normal and fruit is of good quality and texture. Droppage as a result of Hurricane Cleo has ceased. There is some splitting of oranges but this condition is not severe. The insect problem is not severe and spraying, cultivation, etc. are on schedule.

No damage to trees or fruit is reported as a result of Hurricane Isbell. The storm passed south of the citrus area in Martin County and north of the citrus area in Palm Beach County."

Broward County: Tree condition in this area is not good at this time as groves have not yet recovered from the effects of Hurricane

Cleo. Some groves have not been cleaned up because of high water. Fruit droppage caused by Hurricane Cleo had about stopped but Hurricane Isbell spawned tornadoes were close enough to blow off more of the weakened fruit. The remaining fruit is small and badly wind-scarred.

The Davie area had 13 inches of rain in seven days prior to Isbell. Many groves are under water up to and on tree trunks and almost all have water standing in the middles. The drainage canals are filled to overflowing which makes it impossible to pump the water off the groves. Many trees are now in wilt and will very likely lose fruit and foliage as a result of water covering the tree roots. High water covers the Everglades west and north of Davie and with all pumping facilities running, they can lower the water only about two inches per day. With no more rain in the Everglades area, it will take at least a week to give the Davie citrus area any relief from flood water. There is more water in citrus groves now than has been present since the 1947 flood that killed many trees in this area.

St. Lucie County: Tree condition is satisfactory at this time. Trees that lost foliage to Hurricane Cleo are putting out new growth and filling in bare limbs rapidly. Fruit dropping has practically ceased and fruit is showing some growth again. Orange sizes are quite good, but grapefruit are small. A large percentage of the larger grapefruit was lost to the hurricane. There have been good rains in the past few weeks but not enough to damage citrus trees. Rain has delayed cultivation and spraying operations.

There was no damage in this area from Hurricane Isbell. The storm passed to the south through lower Martin and upper Palm Beach counties.

Polk County: Growers were hoping that Hurricane Isbell would not come close enough to bring winds but would come close enough to bring some rain, since rain would be most welcome in all groves. There was no wind and very little rain in this county from the hurricane. Moisture conditions in the county are mostly adequate—but just barely so. There has been some irrigation going on, mostly for young trees or in an effort to "size" fruit.

Personal Observations

Agricultural damage from the hurricane exceeded $5 million (Dunn and Staff, 1965). At the author's house in Winter Haven, well to the north of Isbell, a complete calm prevailed all night. It was hard to believe that a hurricane was crossing the peninsula.

1965

After hurricanes Cleo, Dora and Isbell in 1964, the 1965 Atlantic Basin season would have been relatively uneventful, with only a total of six systems, except for Hurricane Betsy, which dealt a severe blow to south Florida and the Florida Keys on the eighth of September. Later in the year, there was tropical storm Debbie which came ashore in west Florida without reaching hurricane force.

Hurricane Betsy—September 8, 1965

Meteorological Considerations

The system which generated Hurricane Betsy was first noted by the TIROS weather satellite on August 23 in the western Atlantic Ocean near 7.5°N, 29.5°W. By August 27, a tropical depression had formed near 11°N, 51°W or about 350 miles east southeast of Barbados, with a central pressure of 29.74 inches (Sugg, 1966). The system moved on a westerly course for two days crossing Martinique on the twenty-eighth at which time it shifted to a north northwest course along the Leeward Island chain emerging into the open Atlantic at 19°N, 63°W on August 29 as a tropical storm. Betsy attained hurricane force on August 30 as she crossed 22°N, 65°W, then slowed to a crawl on August 31 before beginning a very slow move to the west crossing 70° west longitude on September 2. Betsy then intensified rapidly as the central pressure fell to 27.82 inches, after which she turned to the northwest for the next two days before halting abruptly on September 5 at 29°N, 75°W. Betsy then looped to an unusual south southwest course into the northern Bahamas and was off Great Abaco Island on September 6 where she again made an abrupt shift to a westerly course on September 7 passing just north of Nassau and across the northern tip of Andros Island. Florida was placed first on a hurricane watch and then a hurricane warning on September 6 as the hurricane crossed the gulf stream along 25°N latitude. In the early morning hours of September 8, the eye transversed the Key Largo area bringing hurricane force winds as far north as Jupiter and Port Everglades (Table 9-10) before exiting the state into the Gulf of Mexico (Figure 9-5) and setting a course toward the Louisiana delta. The observer at Tavernier reported that he was in the eye of the storm from 4:30 A.M. until 7:00 A.M. and the barometric pressure reached a low of 28.12 inches at 7:00 A.M. with northwest winds of 140 mph (Butson, 1965). Flooding was severe in south Dade County with land under water south of Homestead Air Force Base and east of U.S. Highway One. Homestead Air Force Base recorded 10.89 inches while the National Hurricane Center measured 6.21 inches and south Miami accumulated 5.16 inches. Meteorological data for key stations are shown in Table 9-10.

TABLE 9-10. METEOROLOGICAL DATA FROM HURRICANE BETSY AS IT CROSSED THE SOUTHERN TIP OF FLORIDA FROM EAST TO WEST, SEPTEMBER 8, 1965.

Location	Minimum pressure (inches)	Maximum sustained winds (mph) direction	Wind gusts (mph) direction	Rainfall (inches)
Key West	28.97	81 SW	88	3.97
Big Pine Key	28.68	125 SW[1]	165[1]	10.52
Homestead AFB	28.71	—	140	10.89
Tavernier	28.12	120 NE	140[1]	
Plantation Key	28.14	100+ ENE	130-140[1]	11.80
Miami (NHC)[2]	28.99	—	104 NE	6.21
Ft. Lauderdale	29.17	—	120	4.32
West Palm Beach	29.46	55 NNE	60	0.45
Vero Beach	28.69	—	65	0.65
Clewiston	29.50	50	68	2.51
Everglades City	29.19	—	140[1]	2.63
Naples	29.24	—	61	2.70
Fort Myers	29.43	46 ENE	56	0.37
Tampa	29.64	31 E	—	0.42
Pensacola	29.78	43 E	60 SE	0.35

[1]Estimated.
[2]NHC = National Hurricane Center.
Source: Sugg, 1966 and Butson, 1965.

FIGURE 9-5. HURRICANE BETSY GRAZED THE SOUTHERN TIP OF FLORIDA ON SEPTEMBER 8, 1965 AS IT MOVED THROUGH THE NORTHERN KEYS. HAD IT PASSED 50 MILES FURTHER NORTH, IT WOULD HAVE BEEN A MAJOR DISASTER. SOURCE: CLIMATOLOGICAL DATA, SEPTEMBER 1965.

The Damage

Wind and rain from Hurricane Betsy were damaging to tomatoes, squash, eggplant, peppers, and the local cucumber crop (Florida Agricultural Statistics Service, 1965), particularly in Dade County where the rains totaled five to six inches with up to ten inches in some locations. The most severe damage was to older tomato plants in Dade County. Of 3,400 acres of tomatoes planted, 700 to 800 acres would require replanting with spot resetting in many fields. Damage was less severe in Martin County where the rains totaled only one to three inches and the wind velocity was much less than in Dade County. However, wind burn from blowing sand was a significant factor with an estimated loss of 20 to 30% of vineripe tomatoes.

Tropical Fruits: Fruit damage in Dade and Broward counties was severe with an estimated loss of 90% of the avocados and 25 to 50% of the citrus, mostly limes (Sugg, 1966). Limb loss of avocados was very heavy but somewhat lighter than during Hurricane Donna in 1960 (Weather and Crop News, 1965).

The damage to tropical fruits from Hurricane Betsy is well-described in the following USDA memo, dated October 4, 1965, from J. W. Todd to P. E. Shuler regarding Mr. Todd's inspection survey conducted September 22 and 23, 1965:

> The following is a factual account of my trip to the south Dade area. On your advice, I had talks with John Campbell and Mr. Cox at the county agent's office, then Pal Brooks, and later Bert Coburn at Kendall's office.
>
> I found that the northern edge of the eye passed just north of Florida City. Greatest havoc to fruit crops was at that location and severity of damage declined to only moderate tree damage along northern edges just south of Miami.
>
> *Avocados*: The bulk of this crop is located close around Homestead and suffered heaviest damage. Many of the old groves in the area had 75 to 90% trees blown over. Younger trees had lesser numbers blown over and trees under 5 years old were all left upright. Most people I talked with believed that possibly 30% of all trees were blown over or left leaning. Limbs on trees still standing were badly broken and leaves stripped off. Looked to me as though these trees have lost half their bearing surface.
>
> Fruit loss was even greater, all concerned report loss at 90 to 95%. Of the fruit remaining, much of it is on downed trees and maturity is somewhat dependent on moisture supply, which is short now. They certainly won't need picking hooks this year, I didn't see any fruit hanging higher than 5 feet. All will be scarred but the committee has agreed to permit shipments of Grade 3 fruit, so all can be moved. Some downed fruit was salvaged and sent to processors but volume was light, because most would not pass maturity.

The task of righting trees is well under way, with labor a restricting factor. The growers I talked with said they certainly had no plan to abandon any groves. John Campbell made the comment that the market structure and future prospects are such as to justify cost of reclaiming them. Most felt that in two years, they can be back. Growers I talked with stated that most trees would live even after being blown down. They are pruning tops back on real large trees before righting them. The future plans appear to include a regular program of cutting back to possibly reduce the amount of damage in event of other storms.

Mangos: The greatest concentration of producing trees, I believe, are located north of Homestead, closer to south Miami. For this, they can be grateful, as the degree of tree damage was much less there than at Florida City. At Florida City and Homestead areas, where most trees are of older groups and many outside hedge rows, damage was severe. More than half of these trees look to be on the ground. They are being set up again, as I am sure they have been in the past. I understand that they expect most all to live and not likely many will be abandoned. The market prospects aren't as bright for these as for avocados, many growers think they are a bother and don't pay too well. Limb damage wasn't too great on trees left standing, as compared to pears. As one travels north, tree damage was only nominal. In the ten acre groves at the corner of Krome Avenue and Eureka Drive there wasn't a tree down. Again, the ones in this area were younger and not as large trees. All fruit had been picked before the storm hit (1965 crop was very small).

Limes: Somewhat the same story is true for limes as for mangoes. The volume of fruit is harvested north of the badly damaged area. Around Homestead where you find the old lime groves that aren't really bearing too much, 25 to 30% are either blown over, leaning or twisted around. They are being propped up as rapidly as possible. Limbs on the east sides of trees were twisted and broken pretty badly. Foliage was stripped off. These trees are very thick and wind didn't filter through, instead they went over. North of there around Quails Roost where many of the more productive groves are located, tree damage was mainly in the outside rows. A lot of these outside trees were left leaning and had much limb damage. Inside trees show only light limb damage.

Growers appear to be holding off on setting up trees and working on avocados instead. In the case of limes, most trees are only leaning instead of blown over and not suffering. Some of the real old non-producing trees may be replaced with younger ones.

Fruit loss varies with location, with 75% of fruit at the time of the storm to 25%. As a whole, I would state that about 30% of all fruit blew off. Some of this was picked up afterwards, but the ground is still covered. Fruit on trees is badly scarred from the many thorns and wind. Most trees are void of leaves on the east side and already these are covered with bloom buds on those defoliated limbs. Growers anticipate they will have a crop of fruit in February, March and April.

Other Citrus: Tree and fruit damage was quite great and I really wonder if a lot of the old half abandoned groves will be maintained.

Hedge Rows: They did provide protection from the winds, but many went down, and some in fruit trees causing much damage. I really was surprised at the amount of Australian pines that blew down. The grafted ones broke off at the bud union.

Citrus: The following report which appeared in the September 10, 1965 issue of the Florida Citrus Mutual Triangle summarized the effect of Hurricane Betsy on the citrus industry:

As the Triangle goes to press, Hurricane Betsy has ripped across the southern tip of the state and roared off into the Gulf. Analysis of preliminary reports indicates very little damage to citrus as a result of this huge storm.

Weather Bureau reports say that Betsy was a very dry storm and consequently the resulting rainfall in the citrus belt was very light. Rain in the extreme southern tip of the belt measured only about 1½ to 2 inches with the central portion of the citrus belt getting an average of ½ to 1 inch.

However, some areas did get slightly more rainfall . . . all of it considered most welcome. In Broward County, in the Davie area west of Ft. Lauderdale, Mutual's field staff reported rains of 4 to 5 inches. While growers welcomed the rain, groves there took some heavy winds which weren't so welcome.

Winds were estimated at 75 to 100 mph and will result in some fruit loss to the area. Immediate fruit losses around Davie were estimated by growers and Mutual's field staff as follows: grapefruit, 30-40% loss; navels, 25% loss; and round oranges, 10 to 15% loss. Some additional loss on all varieties is expected due to bruising and lashing from the winds.

Tree damage in this area was only minor and was confined mostly to groves which were damaged last year when Hurricane Cleo roared through Davie.

Citrus producing areas north of Broward County suffered almost no loss of fruit and no tree damage was reported at press time. Interior areas of the southern part of the citrus belt apparently came through with very little damage; however, Mutual's field staff will continue checks this week and subsequent reports will be made in next week's Triangle.

There is, however, concern that some interior citrus might have been scarred as a result of gusty winds. Some growers have also reported tender leaf damage from wind-blown sand.

But, by and large, the major portion of the citrus industry can count its blessings for another escape from the bad side of nature. Ironically, growers in some areas are actually concerned that the far reaching winds of Hurricane Betsy will remove more water from the soil than was dropped.

Vegetable Damage by Commodity in South Florida was reported by the Florida Agricultural Statistics Service (1965) in Orlando as follows:

The winds from Hurricane Betsy affected vegetable growing areas from Dade and Collier counties in the south to as far north as Martin and the southern portion of St. Lucie County. However, as the storm was relatively dry in most areas, rainfall damage was not as severe as might have been expected.

Lettuce. As early seeding had begun in the Everglades in late August, part of the acreage in this area was up and replanting became necessary.

Green Peppers: At the time of Hurricane Betsy, the Pompano area of Palm Beach County and Martin County were producing 48% of the state's crop. Wind and rain damaged some fields and these were replanted.

Irish Potatoes: Prior to any seeding, Hurricane Betsy brought in a tidal wave which saturated some Dade County fields with salt. An estimated 300 acres were lost and the yields were lowered in many other fields.

Squash: Plantings in Palm Beach and Dade counties did not begin until after Hurricane Betsy and were unaffected by the storm. However, torrential rains in mid-October, unrelated to the hurricane, drowned many plants in the Pompano area.

Tomatoes: Vine-ripe tomatoes grown on stakes or trellises in the Pompano and the Immokalee-Naples areas were severely damaged by wind blown sand. Recovery at Pompano was slowed due to salt spray spread by the hurricane, but Immokalee fared better and harvest was underway in late November. Early fall plantings of mature green tomatoes in Dade County were severely damaged by the hurricane.

1966

Activity in the Atlantic Basin increased in 1966 with seven hurricanes and four tropical storms. Two of the hurricanes affected Florida, Alma which struck the Panhandle region on the ninth of June, and Inez which crossed the northern Keys and south Dade County on October 4-5.

Hurricane Alma—June 8-9, 1966

Meteorological Considerations

Hurricane Alma originated from a low pressure system which formed over Nicaragua on June 4 and moved northward over Honduras into the Caribbean Sea on June 5, after which it moved on a north northeast track and reached hurricane intensity on June 6 and crossed Cuba early on the eighth. Alma was one of the very few tropical storms to reach hurricane intensity before mid-June and was the earliest hurricane ever to cross the Florida coastline.

After crossing western Cuba, Alma battered Key West and the Dry Tortugas with hurricane force winds near mid-day on June 8 (Table 9-11). The highest wind speed recorded during the life of the hurricane was 125 mph from the north at 1:22 P.M. on the eighth at the Dry Tortugas. The lowest minimum pressure of 28.65 inches at 1:15 P.M. on the eighth was also recorded at the Dry Tortugas.

Moving north, Alma paralleled the west coast of Florida, 20 to 60 miles offshore (Figure 9-6). With the entire west coast on the storm's strong side, gale force winds occurred across the entire state (Table 9-11) before Alma slammed ashore at Alligator Point, about

TABLE 9-11. METEOROLOGICAL DATA FOR HURRICANE ALMA AS IT MOVED NORTH ALONG THE FLORIDA GULF COAST, JUNE 8-9, 1966.

Station	Minimum pressure (inches)	Maximum sustained winds (mph) direction	Wind gusts (mph) direction	Rainfall (inches)
Key West	29.42	60 SE	70 SE	3.35
Dry Tortugas	28.65	125 N	—	—
Miami NHC	29.75	—	61 E	7.65
West Palm Beach Airport	29.96	39 E	52 ESE	5.18
Ft. Myers Airport	29.46	46 ESE	64 ESE	2.36
Tampa Airport	29.43	46 ESE	68 E	3.38
Lakeland	29.56	34 NE	45 E	1.97
Orlando Airport	29.67	38 E	53 E	2.20
Daytona Beach Airport	29.67	37 E	55 E	3.29
Gainesville	29.65	25 S	48 E&S	2.97
Apalachicola	29.34	42 N	52 N	3.78
Tallahassee Airport	29.15	43 NE	62 NE	6.90
Alligator Point[1]	29.06	—	75-90[2]	—
Pensacola Airport	29.76	36 N	—	0.01
Jacksonville Airport	29.63	48 E	—	1.43

[1]Point of landfall.
[2]Estimated.
Source: Butson, 1966.

FIGURE 9-6. HURRICANE ALMA, AN UNUSUAL EARLY SEASON STORM, PRODUCED HEAVY GALES ALL ALONG THE FLORIDA WEST COAST BEFORE MOVING INLAND EAST OF APALACHICOLA ON JUNE 9, 1966. SOURCE: CLIMATOLOGICAL DATA, JUNE 1966.

25 miles east of Apalachicola at a little after 4:00 P.M. on June 9. Losing intensity over land, the storm veered northeast toward Tallahassee and into south Georgia in the early evening. Rainfall from Alma was relatively light in most areas as the storm sped north

except for five to eight inch totals in Dade, Broward and Palm Beach counties, and five to seven inches along a narrow path east of Tallahassee. No flooding occurred in peninsular Florida and the rains were thought to benefit the citrus groves.

The Damage

There was no appreciable damage to the Florida citrus crop, but according to Keith Butson (1966), State Climatologist, the late season vegetable crop in the gulf coast counties suffered some damage and the north Florida tobacco crop, both flu cured and shade grown, caught in the midst of the harvesting season, was damaged by high winds when plants were blown over and tobacco leaves shredded.

The most serious vegetable damage was incurred by the gulf coast tomato crop in the Ruskin area of Hillsborough County. The harvest was already past its peak when the storm struck, but the gale winds and moderately heavy rains significantly reduced volume and brought the season to an early end (Florida Agricultural Statistics Service, 1966).

The Florida Citrus Mutual field staff made the following report in the Triangle:

> A spot check of grove conditions shows that Hurricane Alma did more good than harm to Florida's citrus belt. Welcome rains in good quantity covered the area, supplying needed moisture for the present and a backlog of water for future irrigation. Some young fruit was lost to Alma's gusty winds, but not enough to cause any appreciable loss in yield for next season.
>
> As for the remaining Valencia crop of this season, some fruit was blown to the ground, but with only about 4% of the year's crop left, there was no measurable economic loss.

Hurricane Inez—October 4-5, 1966

Meteorological Considerations

Hurricane Inez developed from a low which formed in the eastern Atlantic near 10°N, 35°W about September 21. The budding storm moved on a west northwest course for five days and finally attained hurricane intensity east of Guadeloupe on the twenty-fifth (El Mundo, 1966). Passing over Martinique and Guadeloupe at noon on the twenty-seventh, the hurricane continued its west northwest march toward Puerto Rico. At 9:00 P.M., it was located at 17.1°N, 68.8°W or about 105 miles southeast of Cabo Rojo, Puerto Rico and was continuing to intensify. Fortunately for the south coast of Puerto Rico, Inez passed 75 miles south of Ponce and Cabo Rojo, but she directed her full fury toward the Dominican Republic and Haiti and struck on September 29 with winds estimated at 150 to 160 mph resulting in massive damage and loss of life.

After dealing Port-Au-Prince Haiti a savage blow and slashing along the southern peninsula of Haiti, Inez exited into the windward passage and directed its wind and rain into the Oriente Province of Cuba. Moving along the south Cuban coast to longitude 74°W, the hurricane turned north on October first and crossed the island, losing intensity and entered the Atlantic Ocean as a tropical storm then skirted the southeast Florida coast producing wind gusts of 40 to 60 mph (Figure 9-7).

The Damage

On October 7, 1966, the Florida Crop and Livestock Reporting Service issued a special damage report for south Florida. The report indicated that as Hurricane Inez moved along the Dade County coast, winds of hurricane force were felt over most farming areas, particularly in south Dade County. In addition, the precipitation accompanying the storm was described as "largely sea water fallout" leading to damage from its high salt content (Weather and Crop News, 1966).

The most severe vegetable damage occurred in south Dade County. Approximately 5,500 acres of tomatoes were in the ground and defoliation was severe except in fields irrigated with fresh water immediately after the storm. Damage estimates were as high as 80%.

Pole beans and squash in south Dade also suffered significant damage. The heaviest losses were to older squash plants which showed 40 to 50% damage. Okra and other vegetables were badly defoliated and cucumber plants suffered leaf burn. The winter potato and bush bean crops had not been planted when the hurricane struck.

FIGURE 9-7. THE PATH FOLLOWED BY INEZ AFTER LOSING FORCE AND CROSSING THE FLORIDA KEYS AS A TROPICAL STORM ON OCTOBER 4-5, 1960. SOURCE: CLIMATOLOGICAL DATA, OCTOBER 1966.

From Pompano northward, the principal damage was due to wind burn from blowing sand and some salt water spray. Although the Everglades area reported no damage, tomatoes and cucumbers in the Fort Myers and Immokalee areas experienced some leaf burn and fruit damage.

1967

There were six hurricanes and two tropical storms during the 1967 Atlantic Basin season, but none of these systems affected Florida agriculture. Only Hurricane Beulah which hit the Brownsville, Texas area on September 20 made landfall in North America.

1968

The Atlantic Basin produced four hurricanes, three tropical storms and one subtropical storm in 1968. Three hurricanes-to-be, Abby, Brenda and Dolly crossed Florida in their tropical storm stages and only became hurricanes in the open Atlantic, thus sparing the Sunshine State from their fury. Abby produced local flooding in the Titusville-Cocoa area where some stations recorded up to 14 inches or rain. As a result, tomato harvest was practically terminated in late May (Rose, 1973). However, Hurricane Gladys, a late October storm, crossed the Florida coastline just north of Tampa with hurricane force winds, and passed over the north central peninsula and entered the Atlantic Ocean near St. Augustine.

Hurricane Gladys—October 16-19, 1968

Meteorological Considerations

Hurricane Gladys formed from a low pressure system in the southwestern Caribbean Sea near 14°N, 81°W. The tropical storm phase was first noted on October 13, and was tracked on a north northwest course achieving hurricane intensity just before reaching the south coast of Cuba early on the sixteenth. Gladys exited the north coast of Cuba about noon on the sixteenth, after which she drifted slowly to the north and passed just west of the Dry Tortugas with maximum sustained winds of 64 mph and gusts to 86 mph. Minimum pressures, sustained wind speeds, peak wind gusts and rainfall along the path of Hurricane Gladys are shown in Table 9-12.

TABLE 9-12. METEOROLOGICAL DATA ALONG THE PATH OF HURRICANE GLADYS ALONG THE FLORIDA COAST OCTOBER 16, 1968 AND ACROSS THE FLORIDA PENINSULA OCTOBER 18-19, 1968.

Station	Minimum pressure (inches)	Maximum sustained winds (mph) direction	Wind gusts (mph) direction	Rainfall (inches)
Key West	29.62	49 SE	55	2.95
Dry Tortugas	—	64 E	86	—
Ft. Myers	29.68	30 SE	37	6.09
Clearwater	29.62	90 SSW[1]	100[1]	2.70
Tampa	29.53	32 SSE	55	3.12
Homosassa	—	— NE	100[1]	—
Brooksville	29.39	—	68	6.28
Lakeland	29.66	33 SE	48	2.84
Orlando	29.66	38 SSW	40	3.40
St. Augustine	29.32	60 SE[1]	70[1]	3.55
Jacksonville Beach	—	56 ESE	74	5.42

[1]Estimated.
Source: Mickelson, 1968.

Gladys continued to the north northwest on October 17 as a minimal hurricane, stalling briefly about 75 miles west of St. Petersburg early on October 18, then shifted to a northeasterly course and crossed the Florida coastline near Homosassa in southern Citrus County. Top wind gusts as Gladys came ashore were estimated at 100 mph along the beaches and near the coast. Continuing to move on a northeast track at 15 mph, Gladys passed just south of Ocala in Marion County and finally moved off the Florida east coast near St. Augustine (Figure 9-8). Gale force winds prevailed as the storm passed across the state, approaching hurricane force near the center. Two tornados formed during the storm's jaunt across the peninsula, one near Boca Raton in Palm Beach County and the other at Whitfield Estates in Manatee County.

The Damage

Citrus: The citrus variety most susceptible to hurricane damage is grapefruit, and as less than 10% of the Florida grapefruit crop is

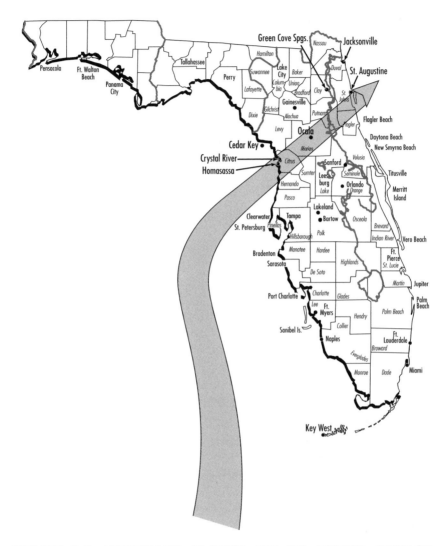

FIGURE 9-8. HURRICANE GLADYS DID ITS WORST DAMAGE ALONG THE WEST COAST NORTH OF TAMPA BEFORE PASSING THROUGH MARION COUNTY AND EXITING THE STATE NEAR ST. AUGUSTINE, OCTOBER 18-19, 1968. SOURCE: MICKELSON, 1968.

grown along the path of Hurricane Gladys, the state's overall grapefruit crop was not seriously damaged (Triangle, October 23, 1968). However, some individual growers in Pinellas County and the coastal areas of Pasco County were very hard hit. Florida Citrus Mutual carried out an area by area survey and reported the following:

West Pasco County and Pinellas County: This was the most severely damaged area in the citrus belt. Grapefruit groves suffered the greatest loss with growers estimating from 25 to 75% loss of fruit. Trees with thick foliage, or close set trees with mostly inside fruit, were reported to have lost less fruit than more exposed trees.

Growers reported orange losses much lighter with estimates of around 10% of fruit on the ground. Tree damage was described as minor and much less than was expected.

Marion, Putnam and Volusia counties: South Marion County, in the area around Weirsdale, suffered the heaviest damage in this area. Some growers estimated their grapefruit loss as high as 25%. There is expected to be some loss of Navels due to thorning and bruising but it is much too early to determine the extent of this damage. Tree damage was limited to breakage of cold damaged limbs plus some tangerine limb breakage.

Lake County: The majority of this county got by with little or no loss of fruit or tree damage. Wind scar damage will have to be determined at a later date. The northern part of the county (from Lady Lake north) did suffer some loss of grapefruit on exposed groves but no loss of oranges is reported. Some reports indicate a loss of oranges in Citrus and Sumter counties but this is not severe.

Polk, Manatee counties, South: There was no loss of fruit in this area except for very minor loss in Manatee and Sarasota counties. Rainfall ranged from five inches on the gulf coast down to one to two inches in the interior. Some fruit scarring is expected. No tree damage reported.

East Coast Area: Growers in Brevard County report that Hurricane Gladys and the rains that accompanied her were more beneficial than harmful. There was no loss of fruit and little wind scarring.

Rainfall along the east coast was general throughout the week from Broward County to Brevard County with the heavier amounts ranging from six to seven inches. These amounts were not harmful.

It should be noted that the citrus growing districts most affected by Hurricane Gladys in 1968 were severely damaged by freezes in 1981, 1982, 1983 and 1985 (Attaway, 1997). Consequently, a hurricane following the same path in the 1990s and 2000s would be less damaging to the overall citrus crop than it was in 1968 because many groves were damaged by the freezes and are no longer producing.

Vegetables: According to Rose (1973), vegetables were adversely affected, but tomatoes had been retarded even more by heavy rains in September than by the hurricane. Insect and disease control was virtually impossible. The wind did not blow over staked tomatoes to any great extent (Weather and Crop News, 1968). Considerable resetting of cabbage was necessary in the Flagler County area, and planting of sweet corn in the Pompano area was delayed by the rains associated with the hurricane. Eggplant in the west coast counties sustained only minor damage and some pepper plants were blown over but most stood up well.

1969

With ten hurricanes, four tropical storms and one subtropical storm, the Atlantic Basin was frantic with activity in 1969. However, none of these systems affected Florida during their hurricane stage. Hurricane-to-be Gerda traversed the Indian River area as a tropical storm on September 6-7 before later becoming a hurricane in the open Atlantic. Tropical storm Jenny passed over peninsular Florida on October 2 and an unnamed subtropical storm affected the Florida panhandle on the first of October. The most important feature of the 1969 Atlantic Basin season was category 5 Hurricane Camille which struck the state of Mississippi with devastating force.

Chapter 10

Caring for Hurricane Damaged Groves

The following information was prepared by Dr. Herman J. Reitz, Director of the Citrus Experiment Station, Lake Alfred, Florida after severe Hurricane Donna struck the Florida citrus belt September 10-11, 1960. It appeared in the Florida Citrus Mutual Triangle dated September 19, 1960:

In the wake of the September 10-11 hurricane, citrus growers will be concerned with returning damaged groves to productive healthy condition in the shortest length of time. The advisable course of action depends upon the condition of the individual grove, but the following generalizations may be of some interest and guidance in making the necessary decisions. The degree of damage varies from essentially none to almost complete loss of fruit and 50% or more defoliation. In many groves, an occasional tree has been blown over, and in some, large numbers of trees are reported blown out of the ground. The response the trees will make in the next few weeks depends on the extent of damage. Trees which have not been noticeably defoliated will probably not put on an unusual fall flush of growth. Those very severely defoliated may be expected to put out an extensive flush of growth in the near future. This is because at least one month or more of warm weather can be anticipated and soil moisture is good.

There will also be additional drop of fruit damaged by weakened stems, bruises, cuts, or thorn pricks during the wind storm. The extent of this droppage is impossible to determine at the present moment, but must be expected.

Growers will be concerned primarily with the fall fertilizer program, as it is the next major operation to be conducted. A very important fact influencing this decision is that trees were in much better nutritional conditions before this hurricane than before any previous one.

In undamaged or moderately damaged groves, the usual fall fertilizer practice should be followed. This application will (except in the relatively undamaged areas along the Indian River) be initiated within the next month, and will in most cases be quite adequate. This course has the advantage of minimizing the danger of coarsening and delayed degreening of any fruit remaining on the trees now.

In severely damaged groves, that is, where fruit loss is severe, where twigs have been frayed by the wind, and where defoliation to the extent of 50% or more has occurred, the fall application of fertilizer should be made as soon as possible. This is in recognition that severely damaged trees will almost certainly put out a flush of growth, regardless of the immediate fertilizer treatment, and this growth should be encouraged.

In no case would it appear necessary to plan now to increase the total annual amount applied, but rather merely to move up the time of application. Later in the fall, after tree response can be observed, a better judgement can be made as to whether additional applications will be necessary. If the first flush to come out does not mature into normal, deep green foliage, additional applications may be advisable. Not many groves will be in this category.

Additional nutritional sprays will not be necessary in most groves. Defer application of these sprays until the need for them appears in the form of visible symptoms. A few groves may show a scattering of zinc or manganese deficiency, but most of these symptoms can be expected to disappear without special treatment.

Insect and mite control will not be drastically modified by the storm. Those who follow mite populations closely will probably find that the storm caused a reduction in the population of both rust mite and purple mite. The reduction in adult mites will almost certainly be temporary, and in a short time, a new population of adults may be expected to hatch from eggs remaining on the trees. Continued checking of groves is very advisable, and plans should be made to spray as soon as necessary.

The only effect of the storm on scale population was the proportionally larger loss of heavily infested fruit and leaves. This may make it appear that the storm has reduced the scale population; however, its potential to cause damage has not been lessened and continued vigilance is advised.

A most perplexing problem is what to do with the trees blown over by the storm. Past experience with the erection of these trees has been variable. Some growers have reported considerable success while others have reported complete failure. At best, it is of doubtful long-range value. Probably no tree in which there is heart rot or foot rot should be set into place again. These trees will almost certainly be short-lived due to wood decay and will continue to be extremely susceptible to additional wind storms, or trunk splitting if they do occasionally set a full crop. As much of the crop as possible should be salvaged, and then these trees should be removed and replaced. Trees with perfectly sound trunks, less than 20 years old, and which have not lost many of their major roots, may be worth the expense of resetting.

Whenever the work is attempted, all broken roots and all broken branches should be removed by making clean cuts with pruning tools. All cut surfaces one inch or more in diameter should be painted with tree-pruning paint. The tree should be placed as nearly as possible at the same level relative to the soil as before and propped securely against further movement in ordinary wind storms. Relatively frequent but moderate applications of fertilizer would be advisable. Special irrigation probably will be necessary during droughts for a year of more. The difficulty of providing proper care is one of the discouraging factors about erecting blown-over trees.

The alternative to resetting blown-over trees is to plant young, vigorous nursery trees. These may be expected to have a longer productive life and ultimately return to greater profit.

Disease control problems may be accentuated to some extent. New flushes of growth on severely defoliated trees may especially need copper sprays next spring for protection against melanose and greasy spot during the recovery period. Also, reset trees and many other trees in seriously damaged groves may have become infected with Phytophthora foot rot. Growers should make a tree-by-tree inspection within the next few months to detect foot rot lesions so that tree surgery may be employed to reduce the loss from this disease.

Irrigation can be expected to be necessary in the damaged groves after less dry weather than before the storm. This will be because of damage to root system.

Additional cultivation beyond that customary at this time of year will not be beneficial.

In summary, the grower caring for a hurricane-damaged grove will need to be unusually vigilant to protect his trees from improper nutrition, attacks by pests, and from unfavorable environmental conditions. These trees have been weakened and special attention to details in their subsequent care will pay good dividends.

The following is from an extension service memo (Lawrence and Mathews, 1960) also written to guide growers after Hurricane Donna in 1960.

Special care and management are needed for citrus trees damaged by the recent hurricane which swept through Florida's vast citrus belt.

Trees partially blown over by high winds should be checked immediately for soundness. If found to be *unsound*, they should be replaced (now or later depending on conditions). When the trees are sound, the root damage isn't severe, leave the trees alone until the crop is harvested. If the root system is damaged severely enough to cause ultimate defoliation, the trees should be reset immediately.

Remove sufficient soil to create a hole for the uncovered roots. If the roots are partially pulled out of the soil, but not broken, it may be necessary to cut them in order to straighten the tree. All torn or severely damaged roots should be pruned to remove the ragged edges. Cuts in excess of ¾ inch should be covered with pruning paint. In those cases where root damage was severe, the top of the tree should be cut back proportionately.

When the tree is raised, soil should be filled in around the roots. Flood the soil in with plenty of water to prevent the formation of air pockets around the roots. It may be necessary to brace the tree after it is righted.

A close check should be made for spider mites and eggs on trees which are partially defoliated. If many of either are found, apply a miticide while the population is still low. Mite damage could cause the remaining leaves to drop or seriously weaken the new flush.

With trees partially defoliated, it is extremely important that the fall flush be kept as healthy as possible. Where greasy spot has been a problem in recent years, a copper spray should be applied when the new flush is out and the leaves fully extended. This precaution will tend to prevent the occurrence of greasy spot lesions and leaf drop during the late winter and early spring.

Water damaged trees will need special care during the next several months. Remove the excess water from the soil as quickly as possible.

Where there is evidence of damage to the root system, large amounts of soluble fertilizers should be avoided and supplemental irrigation should be provided when needed. Groves will recover faster, barring freezes, if they receive light applications of soluble fertilizer and are irrigated as soon as they show signs of wilting during dry weather.

Split the normal fall fertilizer application into two, the second being applied six to eight weeks after the first. Fertilizers should never be applied to wilted groves.

Unflooded trees showing a definite foliage loss should receive the regular fall fertilizer application immediately. This is particularly true if the grower suspects a new flush of growth.

The following information was provided by Dr. A. F. Camp, who was Vice-Director in charge of the Citrus Experiment Station at the time of the 1949 hurricane. It appeared in the October 1949 issue of Citrus Magazine. It presented a different set of circumstances as it occurred in August, as opposed to Hurricane Donna in early September.

Storms and freezes always bring on perplexing problems for citrus growers, and a decision has to be reached as to how to treat the injured groves and at the same time save as much fruit as possible. Unfortunately, no fixed pattern of treatment can be laid out because the timing of the freezes or storms determines to a large extent what may be done and what type reaction may be expected. The storm this year, for instance, came early enough so that it preceded the normal fall growth and the trees are now responding vigorously and rapidly covering up the defoliated twigs. On the other hand, the hurricane of October 1944 occurred subsequent to the normal fall growth period and only a small abnormal growth occurred after the storm period with a consequent weakening of the trees and a good deal of dead wood by the following spring. Our particular concern at the present time is how to retain as good quality in the fruit that remains as is possible while at the same time restoring the trees to enough vigor so that they may set a crop of fruit during the coming spring. There will certainly be a lot of fall growth in storm injured groves and it has already started but its extent and character may be influenced considerably by the treatment given the grove by the owner plus the kind of weather that results following the storm. This is in contrast to 1944 when the damage occurred so late that little rebuilding could be done until the following spring.

In trying to formulate some recommendations that may be helpful to growers it is necessary to set up some sort of classification of storm injured groves since the treatment ought to be keyed to the grove condition primarily and blanket recommendations cannot be made which will be equally good for both severely injured groves and slightly injured groves. For convenience, the following classifications may be useful.

Groves Classified According to Extent of Storm Damage

The first category would include groves which have lost most of their fruit and 50% or more of their leaves. Groves of this type will occur mostly in the coastal district but the category might include a good many grapefruit groves in exposed locations on the ridge. The trees in addition to losing their fruit have been heavily defoliated and in many cases the bark on the small twigs has been almost stripped off or if not stripped off it has been badly beaten up so that the twigs may die unless foliage is produced immediately. Trees in this type of grove will have also suffered very heavy root damage due to the wrenching around of the trees and many large lateral roots will be broken. In fact, the root damage in many cases will have a more lasting effect than the top damage.

In the second category would be groves that have lost a lot of fruit, and considerable foliage in the tops but are not otherwise badly damaged. It would include a lot of groves in the ridge district which have 50% or more of the fruit left on the trees.

The third category would include those groves which have lost more or less fruit but have sustained very little or no foliage damage. Most of these groves in the case of grapefruit may have lost anywhere up to 50% of their crop without being seriously defoliated while in orange groves, the loss of fruit may be no more than 15 or 20%.

In considering the first category, we are not concerned with the problem of fruit quality because on such badly injured trees very little fruit is left, and the main problem is to restore the top as much as possible during the fall period of growth, as twigs that remain defoliated all winter will be dead by spring and a source of melanose spores. It is well to remember also that groves in this category will have heavy root injuries and a vigorous root growth will be very desirable. Groves in this category will thus have to replace both top and root growth and will need all the help that can be supplied. With a depleted root system, moisture will be as important as fertilizer. These groves should probably be fertilized immediately and cultivated so as to produce as much good growth as possible during the early fall and get it hardened up as quickly as possible. The earliest that we have had a severe frost in the central part of the state is about November 12th to 14th and this has only occurred once in a great many years so that we can count with some safety upon the period lasting until the middle of November for a new growth and its hardening. If the weather turns dry in the fall as it frequently does following storms, such groves will need irrigating because they will be very short of roots and will have difficulty in absorbing water from the soil. Generally speaking, it would probably be better to irrigate before the soil gets really dry during September and early October and then slack off on the irrigation to help them harden up by mid-November. Unless they respond with a strong growth this fall, it is likely that they will set a small crop the coming year and that the fruit will be undersized. It should be remembered also that twigs killed by the storm will likely be full of melanose spores by spring, and melanose spraying may be of the utmost importance not only to protect the young fruit but also to protect the spring growth which will be needed to completely cover up the top. Rains have been rather scattered since the storm and many groves are already becoming dry. The main thing in such groves is to get growth back on the tops as quickly as possible and as strong a growth as possible and at the same time to rebuild the root system so that the root rot fungi will not infect the broken roots and cause leather damage to the tree.

We Can Expect Additional Droppage of Fruit

Recommendations for groves in the second category are complicated by the desire of growers to protect the fruit quality as much as possible. Therefore, some decision will have to be made by the owner as to whether he wishes to protect the quality of the remainder of the current crop as much as possible or whether he wishes to help the trees along as much as possible in order to improve his chances for a good crop for the coming year. This will be determined very largely by the crop that is left on the trees in such groves and this should be taken into careful consideration before making any decision. Certain considerations concerning the fruit left on storm damaged trees should be taken into consideration also. The drop that occurs during the storm is not the only drop that results from hurricanes. Our experience following the storm of 1944 showed that there were still drops from the storm as late as the first of January and even thereafter and our counts so far indicate that the same thing is occurring now. We can expect additional droppage of fruit all through the fall and this will further reduce the crop remaining. In addition, the quality of fruit remaining on the trees will not be as good as it would have been otherwise. The fruit has been tossed around by the wind, hit against twigs and limbs and brown spots, resulting from ruptured oil cells, are already developing but are not easily observed because of the green color of the fruit. These will be very noticeable when the fruit ripened as irregular brown sunken patches on the peel. It can be presumed in most of the groves on the ridge that the fruit quality has already been very adversely affected and a high percentage of fruit will probably have to go to the canneries and concentrate plants unless fresh fruit regulations are greatly modified over what has been the custom the last few years. With greatly increased capacity for making frozen concentrate, this may not be as important as might appear at first glance. In addition to the peel markings, the damage to the trees, including defoliation and root damage will probably have the effect of delaying the development of solids and this will be accentuated by the heavy vegetative growth resulting from the shock. This growth cannot be prevented since the trees have received such a shock that they are certain to put on heavy fall growth; in fact, this is already in progress. Vegetative growth is always an enemy of fruit quality when fruit has nearly matured and some puffing may result. An attempt to bring the trees back quickly by early applications of fertilizer will probably increase the puffing and may hold back the solids to some extent though we have no data on this particular point. A choice here is going to have to be made by the owner as to whether he wants to bring back the trees quickly and put them in good shape to resist the winter and to set a heavy crop next year. In the former case and particularly with early varieties, it may be desirable to delay fertilization until the normal time and not to irrigate heavily unless the trees have shown signs of suffering, but if the crop loss has been heavy and it does not seem worth while to try to retain high quality, particularly in mid-season and late fruit, better results as far as the trees are concerned will probably be obtained by earlier fertilization and cultivation so as to help the growth as much as possible and try to get it hardened up before mid-November. It is in this category that the difficulty will come. Groves that have been well-fertilized will probably produce a normal fall growth and carry on satisfactorily until the normal time for fertilization and the quality of early fruit may not be badly hurt. The loss of leaves will hold back the development of solids in the fruit and this can hardly be prevented. Fundamentally, a decision will have to be made between fertilizing early so as to help the fall growth or waiting for cool weather and a hardening up of the fall growth before applying fertilizer. In the case of groves where there is sufficient fruit to justify the maintenance of quality, the latter program which is the normal one followed in central Florida may be desirable, but if the trees have been badly battered and particularly if they are carrying mid-season or late fruit, it may be desirable to help along the fall growth as much as possible with early fertilization and cultivation.

Melanose Spraying Next Year Will Be Important

In most of these groves, there will be a lot of dead twigs, particularly in the tops and melanose spraying next year will be very important.

In the third category where we are dealing only with a loss of fruit and very little visible damage to foliage and no apparent shaking up on the trees in the ground, the normal program should be followed as the reduced crop could be easily puffed by an excessive fall growth, and this could be particularly true in the case of early varieties. There are many grapefruit groves that have lost more than 50% of their fruit with little or no tree damage. If there is an excessive vegetative growth on these this fall, there will naturally be a tendency for fruit puffiness due to the reduced crop that the trees are carrying. This appears to be unavoidable and probably such fruit should be moved fairly early before it has a chance to fall off in quality too much. In the case of orange trees that have not been visibly hurt, the loss of fruit is usually so small that puffiness or coarseness would probably not occur, but there will be a falling off in grade due to external injuries which occurred during the storm but which are not readily visible as yet.

From our past experience, we are inclined to believe that the most serious damage will be that suffered by the root system. A careful examination of many trees following the 1944 hurricane revealed a great many broken roots out around the periphery of the trees due to the trees being moved about in the soil. These broken roots are often a point of infection for root rot fungi and trouble may develop in trees several years after the storm as the infection becomes progressively worse. This was probably aggravated in 1944 by the lateness of the storm which occurred too late for much root growth during the fall so that the roots probably dried out in the succeeding drought and then got infected before spring rains and spring growth set in. It is hoped that this year between the vigorous condition of the trees and the earliness of the storm, proper root growth will occur along with top growth in September and October and much of this infection may be avoided. Special attention in this connection should be given all trees that have had to be set up. If the fall turns off dry, these trees will need to be watered at fairly frequent intervals as otherwise there is likely to be a heavy loss of roots and the starting of root decay. Too often, trees that have been set up following a storm are forgotten after initial watering and left to shift for themselves. This is all right if regular rains occur but if the weather turns dry, it is likely to result either in the immediate death of the tree or its death a few years hence due to root rot which has set in in the broken and drought damaged roots.

Chapter 11

The 1970s

"When Hurricane David arrived, I was living in a second floor apartment in Vero Beach with a sliding glass door on the west side. I spent the entire storm curled up in a bean bag chair reading Michener's *Centennial* by the light of a Coleman lantern with the door wide open. The wind never came out of the west at all."—Pete Spyke, Vero Beach, FL, 1998.

While there were serious freezes in 1970 and 1971 and a near impact freeze in 1977 (Attaway, 1997), Florida was little troubled by hurricanes in the 1970s. Only Hurricanes Agnes and Eloise, which struck the Panhandle in mid-June 1972 and September 1975, respectively, and Hurricane David, which sideswiped the Florida east coast in early September 1979, were of any consequence.

1970

Fortunately in 1970. Florida was free of hurricanes and experienced only three tropical storms, Alma, Becky and Felice. Alma was very unusual in that it occurred during the month of May.

1971

1971 was an active tropical season in the North Atlantic with 13 tropical storms or hurricanes. However, neither west Florida or peninsular Florida was affected by any of these storms. Hurricanes included Beth, Edith, Fern, Ginger and Irene plus an unnamed storm which did not reach hurricane force until it crossed latitude

45°N. Tropical storms included Arlene, Chloe, Doria, Heidi, Janice, Kristy and Laura. The most interesting hurricane of the season was Ginger which formed on September 5 and meandered first east and then west across the Atlantic for 25 days before striking the North Carolina coast on September 30, 1971.

1972

The 1972 season was relatively inactive with only four named storms: Agnes, Betty, Carrie and Dawn. Of these four, only Agnes, which struck west Florida on June 19, brought hurricane force winds to the state. Dawn brushed the southeast Florida coast as a tropical storm on September 5, but its winds were not damaging.

Hurricane Agnes, June 19, 1972

Hurricane Agnes had an interesting path in that it formed from a low pressure area which moved eastward from the Yucatan peninsula on June 15, then turned 90° due north into the Gulf of Mexico where it attained hurricane force near 24°N, 86°W on June 18. It then moved rapidly north going ashore near Port St. Joe on June 19. Agnes was only a minimal hurricane, category 1 on the Saffir-Simpson scale (Table 15-4) with surface winds reaching 85 mph over the open gulf on June 18 and with a minimum pressure of 28.88 inches on the nineteenth. The eyewall was never fully developed, but destructive storm tides occurred along the west coast of Florida with above normal tides of three feet at Fort Myers, four to five feet at Tampa Bay, seven feet at Cedar Key, and 6.4 feet at Apalachicola. Agnes actually moved ashore as a tropical storm with 40 to 45 mph sustained winds, but rainfall was heavy across Georgia, the Carolinas to upstate New York. In addition, Agnes spawned 15 confirmed tornadoes in Florida and two in Georgia.

The heaviest blow came from the floods generated by Agnes, which caused nine deaths in Florida, two in North Carolina, 13 in Virginia, one in Delaware, 21 in Maryland and the District of Columbia, one in New Jersey, 50 in Pennsylvania and 25 in New York. At the time with $3.47 billion in damages, Agnes was considered the costliest hurricane in United States history (DeAngelis and

Hodge, 1972). Damage from tides, winds and tornados accounted for $41 million in damages in Florida. All Florida gulf coast counties from Monroe to Bay were declared disaster areas. Other disaster counties included Brevard, Hardee, Hendry and Okeechobee. Near LaBelle, a tornado damaged several mobile home parks and ripped up citrus groves.

As Agnes was a June hurricane, it did not catch cotton in the fields ready for harvest. However, a similar hurricane in September or October today would find the cotton crop of west Florida susceptible to heavy damage with the annual crop topping 100,000 bales (Table 11-1).

1973

The 1973 season produced four hurricanes and three tropical storms, none of which struck the State of Florida. Hurricanes were Alice, Brenda, Ellen and Fran. Tropical storms were Christina, Delia and Gilda.

1974

Although two subtropical storms affected the Florida peninsula in 1974, neither of the season's three hurricanes or two tropical storms were a threat to the state at any time. Hurricanes were Becky, Carmen, and Fifi. Tropical storms were Alma, Dolly and Elaine. Carmen went ashore in Louisiana.

1975

Peninsular Florida was again free of hurricanes and tropical storms in 1975, but the western panhandle felt the brunt of Hurricane Eloise which stormed ashore midway between Fort Walton Beach and Panama City after 8:00 A.M. EDT on September 23, 1975.

Hurricane Eloise, September 13-24, 1975

Meteorological Aspects

The disorganized system which spawned Eloise left the African coast on September 6 and moved west at about 15 mph finally form-

TABLE 11-1. COTTON PRODUCTION IN NORTHWEST FLORIDA FROM 1960-1997.

Year	Planted acres	Harvested acres	Pounds of lint per acre	Bales
1960	21,350	20,220	346	14,610
1961	20,910	19,830	286	11,840
1962	17,760	17,020	396	14,012
1963	21,260	20,180	406	17,050
1964	—	20,205	344	14,534
1965	—	16,355	346	12,545
1966	—	12,375	352	9,075
1967	—	8,640	353	6,346
1968	—	8,370	515	8,983
1969	—	8,070	508	8,540
1970	—	7,290	464	7,044
1971	—	8,520	633	11,249
1972	12,500	11,300	572	13,500
1973	11,890	11,010	533	12,220
1974	11,927	11,538	508	12,237
1975	4,000	3,700	350	2,700
1976	7,400	7,100	514	7,600
1977	6,200	6,100	425	5,400
1978	3,800	3,600	506	3,800
1979	3,400	3,400	565	4,000
1980	6,000	5,900	610	7,500
1981	14,500	13,700	660	18,840
1982	13,920	13,010	640	17,358
1983	10,735	10,245	650	13,873
1984	15,300	14,810	885	27,292
1985	22,100	20,300	724	30,625
1986	18,400	18,000	715	26,802
1987	28,800	28,300	648	38,200
1988	32,100	28,200	565	33,200
1989	25,100	24,650	557	28,600
1990	36,450	35,500	641	47,400
1991	49,400	48,400	721	72,700
1992	48,900	48,450	702	70,900
1993	52,300	51,900	703	76,000

Source: Florida Agricultural Statistics Service, 1998.

TABLE 11-1. COTTON PRODUCTION IN NORTHWEST FLORIDA FROM 1960-1997.

Year	Planted acres	Harvested acres	Pounds of lint per acre	Bales
1994	65,000	64,050	745	99,400
1995	100,100	99,400	454	95,100
1996	87,100	87,100	644	116,900
1997	90,100	89,100	585	108,500

Source: Florida Agricultural Statistics Service, 1998.

ing a weak tropical depression about 600 miles east of the Virgin Islands on the thirteenth. Slowly strengthening, it reached tropical storm intensity on the sixteenth and the first advisory was issued by the San Juan Hurricane Warning Office. Late on the sixteenth, Eloise struck the northeast coast of the Dominican Republic as a minimal hurricane, after which it tracked westward across southeastern Cuba and into the Caribbean north of Jamaica, losing strength from its encounter with the mountains of Hispaniola and Cuba.

A poorly organized Eloise crossed the northeast coast of the Yucatan peninsula on September 20 and then regained hurricane force in the Gulf of Mexico as it moved on a north northeast course at 25 mph toward its date with west Florida on the twenty-third. The highest winds were estimated at near 115 mph at landfall. Winds of hurricane force were felt from Fort Walton Beach to Panama City and north into Alabama, with rainfall from four to eight inches from southeastern Louisiana to Panama City, Florida. The greatest rainfall total was 14.9 inches at Eglin Air Force Base, Florida.

The Damage

The most severe damage from Hurricane Eloise was to property along the waterfront where losses in the millions were reported. However, the high winds and heavy rains also dealt a destructive blow to the corn and cotton crops of the ten-county panhandle agricultural area, although cotton plantings were at a very low point at the time of the hurricane (Table 11-1).

Other 1975 hurricanes included Doris, Faye and Gladys, which remained at sea, Carmine which made landfall in Texas and Blanche which came ashore in Nova Scotia.

1976

The Florida Keys and the lower east coast were brushed by tropical storm Dottie in August, but none of the season's six Atlantic Basin hurricanes affected the state. Hurricane Belle went ashore on Long Island, the other five hurricanes, Candice, Emmy, Frances, Gloria and Holly remained well at sea.

1977

The Atlantic Basin produced only five hurricanes and one tropical storm in 1977, none of which affected the State of Florida. The hurricanes were Anita, Babe, Clara, Dorothy and Evelyn. The tropical storm was Frieda.

1978

There were five hurricanes and six tropical storms in the Atlantic Basin in 1978, but none of the hurricanes made landfall in Florida. Only Hurricane Greta, which struck Belize in mid-September, made landfall in North America. Other hurricanes included Cora, Ella, Flossie, and Kendra.

1979

The 1979 Atlantic Basin storm season was slightly below average in numbers with four hurricanes (Bob, David, Frederic and Gloria) (Table 11-2) and four tropical storms (Ana, Claudette, Elena and Henri). This was the first year that men's names were used to identify hurricanes. Of these, Hurricane David only affected the fruit and vegetable growing regions of southeast Florida, and Hurricane Frederic only affected the western most Florida panhandle, passing inland along the Mississippi-Alabama border.

TABLE 11-2. 1979 HURRICANES.

Name	Origin	Time span	Landfall
Bob	Bay of Campeche	July 9-16	Louisiana
David	East Atlantic	Aug. 25-Sept. 7	Florida, Georgia, South Carolina
Frederic	East Atlantic	Aug. 29-Sept. 14	Mississippi
Gloria	East Atlantic	Sept. 4-15	None

Source: National Climatic Data Center, 1998.

Hurricane David—August 25-September 7, 1979

Meteorological Aspects

Hurricane David originated from a low pressure area which moved off the coast of Africa in late August and was classified as a Cape Verde hurricane. By August 25, the low had intensified into a tropical depression near 12°N, 36°W, after which it moved directly west and became a tropical storm on August 26 at 12°N, 44°W and reached hurricane strength on August 27 as it crossed over 50°W longitude. A slow curvature began and the storm passed through the lesser Antilles on August 29, south of Puerto Rico on the thirtieth and struck a direct blow at the Dominican Republic and Haiti on August 31 and September 1 resulting in several hundred deaths and causing severe crop damage. As a result of its passage over the mountainous terrain of Hispaniola, David had lost considerable strength when it emerged into the Windward Passage with maximum winds of only 68 mph. The storm then continued across the eastern tip of Cuba and then curved northwestward through the Bahamas toward Florida. It began to reintensify over the warmer Bahamian waters but did not regain its previous strength, crossing Andros Island on September 2 with 70 to 80 mph winds. The Bahamas received up to eight inches of rainfall.

David moved inland over Florida at approximately 12 noon on September 3, 1979 just north of Palm Beach. The eye, 20 to 30 miles in diameter, passed over Stuart, Fort Pierce, Vero Beach, Melbourne, Titusville, and New Smyrna Beach, causing significant damage to the Indian River grapefruit crop, before moving offshore in the vicinity of New Smyrna Beach (Table 11-3). As the storm was moving due north at only about 12 mph, some areas were within the eye for as long as two hours (Figure 11-1).

The minimum pressure as the storm passed was 28.64 inches. Wind gusts of 95 mph were reported at the Ft. Pierce Coast Guard Station, 92 mph at Jupiter, 86 mph at South Melbourne Beach, 80 mph at the Port Canaveral Coast Guard Station, and 77 mph at the Kennedy Space Center. Vero Beach received 8.92 inches of rain (Table 11-3). After moving off the Florida east coast, David made its final landfall near Savannah, Georgia.

TABLE 11-3. METEOROLOGICAL DATA DURING HURRICANE DAVID AS IT MOVED NORTH ALONG THE FLORIDA EAST COAST SEPTEMBER 3-4, 1979.

Location	Minimum pressure (inches)	Maximum sustained winds (mph) direction	Wind gusts (mph) direction	Rainfall (inches)
Key West	29.58	26 N	39 NE	0.03
Miami Beach	—	58 NE	69	—
Miami Airport	29.31	30 WSW	44	—
West Palm Beach	28.87	58 N	75 N	2.17
Jupiter	28.73	—	92	—
Stuart	28.80	—	69 N	—
Ft. Pierce USCG	28.73	70 NE	95	—
Pt. Canaveral USCG	—	60 NE	80	—
Vero Beach	28.36	—	—	8.92
Ft. Drum	—	—	—	8.10
Kenansville	—	—	—	7.30
S. Melbourne Beach	—	61 E	86	—
Melbourne	28.68	31 ENE	70	—
Orlando	—	35 NNE	54 N	—
Kennedy Space Center	—	44 WSW	77	6.89
Ponce Inlet USCG	—	40 ENE	60	—
Daytona Beach	29.89	34 NE	55 NE	3.28
Mayport	29.66	30 ENE	45	—

Source: Herbert, 1980.

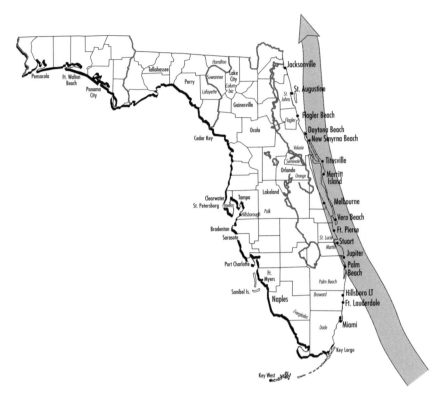

FIGURE 11-1. HURRICANE DAVID BATTERED THE EAST COAST
COUNTIES OF FLORIDA FROM STUART TO NEW SMYRNA BEACH
ON SEPTEMBER 3-4, 1979. SOURCE: HERBERT, 1980.

The Damage

Hurricane David struck the east coast of Florida a full month
before the release of the 1979-80 crop forecast by the Florida Agricultural Statistics Service; therefore, it is not possible to make an
accurate estimate of losses to the citrus crop. As the east coast counties are heavy producers of grapefruit, the type of citrus most vulnerable to wind damage, more grapefruit blanketed the ground after
Hurricane David than did oranges or tangerines. Bob Gibson of
Florida Citrus Mutual (Triangle, Oct. 12, 1979) estimated the grapefruit loss in the five east coast counties, Brevard, Indian River, St.
Lucie, Martin and Palm Beach, to be 5 million boxes or about 25%
of the fruit on tree prior to the hurricane. This included what Gib-

son described as severe secondary drop which continued for two to three weeks after the hurricane had passed. The most severe damage was in a strip 12 to 15 miles from the coastline with most of the losses confined to Martin, St. Lucie and Indian River counties. Orange losses in these same areas were minor. Later in the year (Triangle, Dec. 28, 1979), Mutual described grapefruit sizes as a continuing problem but the crop set was quite good considering the hurricane losses. Table 11-4 compares grapefruit production in a six-county area for the two years prior to the hurricane, the hurricane year (1979-80) and the two years immediately following the hurricane.

Vegetable Damage

When Hurricane David moved up the Florida east coast, the high winds did some crop damage but little acreage was lost. Among commodities affected, snap beans suffered extensive windburn damage and spot resetting of eggplant was required in some areas. Tomato growers in Dade County had begun planting in mid- to late August and the high winds and heavy rains caused some losses estimated at generally less than 10%. Most of the damage resulted from plastic being blown from beds (Florida Agricultural Statistics Service, 1979-80).

TABLE 11-4. GRAPEFRUIT PRODUCTION FROM THE 1977-78 THROUGH THE 1981-82 CROP YEARS FOR COUNTIES MOST AFFECTED BY HURRICANE DAVID.

County	1977-78	1978-79	1979-80[1]	1980-81[1]	1981-82
			1,000 boxes		
Brevard	1,035	904	968	1,065	1,016
Broward	135	97	53	82	55
Indian River	8,388	8,814	8,073	9,064	10,493
Martin	1,556	1,935	1,460	1,953	2,296
Palm Beach	1,556	1,374	1,038	1,312	1,606
St. Lucie	9,102	8,614	7,790	9,094	10,484
Total	21,772	21,535	19,382	22,564	25,950

[1]Hurricane year.
Source: Florida Agricultural Statistics Service, 1983.

Government Response

U.S. Senator Lawton Chiles (Triangle, Sept. 20, 1979) advised that citrus growers in Broward, Martin, Palm Beach, Indian River and St. Lucie counties were eligible for FHA emergency loan assistance as a result of losses from Hurricane David. Growers and foliage nurseries were also declared eligible for loans. Either physical or production losses could be covered. It was advised that applications regarding physical (tree) losses should be made before June 12, 1980, while production (fruit) losses had until September 11, 1980 to make application. These loans were for qualifying farmers who lacked other sources of credit to cover damages from the natural disaster. Loans for the farmer's total loss were made at 5% interest. Amounts in excess of that required to restore production or to make major adjustments were available at 9% interest for real estate purposes and 9.5% for annual operating or chattel purposes.

Personal Observations

Peter Spyke (1998), who was a county agent in Vero Beach in 1979, remembers, "Hurricane David came right over the top of us, but it wasn't a particularly bad storm with maximum winds about 90 mph. It blew off grapefruit in Wabasso, and dropped a fair amount of rain. State Road 60 was flooded at 20-mile bend, which is normal whenever it rains a lot in western Indian River County."

Erroll Fielding (1998) was caretaking the DiGiorgio North Canal Grove east of Ft. Pierce for Orange-Co who owned the grove at the time. They lost an estimated 25% of the grapefruit. That was the only hurricane since the 1960s to blow off a significant amount of fruit.

Chapter 12

From 1980 to 1989

"Florida hurricanes were bad news in the 1920s, the 1940s and the 1960s, so we worried about the 1980s, but no really severe hurricanes came our way."—Clayton White, Okeechobee City, FL, 1998.

The 1980s produced history's most severe weather disasters for Florida agriculture, but those disasters were in the form of freezes, not hurricanes. The 1980s produced serious freezes in 1981 and 1982 and impact freezes in 1983, 1985 and 1989 (Attaway, 1997). However, there were only minor brushes with weather originating in the tropics. The most interesting feature of the decade was Hurricane Kate which became a category 3 hurricane on November 20, 1985 and was a category 1 hurricane when it came ashore between Panama City and Apalachicola on November 22, 1985.

1980

Although there were nine hurricanes and two tropical storms in the Atlantic Basin in 1980, activity was heaviest in the far eastern Atlantic, with seven tropical systems, and in the Caribbean Sea and the Gulf of Mexico with four systems. No hurricane or tropical storm threatened any part of Florida during the entire season.

1981

The 1981 Atlantic Basin hurricane season was average in terms of activity, with seven storms reaching hurricane strength and four remaining in the tropical storm category. Of these, only Hurricane Dennis made landfall in Florida, and then only as a tropical storm. It did not reach hurricane intensity until it was east of the Virginia Capes, and then was classified as a hurricane for only a very short time.

Tropical Storm Dennis, August 17-19, 1981

Meteorological Considerations

Dennis developed as a typical Cape Verde storm (National Hurricane Center, 1981). It moved off the coast of Africa as a well-developed low pressure area on August 5 and became a tropical depression and later a tropical storm on August 7 in the eastern Atlantic near 12°N, 32°W. It remained a tropical storm for three days before encountering upper level wind shear and being downgraded to a tropical depression on August 11, and then weakened further to a tropical wave on August 12. The wave moved steadily westward and stalled briefly near the south coast of Cuba on August 15, at which time it reorganized first into a tropical depression and then regained tropical storm strength. As it continued to strengthen slowly, Dennis moved over Cuba and into the Florida Straits on August 16 and 17 then moved very slowly across the warm waters producing very heavy rainfall and wind gusts to tropical storm force in extreme south Florida (Table 12-1) before moving north across south and central Florida and finally out over the Atlantic ocean north of Cape Canaveral. The minimum pressures recorded were 29.47 inches on August 17 over the Florida Straits and 29.38 inches on August 20 and 21 when the storm briefly reached hurricane strength over the open Atlantic.

Two tornadoes accompanied Dennis in its path across south Florida, one on Plantation Key and another at Haulover Beach in northern Dade County. No major damage resulted from these tornadoes.

The Damage

The greatest damage from Dennis was in Dade County and resulted from heavy rains, not wind gusts. The 20-inch rainfall totals

TABLE 12-1. METEOROLOGICAL DATA AS DENNIS PASSED OVER SOUTH FLORIDA AS A TROPICAL STORM, AUGUST 17-19, 1981.

Location	Maximum sustained winds (mph) direction	Wind gusts (mph)	Rainfall (inches)
Key West	28 W	52	5.04
Homestead	—	—	19.15
Kendall	—	—	20.38
Miami Airport	23 S	39	6.92
Ft. Lauderdale	—	60	5.98
West Palm Beach	25 S	45	5.94
Orlando	20 SW	37	1.20
Daytona Beach	23 NE	32	0.58

Source: National Hurricane Center, 1981.

in the Kendall and Homestead areas of southwest Dade County lead to prolonged standing water resulting in losses to agriculture exceeding $15 million in Dade County alone (National Hurricane Center, 1981). Field work and planting were delayed in south Dade County while farmers waited for fields to dry (Weather and Crop News, August 31, 1981).The rain was welcome in some parts of south Florida as it relieved a prolonged drought. However, very little fell across Lake Okeechobee or its catch basin to the north of the lake which could have provided further drought relief.Other named storms for the 1981 season were Hurricanes Emily, Floyd, Gert, Harvey, Irene and Katrina. Tropical storms were Arlene, Bret, Cindy and Jose. None of these other hurricanes or tropical storms affected either peninsular Florida or the Florida panhandle.

1982

1982 was a very inactive tropical season in the Atlantic Basin with only two hurricanes, Alberto in June and Debbie in September. There were three tropical storms: Beryl, Chris and Ernesto. None of these hurricanes or tropical storms crossed the Florida coastline, although a subtropical storm passed from the gulf north of Tampa to the Atlantic near Jacksonville in June.

1983

The 1983 season was also relatively inactive with only three storms which briefly attained hurricane force, and one tropical storm. The hurricanes were Alicia, Barry and Chantal. The tropical storm was named Dean. Of these, only Barry made landfall in Florida, but as a tropical depression, not a hurricane. It did not reach hurricane intensity until after it crossed the Gulf of Mexico and made landfall on the northeast coast of Mexico (National Hurricane Center, 1983).

Tropical Depression Barry - August 25-26, 1983

Meteorological Considerations

The storm, which would ultimately become tropical storm and later Hurricane Barry, developed from a low pressure disturbance which moved off the northwest coast of Africa on August 13. However, due to strong upper level winds it did not show signs of development until it approached the Bahamas on August 22 and as upper level winds decreased it became a tropical depression on the evening of August 23. The depression quickly strengthened into Tropical Storm Barry and turned west toward the central Florida east coast on August 24. At this point, a return of strong upper level winds prevented further development and Barry was downgraded to a depression as the center crossed the coast just south of Melbourne on the morning of August 25 with only 35 mph winds, and crossed the Florida peninsula exiting into the Gulf of Mexico near Bradenton. After leaving Florida, Barry again became a tropical storm in the central gulf and reached minimal hurricane status before crossing the coast of northeast Mexico just south of Brownsville, Texas.

As Barry neared Florida, the minimum pressure was only 29.85 inches, although winds just east of the center were 55 mph, higher than would be expected for that high a minimum pressure. Throughout the entire track of Barry, most of the heavy convection was southwest of the center which caused most of the rain in Florida to fall along the lower east coast. Rainfall totals in inches along the lower east coast included 1.20 inches at Miami Beach; 1.02 inches at

West Palm Beach; 1,66 inches at Pompano Beach; 1.37 inches at Hialeah and 2.18 inches at Homestead (Climatological Data, August 1983).

The Damage

There was no appreciable damage to agriculture from Tropical Depression Barry.

1984

The frequency and number of storms was more normal in 1984 with five hurricanes and seven tropical storms in the Atlantic Basin. The hurricanes were named Diana, Hortense, Josephine, Klaus, and Lili. The tropical storms were Arthur, Bertha, Cesar, Edouard, Fran, Gustav, and Isidore. However, only Tropical Storm Isidore, in late September, had a direct effect on Florida.

Tropical Storm Isidore - September 26-28, 1984

Meteorological Considerations

Isidore formed as a tropical depression just east of the central Bahama Islands near 23.5°N, 70°W and reached tropical storm strength at about 7:00 A.M. on September 26 (National Hurricane Center, 1984). The lowest central pressure of 29.50 inches was reached when the storm passed near the northern tip of Andros Island. It made landfall near Jupiter after which it turned to the west northwest and crossed central Florida. Hourly surface maps and radar indicated that the center passed between Frostproof and Lake Wales, between Lakeland and Winter Haven then turned north to Dade City, north northeast to near Ocala and northeast to east of Gainesville, just west of Jacksonville after which it reentered the Atlantic Ocean at Fernandina Beach. The central pressure remained constant while Isidore was over the Florida peninsula and weakening was not evident. While the storm was over land, reconnaissance reports documented 1500 foot flight level winds of 58 mph along both the east and west coasts of Florida. The highest gusts recorded were 73 mph near St. Augustine. Gusts of 60 mph were recorded at Patrick AFB and 65 mph on the Skyway Bridge across Tampa Bay (Table 12-2). Sustained winds of 46 mph were felt near St. Augustine and 43 mph at Mayport east of Jacksonville. Highest rainfall totals were 5 to 7 inches in north Florida. Examples were 6.59 inches at Live Oak; 5.67 inches at Jasper; 5.86 inches at Usher Tower and 5.01 inches at Jacksonville.

The Damage

There was considerable concern in the Florida citrus industry as Isidore crossed several major citrus producing counties including St. Lucie, Indian River, Brevard, Highlands, Polk, Osceola, Pasco, Lake, Orange, Seminole, Marion and Volusia at the time of a perceived major citrus canker scare; and it was feared that the winds and rain might spread the canker bacteria.

TABLE 12-2. METEOROLOGICAL DATA AS TROPICAL STORM ISIDORE VISITED THE FLORIDA PENINSULA, SEPTEMBER 26-28, 1984.

Location	Minimum pressure (inches)	Wind gusts (mph)	Rainfall (inches)
Homestead AFB	29.70	29	—
Miami Airport	29.68	32	1.26
West Palm Beach	29.62	45	2.80
Vero Beach	—	56	1.62
Cape Canaveral	—	54	—
Patrick AFB	—	60	—
Melbourne	—	52	1.72
Sanford	—	58	2.44
Daytona Beach WSO	29.71	48	4.06
Tampa Airport	29.64	45	2.50
Ruskin	29.62	—	—
Skyway Bridge	—	65	—
St. Petersburg	—	40	2.11
Cedar Key	—	48	—

Source: National Hurricane Center, 1984 and Climatological Data, September 1984.

However, these concerns were relieved when the canker was found not to be the serious "A" strain (Graham and Gottwald, 1991).

Rains from Isidore improved soil moisture in all areas other than the panhandle (Weather and Crop News, October 1, 1984).

1985

With seven hurricanes and four tropical storms, 1985 was an average year in the Atlantic Basin. Hurricane names were Bob, Claudette, Danny, Elena, Gloria, Juan and Kate. The tropical storms were Ana, Fabian, Henri and Isabel.Fortunately, only Bob and Kate affected Florida, Bob crossing south Florida in July as a tropical storm, and Kate going ashore near Port St. Joe in November. Bob did not reach hurricane force until after crossing Florida, and then only briefly before going ashore in South Carolina.

Tropical Storm Bob - July 23, 1985

Meteorological Considerations

Bob developed in the eastern Gulf of Mexico on July 20 near 26°N, 85°W from the remnants of a tropical wave (Laurence, 1985). It moved slowly eastward and attained tropical storm status at 10:00 P.M. on July 22. The center moved ashore between Naples and Fort Myers at noon on July 23. Most of the heavy rains and strongest winds were located to the south and east of the center, with Naples reporting sustained winds of 40 mph from the west as the center passed north of the city. Bob's forward speed increased to 12 mph as the storm turned sharply to the north and moved off the Florida coast near Vero Beach at 7:00 P.M. on July 23. It attained minimal hurricane status in the Atlantic before reaching the South Carolina coast at 10:00 P.M. on July 24. South Florida received heavy rains from Bob with Naples recording a three day total of 12.07 inches to 7.3 inches at the Oasis Ranger Station; 2.41 inches at the Belle Glade Experiment Station; 7.13 inches at Ft. Lauderdale; 4.46 inches at Pompano Beach; 3.56 inches at West Palm Beach and 1.25 inches at Vero Beach.

The Damage

No major agricultural damage to crops in the ground resulted from Tropical Storm Bob, but fall planting was interrupted in the vegetable areas of south Florida.

Hurricane Kate - November 21, 1985

Meteorological Considerations

Kate was a very unusual hurricane in that it formed in the western Atlantic in mid-November and held together until it struck west Florida on the twenty-first. Kate was first detected on November 15 as a tropical storm northeast of Puerto Rico near 21.1°N, 63.8°W. The storm's initial track was to the southwest where its winds reached minimal hurricane force at latitude 20.4°N early on November 17. Kate then moved along a west northwest path for the next two days with winds of 85 to 95 mph, category 1 on the Saffir-Simpson scale (Table 15-4). After skirting the north coast of Cuba on November 19, Kate emerged into the open Gulf of Mexico on the twentieth and took a more northwesterly course as its winds increased to 105 mph, category 2 on the Saffir-Simpson scale (Table 15-4). Fortunately, Kate began to lose intensity as it approached the Florida coastline late on November 21, and as it came ashore winds had diminished to near 80 mph. Rainfall amounts in the affected area ranged from 4.20 inches at Apalachicola as the hurricane came ashore to 4.25 inches at Chipley and 4.40 inches at Tallahassee as the center moved inland as a tropical storm.

The Damage

According to Weather and Crop News for the week ending December 1, 1985, both quality and yield of late maturing soybeans were reduced by the hurricane. Unharvested cotton fields also suffered significant damage. The loss of ungathered pecans may be assumed but no report was made.

1986

The 1986 Atlantic Basin hurricane and tropical storm season was relatively inactive with only four hurricanes: Bonnie, Charley, Earl and Frances and two tropical storms, Andrew and Danielle. Of the six, only Charley, which moved inland from Apalachee Bay as a subtropical depression had any affect on Florida, and its affects were minimal.

1987

For the second consecutive year, the Atlantic Basin season was relatively inactive with three hurricanes and four tropical storms. The only storms affecting Florida were Floyd, a minimal hurricane which passed from the Gulf of Mexico through the Florida Keys to the Atlantic in mid-October, and an unnamed tropical storm which affected the Florida panhandle in August. Other 1987 hurricanes were Arlene and Emily. Tropical storms were Bret, Cindy and Dennis.

Hurricane Floyd - October 12, 1987

Meteorological Considerations

Floyd was the first storm since 1968 to become a hurricane in the northwestern Caribbean in October (National Hurricane Center, 1987). It formed from a broad low pressure area first noted over the Gulf of Honduras on October 5. The system drifted east and then south to a position off the northeast coast of Nicaragua where it was classified a tropical depression on October 9. After drifting south and then north, it strengthened to tropical storm force on October 10 and moved northwest and then north and crossed western Cuba on October 11. Early on October 12, a reconnaissance plane measured 80 mph winds and Floyd was upgraded to a hurricane as its direction shifted east northeast, a turn not forecast by any of the guidance models, toward the Florida Keys. Floyd passed south of Miami the evening of the twelfth, became disorganized and weakened, gradually becoming extra-tropical and merging with a large low pressure system between Florida and Bermuda.

It had remained a hurricane for only about 12 hours. Reconnaissance aircraft reports indicated that the area of maximum winds remained south of the center throughout the storms passage through the Florida straits. None of the reporting stations in south Florida measured sustained winds of hurricane force. North of the Keys, sustained winds were only 35 to 40 mph with gusts in the 45 to 55 mph range (Table 12-3).

Rainfall directly attributable to the area around Floyd's center was quite low. However, interaction of Floyd's circulation with a frontal trough across central Florida resulted in generally two to four inches of rain across southern Florida except for a band of five to eight inches between Naples and Lake Okeechobee.

The Damage

Crop damage in south Florida could not be precisely estimated, but was considered to be less than $100,000 (Climatological Data, October 1987).

TABLE 12-3. METEOROLOGICAL DATA ASSOCIATED WITH HURRICANE FLOYD AT SELECTED STATIONS IN SOUTH FLORIDA, OCTOBER 12, 1987.

Location	Minimum pressure (inches)	Maximum sustained winds (mph) direction	Wind gusts (mph)	Rainfall (inches)
Key West WSO	29.37	45 N	67	3.28
Miami WSO	29.51	35 SE	38	2.39
Nat. Hur. Ctr.	—	36 N	40	—
Naples	—	—	—	5.20
Palm Beach WSO	—	31 N	50	2.92
Vero Beach	—	—	40	3.77
Melbourne	—	25	37	3.10
Fort Myers	29.60	25	51	4.83
Tampa WSO	29.68	16	24	1.28

Source: National Hurricane Center, 1987.

1988

Hurricane and tropical storm activity returned to normal levels in 1988 with five hurricanes and seven tropical storms being monitored in the Atlantic Basin. Fortunately, of the 12 systems, only tropical storm Keith, which crossed the Florida peninsula, and tropical storm Chris, which paralleled the Florida coast just to the east of the Fort Pierce area had any effect on the Sunshine state.

Tropical Storm Keith - November 23, 1988

Meteorological Considerations

Keith was an unusual late season (November) storm which almost attained hurricane force. The system which became Keith came off the coast of Africa as a tropical wave on November 5, but did not become a tropical depression until it reached the central Caribbean on November 17 (National Hurricane Center, 1988). It moved to the west as a poorly organized depression, and by November 20 as deep convection developed, the storm gradually strengthened and was upgraded to a tropical storm. Keith turned to the northwest and north and as it passed the northeast tip of the Yucatan peninsula it reached near hurricane force, but weakened due to westerly shear and turned to the northeast and made landfall on the Florida west coast near Sarasota the morning of November 23. The central pressure at landfall was 29.38 inches, the maximum sustained winds were 63 mph with a gust to 80 mph at an elevation of 250 feet. Keith maintained tropical storm strength as it crossed central Florida and exited north of Melbourne later in the day on November 23.

Rainfall along the central Florida west coast was very heavy (Table 12-4) with 11 inches reported at Largo, Florida. Elsewhere to the north of the track, amounts ranged from four to seven inches, but rainfall to the south of the track was generally less than one-half inch.

The Damage

Heavy rains resulted in low level flooding but no specific information on damage is available.

Table 12-4. Meteorological data associated with Tropical Storm Keith at selected Florida stations, November 23, 1988.

Location	Minimum pressure (inches)	Maximum sustained winds (mph)	Wind gusts (mph)	Rainfall (inches)
Tampa	29.55	31	47	4.48
Brandon	29.51	—	—	4.45
Indian Rocks Beach	29.47	40	54	—
St. Leo	—	—	—	10.25
Largo	—	—	—	11.00
Safety Harbor	—	—	—	10.00
Bushnell	—	—	—	5.50
Wauchula		—	—	3.10
Sarasota		—	—	0.33

Source: National Hurricane Center, 1988.

1989

The 1989 season was again average in numbers, with seven hurricanes and four tropical storms, none of which crossed the Florida coastline. However, it was the year of Hurricane Hugo which struck a devastating blow to the state of South Carolina. Hurricane names were Chantal, Dean, Erin, Felix, Gabrielle, Hugo and Jerry. Tropical storms were Allison, Barry, Iris and Karen.

Chapter 13

Devastating Hurricane Andrew - The Most Powerful Storm to Strike Florida Since the Keys Hurricane in 1935

Hurricane Andrew - August 24, 1992

"It is as if we had been transported, under conditions of incredible violence, to an entirely different land"—William H. Krome (1992), Homestead, Florida

Hurricane Andrew was a "Cape Verde hurricane," one of the strongest ever recorded on the U.S. east coast. It was one of the two most destructive hurricanes to strike Florida in this century, if not for all time. No other hurricane has ever so devastated south Dade County, and agricultural losses were said to be the greatest in U.S. history, second only to a Midwest drought (FEMA, 1992). For that reason, it has been given its own chapter.

Meteorological Considerations

Hurricane Andrew formed from a tropical wave which moved off the coast of Africa on August 14 and passed to the south of the Cape Verde Islands on August 15 (Mayfield and Avila, 1992). It became a tropical depression near 12°N, 37°W on August 16 and a tropical storm on the seventeenth after which it moved rapidly on a west northwest course at 20 to 25 mph toward the Lesser Antilles. However, after encountering a strong upper level low pressure system, Andrew turned to the northwest on August 19 and passed east of the Lesser Antilles and was weakened to below hurricane strength by a strong southwesterly vertical wind shear. Then on August 22, the environment changed with the formation of a high pressure ridge eastward from the southern U.S. coast and as an eye formed, Andrew rapidly intensified to hurricane strength on the twenty-second and became an extremely dangerous category 4-5 hurricane on the twenty-third as it moved west through the Bahama Islands along the 25th parallel. The central pressure in the storm dropped over 2.5 inches in 36 hours as maximum sustained winds reached 155 mph. After leaving the Bahamas, Andrew weakened somewhat as it passed over the Grand Bahama Bank and the central pressure rose to 27.79 inches. However, there was to be no escape for south Florida as Andrew rapidly re-intensified as it crossed the warm waters of the gulf stream at 18 mph and slammed into the Florida City/ Homestead area with a fury never before experienced by long time residents, with official estimates of maximum sustained winds of 145 mph and gusts to 175 mph (Mayfield, et al., 1994). Unofficially, a gust of 178 mph was recorded at Princeton and a gust of 211 mph at Perrine (FEMA, 1992) and the National Hurricane Center noted that, "winds in excess of the officially recorded values occurred in the northern eyewall, a little to the left or right of the flight path." Its central pressure at landfall near Homestead Air Force Base at 5:05 A.M. on August 24 was estimated at 27.23 inches (922 mb), category 4 on the Saffir-Simpson scale (Table 15-4), a value eclipsed in Florida only by the Labor Day Hurricane in 1935 with 26.34 inches.

Meteorological data for Hurricane Andrew is presented in Table 13-1. However, it should be noted that as Homestead Air Force Base and Tamiami Airport had discontinued routine meteorological observations prior to Hurricane Andrew, Miami International Airport, located 6 miles north of the eyewall, was the closest official station. Consequently, wind speeds from the northern and southern eyewalls were either calculated values or unofficial values from residents of the area who frequently risked serious injury by continuing to make observations at the height of the storm.

It should also be noted that while a storm with unparallel ferocity, Andrew was such a small hurricane that it has been described as, "like a big tornado." This can be seen from the wind speeds (Table 13-1) which decreased rapidly a few miles from the center.

Andrew continued its westerly direction across the extreme southern portion of the Florida peninsula and emerged into the Gulf of Mexico about four hours later between Flamingo and Marco Island (Figure 13-1) after causing catastrophic damage to farms, groves and the agricultural infrastructure of Dade County, where lime and avocado groves were devastated. It was still a major category 3 hurricane with an estimated minimum pressure reading of 28.05 inches (950 mb) when it entered the Gulf of Mexico and turned gradually to the west northwest, and later to the northwest, making a final landfall in south central Louisiana on August 26.

TABLE 13-1. METEOROLOGICAL DATA DURING HURRICANE ANDREW AS IT CROSSED SOUTH FLORIDA ON AUGUST 24, 1992.

Location	Minimum pressure (inches)	Maximum sustained winds (mph)	Wind gusts (mph)	Rainfall (inches)
Fowey Rocks	28.57	141[1]	170[1]	—
Northern eyewall	—	143[2]	172[2]	—
Homestead	27.23	—	—	6.9[3]
Miami (NHC)[4]	29.00	115	164	—
Tamiami Airport	29.18	127[1]	—	—
Miami Airport	29.31	86[3]	115[3]	2.04
Haulover Pier	29.65	67[1]	132	—
Fort Lauderdale	—	—	60	0.40
Palm Beach	29.85	49	58	0.95
Patrick AFB	30.01	25	36	—
Key West	29.83	29	43	0.33
Captiva Fire Station	—	—	72	—
Fort Myers	29.83	34	51	0.55

[1]A higher value may have occurred. Equipment became inoperable shortly after observation.
[2]Estimated on the assumption that for a westward moving hurricane the wind speed in the northern eyewall exceeds the wind speed in the southern eyewall by about twice the wind speed in the southern eyewall.
[3]Estimated.
[4]NHC = National Hurricane Center.
Source: Mayfield, et al., 1994.

318

**FIGURE 13-1. THE PATH OF HURRICANE ANDREW AS IT RAVAGED
SOUTH DADE COUNTY ON AUGUST 24, 1992 WITH WIND GUSTS
ESTIMATED AT UP TO 200 MPH. SOURCE: MAYFIELD, 1974.**

The Damage

"When the eyewall of Hurricane Andrew came ashore, I'm
convinced that the wind gusts were 200 mph"—Phoebe
Krome (Mrs. William Krome), 1997, Homestead, Florida.
"Hurricane Andrew was a storm which took no prisoners. It
showed no mercy to tree or shrub, man or beast"—Anonymous.

Speaking to the Indian River Citrus League in September, Governor Lawton Chiles estimated that damage to the Dade County agricultural base would exceed $1 billion, including infrastructure, with eight out of ten Dade County farms suffering major damage from the hurricane (Florida Grower and Rancher, 1992A). He indicated that the lime industry had been destroyed, although growers were heavily pruning the surviving trees in an effort to restore their groves. However, he estimated that it would take three to six years before a significant crop would come out of the area. Losses to the nursery industry were indicated to exceed $300 million, the $15 million okra crop was lost and he also cited heavy losses to avocados, mangos, papayas and kumquats. Fortunately, the winter vegetables were not yet in the ground (Table 13-2). Chiles indicated that the heavily damaged state farmers' market in Homestead would be rebuilt.

Fall planting of traditional vegetables had been scheduled for the last week of August but was delayed by the hurricane, accumulation of storm debris and damage to farm facilities and equipment (Lamberts and Bryan, 1993).

Losses to Agriculture Recorded by FEMA

Agricultural damage by category is detailed by FEMA (1992) and summarized in Table 13-3. Of the $1.04 billion in loss to Dade

TABLE 13-2. APPROXIMATE INITIAL PLANTING DATES FOR SOME VEGETABLE CROPS GROWN IN DADE COUNTY, FLORIDA.

Vegetable crop	Month
Cabbage	Mid-October
Eggplant	August
Potatoes	Late October to early November
Squash .	Late September
Strawberries	Late September
Sweet corn	Late October to early November
Tomatoes	Transplanted in mid- to late September

Source: Florida Agricultural Statistics Service, 1992.

TABLE 13-3. DAMAGE TO AGRICULTURAL INDUSTRY OF DADE COUNTY, FLORIDA AND ESTIMATED ECONOMIC LOSSES FROM HURRICANE ANDREW, AUGUST 24, 1992.

Type	Damage	Estimated loss
		--- millions ---
Dade County farms	80% of 3,655 farms damaged	$1,040
Employment[1]	Permanent income loss from crop destruction, foliage and nursery industry	$250
Tropical fruits	Includes damage to 1992 and 1993 mango, lime and avocado crops and trees	$128.7
Horticulture	Damage to 10,000 acres of foliage and nursery industry	$280
Vegetable crops	Damage to tropical and specialty vegetables and a few early plantings of traditional vegetables	$12.9
Livestock	Includes damage to quail, aquaculture and horses	$12.5

[1]Dade County agriculture supported 23,000 jobs at the time of Hurricane Andrew.
Source: FEMA, 1992; Lamberts and Bryan, 1993.

County farms, a minimum of $349 million was in crop loss, $580 million in physical loss of plants and equipment and a $125 million loss in aquaculture and livestock. This does not include losses resulting from salt water inundation of approximately 5,200 acres.

Overall Effect of Hurricane Andrew on Dade County Nurseries

According to Earl Wells, Executive Vice President of the Florida Nurserymen's Association, Dade County was home to 842 ornamental and foliage nurseries embracing 7,200 acres of high volume growing operations. Immediately after the hurricane, he estimated a virtually total loss of more than $145 million in plants alone, with even heavier capital losses.

The nursery industry in Dade County was a major victim of Hurricane Andrew because it is a year round industry as opposed to traditional vegetables which are highly seasonal and not yet planted in August when the storm struck (Citrus and Vegetable Magazine, 1992C) (Table 13-2). There were over 7,000 nursery acres producing foliage, woody ornamentals, vegetable transplants and nursery stock with an estimated value of over $150 million, most of which was either destroyed or severely damaged. Nurseries in adjacent counties including Broward, Collier and Palm Beach also suffered, but the damage was not comparable to that in Dade County. Hull and Hodges (1993) studied the impact of Hurricane Andrew on eleven commercial nurseries in Dade County. A review of the financial records of these nurseries showed that the net income fell from a three year (1987-1991) prehurricane average of $130,000 to a *minus* $234,000 in 1992. The storm destroyed approximately 1,300 acres of nursery shadehouses and greenhouses and over 4,000 acres of woody ornamental nursery crops. Total losses for the nursery industry were estimated at $206 million including $120 million in plant inventory losses, $22 million in fixed capital losses for buildings and equipment, $34 million for storm cleanup and $29 million for the time value of lost sales, replanting, and capital replacement costs. Lost export sales for the nursery industry were $85 million and were projected to cause additional losses of $155 million in local economic activity for all allied industries in Dade County. Overall, this catastrophic damage translated to a loss of 4,139 jobs.

Effect on the Orange and Grapefruit Crops

"Although the eye of Hurricane Andrew passed less than 50 miles south of our grove (Bob Paul, Inc.) in northern Collier County, we felt winds of only 40 mph. I've seen thunderstorms which were much worse."—Norman Todd, 1998.

According to Rouse (1992) although Hurricane Andrew was the most severe storm to hit the Florida coast since Hurricane Donna in 1960, losses to citrus in the gulf production area of Charlotte, Glades, Collier, Lee and Hendry counties was minimal. The only losses may have been in the southern part of Collier County south of Alligator Alley (Highway 84 & I-75). Wind gusts did not exceed 40 to 50 mph and rainfall varied from 0.5 to 1.75 inches. Fruit loss was limited to a few grapefruit on the ground under trees at the end of rows, although one grapefruit grove was reported to have 5% of the fruit on the ground. No orange losses were reported.

Scientific Studies and Evaluations

In terms of agricultural damage, Hurricane Andrew is undoubtably the most well-documented storm in Florida history, and publications by scientists, economists and statisticians including Jonathan H. Crane, Richard J. Campbell, Carlos Balerdi, Carl W. Campbell, Seymour Goldweber, Robert R. Terry, Mary Lamberts, Herbert Bryan, DeArmand L. Hull and Alan W. Hodges, made enormous contributions to understanding the effects of the disaster.

In the months immediately following the hurricane, Campbell, et al. (1993) surveyed the damage to south Florida's major tropical fruits as: avocados - moderate, limes and mangos - severe. Crops of lesser importance, including atemoya, banana, carambola, coconut, guava, longan, lychee, mamey sapote, papaya, passion fruit and sapodilla ranged from light in the case of carambola, to moderate for bananas and papayas, moderate-severe for lychee and severe for passion fruit.

In a separate study 14 to 15 months after the hurricane, Crane, et al. (1993) reported the percentage of the various varieties of tropical fruit damaged by Andrew, and characterized the type of damage for each variety. Tree condition was described as toppled (tipped over), stumped (reduced to major scaffold limbs), destroyed (empty tree hole or dead tree) or standing upright. Size and anchorage of root system were important. They found that tree age and height were among the most important factors which affected the susceptibility of a tree to damage. Trees which best survived the hurricane included lime, carambola, atemoya, avocado, mamey sapote and guava, while mango, longan and lychee trees were more vulnerable (Table 13-4).

Regarding the overall effect of Hurricane Andrew on the acreage of tropical fruit in Dade County, the Florida Agricultural Statistics Service (Terry, 1993) found limes to be most affected with a 73% loss in acreage, while avocado acreage was reduced 34% and mango acreage was down 42% (Table 13-5). In terms of tree count, the

TABLE 13-4. PERCENTAGE OF VARIOUS VARIETIES OF TROPICAL FRUIT TREES SURVIVING BUT OFTEN HEAVILY DAMAGED BY HURRICANE ANDREW. MANY HEAVILY DAMAGED TREES SURVIVED, BUT REQUIRED MAJOR REHABILITATION, I.E. TOPPLED OR STUMPED TREES.

Tree variety	% Surviving
Lime[1]	95
Carambola	93
Atemoya	90
Avocado	87
Mamey sapote	84
Guava	84
Mango	71
Longan	70
Lychee	60

[1]Only grafted trees were counted. Groves with air layered trees suffered greater damage.
Source: Crane, et al., 1993.

numbers of lime trees were reduced from 963,000 to 290,000, avocados from 892,000 to 596,000 and mangos from 240,000 to 138,000 (Table 13-6).

J. H. Crane and C. F. Balerdi (1996) further surveyed the status of mango plantings and reported that prior to Hurricane Andrew, Florida had produced 10 metric tons of mangos annually with a

TABLE 13-5. ACREAGE OF LIMES, AVOCADOS AND MANGOS IN DADE AND COLLIER COUNTIES IN THE YEARS BEFORE AND AFTER HURRICANE ANDREW.

Year	Limes	Avocados	Mangos
1990	6,700	9,078	2,759
1993	2,235[1]	6,104	1,732
1994	2,618	6,040	1,550
1996	2,792	6,305	1,505

[1]The October 1992 survey was delayed to March 1993 to more accurately reflect the damage from Hurricane Andrew.
Source: Florida Agricultural Statistics Service, 1995 and Terry, 1993.

TABLE 13-6. TOTAL NUMBERS OF LIME, AVOCADO AND MANGO TREES IN DADE COUNTY ALONE IN THE YEARS BEFORE AND AFTER HURRICANE ANDREW.

Year	Limes	Avocados	Mangos
	---------------- thousand trees ----------------		
1990	962.8	891.8	239.5
1993[1]	290.4	596.0	138.3
1994	440.8	594.8	157.2
1996	470.9	625.7	155.5

[1]The October 1992 survey was delayed to March 1993 to more accurately reflect the damage from Hurricane Andrew.
Source: Florida Agricultural Statistics Service, 1993.

value of approximately $4.3 million. However, immediately after the storm, the yield was only 1.25 metric tons with a value of less than $1 million. The authors also provided valuable information on restoration of a mango orchard after a devastating hurricane. In a further evaluation of hurricane damage to young carambola trees, Crane, et al. (1994) reported that the major cause of death and decline of trees at the Tropical Research and Education Center, Homestead, was detachment of the bark at the soil line and major roots.

M. Lamberts and H. Bryan (1993) evaluated non-cultural factors which affected Dade County vegetable production after Hurricane Andrew. They found several adverse factors including delayed planting dates, debris blown into prepared fields from adjacent properties, the appropriation of vegetable land for tent cities and sites for disposal of urban debris, lack of housing for migrant labor and damage to tractors, trucks and other equipment and facilities such as barns and packinghouses. In addition, heavy equipment used in the building of tent cities compacted the soil so densely that extensive plowing was necessary to bring fields back into production.

D. L. Hull and A. W. Hodges (1993) reported on the impact of Hurricane Andrew on the profitability of ornamental nurseries in Dade County. They found that lost nursery sales exceeded $205 million, lost export sales over $145 million, lost personal income almost $92 million and over 7,000 jobs lost (Table 13-7). Estimated direct and indirect losses to sectors of the ornamental industry are shown in Table 13-8.

TABLE 13-7. ECONOMIC IMPACT OF HURRICANE ANDREW ON THE ORNAMENTAL NURSERIES OF DADE COUNTY, FLORIDA.

Sector	Direct and indirect losses
Lost nursery sales	$205,798,954
Lost export sales	$145,088,263
Loss in personal income	$ 91,739,309
Number of jobs lost	7,073

Source: Hull and Hodges, 1993.

The Florida Lime Industry After Hurricane Andrew

Despite the horrendous losses suffered by the Florida lime growers, many in the industry remain optimistic, though the challenges from the North American Free Trade Agreement (NAFTA) and the emergence of Mexico as the major North American producer of limes, may at time seem overwhelming.

The previous major weather event affecting the Florida lime grower was the severe freeze of January 1977 which brought killing cold even as far south as Dade County (Attaway, 1997). This freeze caused the price of a 40 pound carton of limes to increase to a record $70.00 (Roy, et al., 1996), and stimulated the Mexican lime growers to increase their production leading to greater and greater imports from south of the Rio Grande River.

TABLE 13-8. ESTIMATED COSTS TO ORNAMENTAL NURSERY INDUSTRY OF DADE COUNTY BY SECTOR.

Sector	Direct and indirect losses
	----------- millions -----------
Container ornamentals	$72.1
Field ornamentals	66.4
Greenhouse	17.8
Foliage	49.3
Total	205.5

Source: Hull and Hodges, 1993.

In the intervening years of the 1980s, the Florida growers of tropical fruits and vegetables began to lose their market share not only as a result of continuing increasing foreign competition aided by trade liberalization, but due to stricter and stricter environmental laws compounded by urban sprawl from the Miami megalopolis to their north. Nevertheless, prior to August 24, 1992, more than 90% of U.S. Persian lime growers were located in Dade County and the production of limes had held firm despite declining acreage (Roy, et al., 1996). However, Andrew's shrieking winds reduced the south Florida Persian lime acreage from 6,700 to 2,235 acres, 1,668 acres in Dade County and 567 acres in Collier County (Tables 13-5 and 13-6), and the Mexican share of the U.S. lime market and increased from 67% to 90%. The outlook for the south Florida grower seemed bleak, but most growers with 100 acres or more decided to replant using grafted trees. They had learned the hard way that groves with air layered trees suffered much greater losses.

Damage to Agricultural Research Facilities

Both the University of Florida, Institute of Food and Agricultural Sciences (IFAS) and the USDA Agricultural Research Service maintained and operated important research facilities in Dade County at the time of Hurricane Andrew. At the IFAS Tropical Research and Education Center in Homestead, the winds with gusts estimated to 200 mph tore roofs from buildings, blew out windows and doors, demolished greenhouses, shadehouses, trailers and metal maintenance buildings and felled palm trees across the property (Citrus and Vegetable Magazine, 1992A). According to Center Director Dr. Richard Baranowski, the main administration building and the newer laboratory buildings were damaged but remained intact. Damage to the Center was estimated at $2.5 million, however the value of many research projects, which were virtually eliminated, was beyond calculation. The Caribbean fruit fly colonies were gone, groves were defoliated, many trees were uprooted and totally lost as well as fields of vegetables ruined.

As described by Mr. Don Poucher, IFAS Director of Education, Media and Services, "trailers just evaporated. The trailer anchors were still there but they weren't anchored to anything but the foundation" (Florida Grower and Rancher, 1992B). Poucher also described avocado trees as "picked up and blown away, and limes trees uprooted and defoliated, looking as if they had been sandblasted."

Miraculously, no employees at the Tropical Research and Education Center were injured, but some lost their entire homes and all suffered some damage. University of Florida employees across the state donated truck loads of bottled water, canned goods, baby food, clothing and building supplies. According to Dr. James M. Davidson, IFAS Vice President for Agriculture and Natural Resources, these shipments, plus volunteer cleanup crews from Gainesville and other IFAS research centers, provided Homestead personnel the time they needed to take care of their personal needs and responsibilities.

Hugh Popenoe, Director of International Programs for IFAS, described Hurricane Andrew as a blow to international agriculture. "The Homestead Station has been a jewel in our international crown," Popenoe said. "It's provided germplasm for many parts of the world. The plant collection is known worldwide. It is also a major center for biological control" (FOCUS, 1992).

The Tropical Research and Education Center (TREC) has been a major facility for research on banana diseases, and hosted the Fourth International Mango Symposium in July 1992 which drew a record number of participants from throughout the world. Graduate students from dozens of foreign countries have used the TREC facilities to conduct research to benefit their native countries (FOCUS, 1992).

The wind and water destroyed many office records and notebooks at TREC, and Neal Thompson, IFAS Acting Dean for Research, noted that, "we may have lost all the cultures we had as well as insects and other microorganisms." John Capinera, Chairman of the Department of Entomology and Nematology, feared that "much research in biological control was lost, including the repository of genetic material (FOCUS, 1992).

According to Dr. Jennifer Sharp, Director of the USDA 210-acre Subtropical Horticulture Laboratory in Miami, that facility also suffered severe damage, but fortunately the plant germplasm collection, started in 1898, survived (Citrus and Vegetable Magazine, 1992F). At the time the hurricane struck, more than 8,500 tropical and subtropical fruit, nut and ornamental plants were maintained at the Miami location. Crops most severely damaged were cocoa and coffee which were a total loss. The least damaged was sugarcane. There was some damage to avocados, bananas, carambola, sugar apple, mangos, palms, passion fruit and tripsicum grass.

Of the site's 25 buildings, the older structures built from coral rock had the least damage. The Plant Science Building dedicated in 1989, and the administration building, each lost part of their roof. Greenhouses and temporary buildings were lost. According to Dr. Kenneth H. Vick of the USDA, ARS National Program Staff, other USDA laboratories were providing support to the Miami facility.

The Storm Surge

From north to south, the storm surge accompanying Hurricane Andrew ranged from 5.7 feet at NE 74th Street, 7.1 feet at SW 8th Street, 9.8 feet at Dinner Key, 10.0 feet at Matheson Hammock, peaked at 16.6 feet at the Charles Deering Estate and 16.9 feet at the Corporate Headquarters of Burger King east of Old Cutler Road at SW 177th Street; 12.5 feet at Black Point Marina, 8.5 feet east of Homestead Air Force Base and 6.7 feet at Homestead's Bayfront Park (FEMA, 1992). The highest surge, at Burger King and the Deering Estate, was in the center of the path of Andrew's northern eye wall, the strongest part of the hurricane, and pushed seawater as much as three miles inland, almost to U.S. One (Old Dixie Highway). Ornamentals in the path of the storm surge were devastated.

At the Kampong

The storm surge at the Kampong (estate of the Fairchild Family and a major show place for tropical plants) reached 10 feet and, coupled with sodium toxicity, destroyed more plants than did the winds (Schokman, 1993A). The storm surge also contributed to heavy damage to the irrigation system, including underground PVC pipe broken by uprooted trees and risers knocked askew by falling branches.

After the hurricane had passed, an uncanny quiet prevailed at the Kampong as the blocked streets were devoid of normal traffic noise. The collection was a tangle of fallen trees and branches and the ground was covered by a layer of smelly, rotting fruit. The first week was spent simply trying to clear a path through the debris. Fifteen large trees were blown down, five of which were removed and the others put back in place using a special tree crane.

Thirty-two medium sized trees and palms were blown over including mango and pomelo trees which were weighted with fruit. Fortunately, with immediate action, it was possible to put these back in place. The assumption was made that if a tree had been blown over it would have lost 50 to 70% of its root system. To compensate for this loss, a corresponding 50 to 70% of the tree's foliage was pruned and placed on top of the exposed roots to slow dehydration. Large trees at the Kampong had been surface planted and consequently had shallow, 12 to 16 inch, root systems. A lesson learned from this disaster was that when growing large trees in a hurricane prone area, it is important to keep them pruned.

It was interesting to note that many grafted trees were not destroyed by the hurricane, and that totally defoliated trees flowered quickly after the hurricane, either due to the lack of leaves or the generation of ethylene gas from the dead fruit and vegetation. Vines became a serious problem because of the increased sunlight.

Schokman (1993B) recommends "that in preparing for a hurricane, nurseries should keep trees healthy and properly pruned, and that if a tree is blown down, it will require rapid action to save it. Don't destroy or remove it simply because it was blown over." He also makes the following common sense recommendations:

1. "Have water on hand for drinking as well as irrigating young or tender plants.

2. Have chain saws with spare chains.

3. Have sufficient oil and gas, properly stored.

4. Stock some 2 × 4 timber to prop up fallen trees.

5. When you replant a tree, don't put a $5.00 tree in a 50¢ hole."

Schokman (1993B) concludes with the statement that although "we did not lose many lives because of well-planned evacuation procedures and general hurricane preparedness, we in south Florida lost many valuable trees and plants that could and should have been saved."

Damage to South Florida Vegetable Crops

"It looked like a 50 mile bomb hit Homestead."—David Neill, tomato grower, quoted by S. Krause (1992A).

When Hurricane Andrew struck south Florida, Dade County was producing 18 different traditional vegetables plus 16 different tropical and specialty vegetables with a total value of over $270 million (Citrus and Vegetable Magazine, 1992C). Fortunately, the planting season had not begun (Table 13-2) but the vegetable industry's infrastructure was seriously damaged.

According to George Sorn, Executive Vice President of the Florida Fruit and Vegetable Association (FFVA), despite the fact that many vegetable packinghouses were hard hit by the hurricane, most of the equipment was operating and was expected to be available when harvesting began in November (Citrus and Vegetable Magazine, 1992E). He indicated that "vegetables will come in a little late, but we will have close to a normal year."

Some of the more important crops include bush and pole beans, cabbage, sweet corn, eggplant, okra, pickles, potatoes, squash, strawberries, tomatoes, cherry tomatoes and plum tomatoes. Table 13-9 shows the acres of squash harvested in Dade County from 1990-91 to 1994-95. Major tropical vegetables include bonito, calabaza and malanga.

FFVA indicated that annual agricultural values in south Dade County were vegetables, $293 million; tropical fruits, $74 million; and nurseries $171 million for a total value of $538 million. Of these, tropical fruits and nurseries were virtually a total loss. How-

TABLE 13-9. ACRES OF SQUASH HARVESTED IN DADE COUNTY IN YEARS BEFORE AND AFTER HURRICANE ANDREW.

Season	Acres
1990-91	4,600
1991-92	5,400
1992-93[1]	3,700
1993-94	5,300
1994-95	5,150

[1]Hurricane year.
Source: Florida Agricultural Statistics Service, 1996.

ever, Sorn noted that in southwest Florida, Collier, Hendry and Lee counties, the situation was normal with even citrus unscathed. It would be a full year as far as vegetables were concerned.

Wayne Hawkins, Executive Vice President of the Florida Tomato Committee, pointed out that although 17% of the Florida tomato crop was grown in south Dade County, and considerable damage was done to farm buildings and packinghouses, there was no crop in the ground (Krause, 1992B). He felt that once debris was removed from the fields, it would not affect what was planted in Dade County. Lack of labor, as people were concerned with their housing, would be the problem. He had found no tomato damage in southwest Florida or in the Ruskin and Palmetto areas.

The state's largest tropical foliage area, which stretched from Kendall south through Goulds, Princeton, and Homestead to Florida City suffered severe damage. According to Steve Munnell, Executive Vice President of the Florida Foliage Association, the damage was "breathtaking." It was total and no one was spared (Florida Grower and Rancher, 1992B).

Economic Considerations

"Today, after 80 years of trial and error, the southern half of Dade County produced over a million bushels of avocados, close to two million bushels of limes and 300,000 bushels of mangos yearly."—W. H. Krome and S. Goldweber (1987).

Tropical Fruits: Tropical fruit sales by Dade County growers exceeded $74 million, including 90% of the limes grown in the United States and 90% of over 9,000 acres of domestically grown avocados. With the lime and avocado crops demolished, buyers were desperate for sources of these fruits, as well as mango, passion fruit and carambolas. Some avocados were available in California and limes in Mexico, but there were no alternate sources for carambola and passion fruit and the mango season was over in Mexico (Krause, 1992A). Damage to avocados, limes and mangos, the principal fruits grown in the area of south Dade County, was severe (Tables 13-5, 13-10, 13-11). Florida avocado production dropped from 1,132,000 bushels in 1991-92 to only 176,000 bushels in 1993-94 (Table 13-12). During this same period, the price rose from $11.90/bushel to

TABLE 13-10. ACREAGE AND NUMBER OF LIME TREES IN DADE COUNTY, FLORIDA BEFORE AND AFTER HURRICANE ANDREW, AUGUST 24, 1992.

Year	Acres	Trees
1990	6,288	997,500
1992	6,074	962,400
1993	1,668	290,400
1996	2,618[1]	440,800

[1]Dade County lime planting through October 1994.
Source: Florida Agricultural Statistics Service, 1996.

$20.50/bushel. Despite the higher price, the total value of the crop fell from $13,471,000 in 1991-92 to only $3,608,000 in 1993-94. Fortunately, the recovery of the avocado industry has been relatively swift as the trees had almost completely recovered by the 1996-97 season as demonstrated by the production of 940,000 bushels which, with a price of $13.20/bushel, yielded a total value of $12,408,000. Unfortunately, the lime industry has been much less resilient due to the long time necessary to bring a new lime tree into production. Florida produced 1,600,000 boxes of limes in 1991-92. By 1995-96, Florida lime production had only recovered to 300,000 boxes, about

TABLE 13-11. NUMBERS OF LIME, AVOCADO, AND MANGO TREES SET IN DADE COUNTY, FLORIDA IN THE YEARS BEFORE AND AFTER HURRICANE ANDREW.

Year	Limes	Avocados	Mangos
	----------------- thousand trees -----------------		
1990	29.8	0.2	1.1
1991	9.3	0.4	0.8
1992	48.8	0.5	2.4
1993[1]	80.7	5.6	14.1
1994	108.4	30.6	14.9
1995	42.9	10.1	4.4
1996	28.2	27.3	4.7

[1]Plantings through mid-March.
Source: Florida Agricultural Statistics Service, 1996.

TABLE 13-12. PRODUCTION AND VALUE OF FLORIDA AVOCADOS BEFORE AND AFTER HURRICANE ANDREW.

Season	Bushels	Price per bushel	Value
1989-90	1,340,000	$ 8.30	$ 11,122,000
1990-91	784,000	17.10	13,406,000
1991-92	1,132,000	11.90	13,471,000
1992-93[1]	288,000	14.60	4,198,000
1993-94	176,000	20.50	3,608,000
1994-95	800,000	15.40	12,320,000
1995-96	760,000	14.90	11,324,000
1996-97	940,000	13.20	12,408,000

[1]Hurricane year.
Source: Florida Agricultural Statistics Service, 1997.

1/5 of pre-hurricane production. The price per box in 1991-92 had been $9.12 to produce a total value of $14,589,000. In 1993-94, the price had increased to $12.70/box but the total value of the diminished crop was $2,541,000. In 1995-96, the price dropped to $8.05/box for a total value of only $2,415,000 (Table 13-13). The precipitous drop in price is largely attributed to increasing imports of limes from Mexico (Roy, et al., 1996).

TABLE 13-13. PRODUCTION AND VALUE OF FLORIDA LIMES BEFORE AND AFTER HURRICANE ANDREW.

Season	Boxes (88 pounds)	Price per box	On-tree value
1989-90	1,650,000	$ 8.26	$ 13,634,000
1990-91	1,450,000	13.99	20,289,000
1991-92	1,600,000	9.12	14,589,000
1992-93[1]	1,000,000	1.02	1,017,000
1993-94	200,000	12.70	2,541,000
1994-95	230,000	8.65	1,989,000
1995-96	300,000	8.05	2,415,000

[1]Hurricane year.
Source: Florida Agricultural Statistics Service, 1996.

Mango production has also been very slow to recover from Hurricane Andrew. In 1991, Florida produced 500,000 bushels of mangos which yielded $12.30/bushel for a total crop value of $6,150,000. However, after the hurricane, the production was only 50,000 bushels which, at a price of $19.00/bushel, produced a total crop value of $950,000 or about 1/6 of the pre-hurricane value. Unfortunately, the recovery of the mango trees in Dade County has been disappointing. In 1996, only 100,000 bushels were produced with a total crop value of $1,500,000, still only about 1/4 of the total crop value before Hurricane Andrew (Table 13-14).

Government Response

At the State Level

Commissioner of Agriculture Bob Crawford made a formal request to Governor Lawton Chiles asking that the state seek an assessment of damage to farms, fields, groves and agricultural infrastructure by the U.S.D.A. (Citrus and Vegetable Magazine, 1992B). To produce immediate relief, (1) the Department of Agriculture and Consumer Services (DOACS) obtained tank trucks to transport water into hard hit areas which had lost electric service; (2) the Department's Bureau of Food Distribution arranged for shipment of

TABLE 13-14. PRODUCTION AND VALUE OF FLORIDA MANGOS BEFORE AND AFTER HURRICANE ANDREW.

Crop year	Bushels[1]	Price per bushel	Value
1991	500,000	$ 12.30	$ 6,150,000
1992	400,000	10.70	4,280,000
1993[2]	50,000	19.00	950,000
1994	100,000	15.00	1,500,000
1995	150,000	11.50	1,725,000
1996	100,000	15.00	1,500,000

[1]Estimated.
[2]After Hurricane Andrew.
Source: Florida Agricultural Statistics Service, 1996.

emergency food supplies; (3) DOACS law enforcement personnel joined other law enforcement personnel in maintaining order; (4) DOACS personnel from the Division of Forestry helped in clearing roads blocked by fallen trees, brush and other debris; (5) soil specialists worked to determine if farm fields were contaminated with salt; (6) the reestablishment of fruit fly detection systems was begun; (7) the sourcing of seed, feed, fertilizer and other supplies and equipment was begun, and (8) DOACS personnel were assigned to coordinate efforts with the Federal Emergency Management Agency.

Andrew had left four million cubic yards of vegetative debris in south Dade County. Thanks to efforts by the South Dade Soil and Water Conservation District (SDSWCD), the University of Florida, Institute of Food and Agricultural Sciences (IFAS) and funding from the Federal Emergency Management Agency (FEMA), much of this mulch was delivered to growers. The mulch was used to reduce the use of herbicides by suppressing the growth of weeds, and enriching the soil by providing organic matter (Florida Grower and Rancher, 1993).

At the Federal Level

"The $20 billion estimate of damages resulting from Hurricane Andrew made it the costliest natural disaster in United States history. Within hours, President George Bush declared Florida a major disaster area and designated the counties of Broward, Dade, Collier and Monroe eligible for Federal disaster assistance. Additionally, a full response effort was quickly mounted by representatives of designated Federal agencies and the military (through Joint Task Force Andrew) to support State and local response efforts" (FEMA, 1992).

On September 2, 1992, President Bush released $755 million of emergency appropriations through the United States Department of Agriculture (USDA) to farmers and growers in Dade County who were affected by the hurricane (Florida Grower and Rancher, 1992D).

Funds were processed by the Agricultural Stabilization and Conservation Office (ASCS) for relief offered by the Soil Conservation Service (SCS) and the Farmers Home Administration (FmHA). Other types of relief included additional food stamps, plus low interest loans through FmHA.

The Rural Development Administration (RDA) had funds for rebuilding communities and help was available from the USDA water and waste loan program and the Rural Electrification Administration.

"From September 15-18, 1992, an Interagency Hazard Mitigation Team (IHMT) was convened in Miami by the Federal Emergency Management Agency (FEMA) and Florida Division of Emergency Management (DEM). Over 100 participants, representing a wide range of Federal, State, local and private agencies, gathered to identify mitigation issues and develop recommendations which form the basis for the IHMT report. In order to better focus on mitigation measures most likely to reduce damages from future disasters, IHMT members used a scenario of a hurricane striking the same area of south Florida 20 years in the future" (FEMA, 1992).

Recommendations in the IHMT report were organized in the following functional categories of which agriculture was only a relatively small part:

"1. Building codes, standards and practices

2. Coastal and flood plain management

3. Planning, both pre-disaster and post-disaster

4. Infrastructure

5. Marine

6. Agriculture/secondary effects

7. Environmental issues"

The Executive Summary of the report concluded with the observation that it is hoped that it will assist local governments in designing ways to reduce damages and impacts from future Florida hurricanes.

Looking to the section on agriculture/forestry, the report identified six issues: landscape management, windbreaks, restoring tree canopy, wildland fire management, residential vegetation and agriculture infrastructure. Under these headings, ten recommendations were made, three of which were in the category of agriculture infrastructure. As background, the report reiterated that approximately 3,000 Dade County farms had suffered major damage, with crop losses estimated at $349.1 million, losses to irrigation and drainage systems, packinghouses and processing plants and other structures and equipment at $580 million and losses to aquaculture and livestock at $12.5 million.

The three recommendations were: "(1) to replace damaged irrigation systems with more efficient and environmentally sensitive storm resistant systems; (2) inspect agricultural drainage and irrigation systems to ensure that proper design and construction practices were followed and that maintenance is being performed on existing water courses; and (3) flood proof irrigation and drainage pumping systems." The lead agency for these three recommendations was the ASCS-Soil Conservation Districts with the South Florida Water Management District and the Florida Department of Agriculture and Consumer Services as support agencies.

Under the heading of environmental issues, nine recommendations were made, three of which were particularly important to agriculture. These were: "to conduct an environmental assessment on the impact of dumping, burning, clearing and other damage from post-hurricane activities and develop a plan of action with recommendations for recovery and future post-storm recovery sites and activities." This was of interest to Dade County farmers as many farm fields were used as dump sites for urban debris.

Another important recommendation was to "familiarize disaster personnel with EPA response plans for timely cleanup of environmental hazards, including pollutant and chemical spills, which occur as a result of disasters." And the final recommendation was to "require storm resistant facilities, located above minimum flood elevations, for the storage of agricultural chemicals."

The University of Florida Response

The response by the University of Florida, IFAS was prompt and well-organized. On August 26, Dr. James M. Davidson, Vice President for Agricultural Affairs, notified all IFAS units that a phone call on August 25 from Dr. Richard Baranowski, Center Director of TREC in Homestead indicated that the Center was not totally destroyed but had sustained major damage. Fortunately, there were no serious injuries or loss of life among Center personnel. A report from Mr.

John Carlson, Director of IFAS Facilities and Operations in Gainesville, indicated what could be done to help the staff of the center.

On August 31, Dr. Davidson advised all IFAS personnel that the situation at Homestead was "moving toward recovery." The University family had been generous in providing relief in the form of cash and supplies. Several shipments of food, clothing, water and other needed items had been sent to the Homestead area and that the response of volunteer clean-up crews from Gainesville and other University locations had been impressive. Dr. Walter Kender (1999), Director of the Citrus Research and Education Center (CREC) at Lake Alfred, recalls that "several CREC staff traveled to Homestead with truck loads of furniture, food, clothing, generators, chain saws and supplies. Volunteers from CREC and other centers remained at TREC for two weeks to help with the cleanup." Dr. Davidson appointed an IFAS Hurricane Andrew Task Force co-chaired by Mr. Peter Warnock and Mr. Edwin Heffelfinger to work with industry, local constituents and IFAS personnel to consider the following: (1) the economic effect of Hurricane Andrew on agricultural production, recovery, employment, etc.; (2) health and personnel issues including food safety and nutrition; (3) environmental issues including fuel and pesticide spills and salt water contamination; (4) faculty to assist commodity crops with recovery; and (5) cleanup and rebuilding of the TREC physical plant.

In response to the Vice President's action, Heffelfinger and Warnock notified department heads and other key personnel on September 3, to attend an organizational meeting of the Hurricane Andrew Task Force on September 8, each bringing one to two pages for a preliminary plan of action.

On September 4, Dr. Davidson reported substantial progress in restoring operations at TREC, and that most of the facilities could be reoccupied in two to three weeks. At this point in time, the needs of Homestead faculty and staff were for non-refrigerated milk, men's and women's heavy duty cotton work gloves, clothespins and clothesline, chain saws, insect repellant, large plastic garbage bags, batteries, and last but not least, diapers.

After the organizational meeting on September 8, Heffelfinger and Warnock proposed the following objectives and work groups for the IFAS Hurricane Andrew Task Force:

IFAS HURRICANE ANDREW TASK FORCE
9-15-92

Objectives

▶ Help IFAS and county extension employees put their homes in order and get back on their feet.

▶ Facilitate cooperation and communication between IFAS work groups.

▶ Establish correct record keeping to qualify for subsequent reimbursement from FEMA, DCA, insurance companies.

▶ Keep all IFAS faculty, staff and administrators informed on a weekly basis.

▶ Assist research and extension personnel to return to business as quickly as possible.

▶ Maintain a record of activities and prepare a plan to handle future disasters.

Work Groups (General)

▶ County Operations—Coordinate work of extension agents, staff and volunteers. Reestablish agent and staff well being. Help faculty and volunteers conduct appropriate educational programs and services - M. Cole (Agent contact - D. Holmes).

▶ Analyses of economic impacts in the area to basic agricultural industries. Determine farm business loans, grants and insurance settlement opportunities - L. Libby, R. Clouser (AREC contact - D. Baranowski) (Agent contact - D. Hull).

▶ Food safety, nutrition, health, housing, human services - D. Tichenor, N. Torres (Agent contact - A. Cooper).

▶ Physical repair and reestablishment of IFAS facilities and equipment - J. Carlson (AREC contact - D. Baranowski) (Agent contact - D. Holmes).

▶ Effect of salt water intrusion on soils - G. O'Connor (AREC contact - D. Baranowski) (Agent contact - M. Lamberts).

▶ Petroleum and chemical spills, handling of pesticides and toxic materials - E. Freeman, N. Nesheim (AREC contact - D. Baranowski) (Agent contact - D. Holmes).

▶ Engineering plans for farm and processing facilities, determination of strategies for rebuilding, provide relevant information on farm safety - O. Loewer, B. Becker (AREC contact - D. Baranowski) (Agent contact - D. Holmes).

▶ Animal health and welfare - K. Braun, T. Lane (Agent contact - D. Holmes).

▶ Pets (cage) birds - C. Douglas (Agent contact - D. Hull).

▶ School for Forest Resources and Conservation - M. Duryea (Agent contact - J. Ritter).

▶ Tropical fish - C. Watson (Agent contact - D. Pybas).

▶ Chipping and mulching - W. H. Smith

▶ Mosquito control - D. Baker (Agent contact - M. Lamberts).

▶ Marine related problems - M. Clarke (Agent contact - D. Pybas).

▶ 4-H - S. Fisher (Agent contact - G. Dietz).

▶ Employee support - W. Summerhill (Agent contact - D. Holmes).

▶ IFAS information and publications - D. Poucher, J. Graddy, Susan O'Reilly.

Work Groups (Production)

▶ Turf and Landscaping - John Cisar, Chair, Bert McCarty, Jerry Sartain, Bob Block (Agent contact - J. Ritter)

▶ Tropical and Sub-tropical Fruit - Tim Crocker, Chair, Wayne Sherman, Norm Childers, Jonathan Crane (Agent contact - C. Balerdi)

▶ Vegetables - Bill Stall, Chair, Ken Pohronezny, Allen Smajstrla, Steve Sargent (Agent contact - M. Lamberts)

▶ Ornamentals - Tom Yeager, Chair, Ed Gilman, George Fitzpatrick, Alan Meerow (Agent contact - D. Hull)

Task Force Co-Chairs: Ed Heffelfinger and Pete Warnock

Task Force Coordinators: Bill Rossi in Gainesville and Rod Clements in Homestead

The various work groups (general and production) assigned by the Task Force carried out their assigned responsibilities based on the objectives established by the IFAS Hurricane Andrew Task Force. After meeting their goals, Dr. Peter Warnock and Mr. Edwin Heffelfinger met with Dr. Davidson and gave him an oral report on the accomplishments of the Task Force and recommending that the Task Force be discharged.

As a result of the Hurricane Andrew experience, Dr. Davidson appointed a group of IFAS faculty and charged them with the responsibility of developing a Handbook on Disaster Preparedness and Recovery. The Handbook was published in 1997 and is available through the IFAS Publication Office. Some of the topics covered in the handbook include: disaster preparation, what to do during and after the disaster, stress and coping, home recovery and farm recovery. Types of disasters discussed are: hurricanes, lightning, floods, tornadoes, hazardous waste materials, terrorism, etc.

Efforts by Private Organizations

In support of Hurricane Andrew relief efforts, the Florida Gift Fruit Shippers volunteered its entire fleet of 60 trailers to transport donated items and to support efforts by the relief agencies (Citrus and Vegetable Magazine, 1992B). The Florida Fertilizer and Agrichemical Association provided generators, the Florida Nurserymen and Florida Foliage Associations joined to deliver goods and supplies, the Florida Sugarcane League provided financial assistance, the Sugarcane Growers Coop launched a major relief program, A. Duda and Sons provided water, ice and food, Coca-Cola Foods and Tropicana delivered water and juices. Collier Enterprises helped with centers for workers and furnished equipment and the Florida Fruit and Vegetable Association established a fund to provide relief to workers (Florida Grower and Rancher, 1992C).

Freshwater Resources

Roman (1994) evaluated the impact of Hurricane Andrew on freshwater resources and found that water levels at the Everglades National Park had increased only slightly as Andrew was a relatively dry hurricane, producing much less rainfall than would be normally expected in a major storm. Of two rain gages which survived the fierce winds, one recorded only 1.7 inches and the other 2.8 inches on August 24. However, these values were considered to be too low due to the effect of winds. The effect was also minimal at the Big Cypress National Preserve. Nevertheless, the dense periphyton mat and herbaceous vegetation was disrupted by the hurricane leading to altered habit for fish and macro invertebrates. The reproductive success of alligators was also reduced and wading bird roost and rookery sites were damaged. Fortunately, the quality of water was not significantly affected.

Relandscaping Dade County

Ritter (1993) reported on a survey of ten nurserymen and over 200 homeowners to determine the type of plants used to reland-scape the hurricane-devastated areas. He found that the most commonly planted trees were citrus, mango, avocado, banana and lychee, with homeowners also using some of the minor tropical fruits.

Native plants, such as live oak and sabal palms, have been planted in large numbers with the belief that they would come through future hurricanes better than non-native trees. Pines were used extensively to reforest the pine canopies which had been destroyed by Hurricane Andrew.

Personal Observations

Harry Huffman of Redlands, Florida recounts:

"When it got a little light, I looked out the window and I couldn't see anything. The wind was still blowing pretty hard so I waited a while, then looked out again. I still couldn't see anything. I told my wife, 'there's some kind of fog. I can't see anything. Finally I realized there wasn't anything to see. All the trees were gone"—Huffman (1997).

William H. "Bill" Krome (1997) of the Homestead area noted a number of interesting anecdotes and observations:

"Jack Kates lives in Dick Bromely's poured-concrete house in the middle of a lime grove on Avocado Drive. He left the front door open (it was on the lee side) and spent the storm at the other end of the hallway. When the wind died down and there was light enough to see, he went to the front door and peered out. 'Hey, Evelyn! The trees are gone!' So was a lot of other stuff."

"Bob Eckard, a Viet Nam veteran, rented Casa Rodriguez, a frame house the Krome family lived in in 1919. It was pretty stout—had stood all the storms until this one. When I saw the heap of boards that was all that remained of it, I figured that Eckard and his son, who was living with him, had doubtless been removed to a safer place (and if they hadn't I certainly didn't want to disturb them three days after the storm). Well, about three weeks later Eckard showed up at my office. He and his son had *not* left Casa Rodriguez before the storm. The first half they spent in the house, both men getting into the bathtub when boards started to come off. When the lull came they went out and got under Barney Rutzke's D-8 tractor which was parked by the house. They spent the second half of the storm under the tractor. Eckard said he hasn't been able to find the bath tub since the storm. "Bob," I said, "you missed the ride of your life."

"Winston Mancur lived in the house in KK Grove. It was totally wrecked. I figured that wise old man had taken his five dogs to a safer place, but again, he didn't. When the lull came, Mancur went out and lay down under a hardy avocado tree. He spent the second half clinging to its trunk. Fortunately the tree didn't go over. I don't know what became of his dogs."

"Bill Losner assessed damage to five banks. Generators were furnished for employees, some of whom lived in the bank. The board room was converted into a commissary for two months. The main branch was operating partly by Wednesday, August 26, 1992."

"I got my electricity back on September 18, three weeks after the storm. Others were not so lucky and did not have electricity until October 8, about six weeks after the storm. Phone service was restored November 7 for four days, then off again. Quite a bit on-again off-again."

"On September 4, 1992, Carl Holcomb and a helper, Roger McGowan, came from Charleston, S.C. to help. They repaired the roof and made temporary repairs to the rest of our house. Later, Wally Witzen fixed the windows in our bedroom and the office and did some other good work. Carl and Roger stayed here for several weeks, sleeping in the yellow bedroom and the floor of the den."

"Will and his friend, Keith Adkins, came from Gainesville to help. They brought with them a 3500 watt generator, romex, five chain saws, gas cans, and other equipment and supplies. Keith was very good and when Will was preparing to go home after a few weeks, I offered to pay Keith $20 an hour to remain and supervise the rehabilitation of the groves. He accepted, staying here until he got a house trailer and installed it near the former site of Mancur's house in the KK Grove. He is living there now."

"Work on the groves continues. A lot remains to be done and I estimate it will be four or five years before the groves show an operating profit."

"Most of the native pines south of Kendall Drive have been killed—a combination of hurricane damage and pine bark beetles. Forestry Service says not to replant pines until the epidemic infestation is over. It will be several years."

"The TV Channel 6 tower north of Coconut Palm Drive and east of Krome Avenue had cable to guy it as thick as your arm. I'm sure it was designed to withstand any conceivable wind. It didn't"—Krome (1997).

The Krome family have been involved with Dade County agriculture for generations. Bill Krome was president of the Florida State Horticultural Society in 1989. His father was recognized for his outstanding contributions to the society in 1927.

Erroll Fielding (1998) recalls that, "we were all set for Hurricane Andrew in our southwest Florida groves, but it didn't spread out."

Rosario "Rosy" Strano (1997), tomato grower in Florida City and past chairman of the Florida Tomato Committee, described the scene after the passing of Andrew. "The storm blew all the windows out of the office and then sucked out the filing cabinets. There were over 300 lime trees piled up along side the house even though the closest grove was 3 miles away."

Chapter 14

Other Storms of the 1990s

1990

The 1990 season was unusually busy with a total of 14 tropical systems, eight hurricanes and six tropical storms. Fortunately, only one of the 14, Tropical Storm Marco, made landfall in Florida.

Tropical Storm Marco - October 11-12, 1990

Meteorological Considerations

Tropical Storm Marco began as a tropical depression near Caibarien, Cuba on October 9 (National Hurricane Center, 1990). The depression moved west-northwest along the north coast of Cuba, then turned northwest over the Florida Straits where it was designated a tropical storm on October 10 while centered about 30 nautical miles south-southwest of Key West. After passing midway between the Dry Tortugas and Key West, Marco moved slowly north just off the Florida west coast. It reached its peak intensity near 1:00 A.M. on October 11 with 63 mph sustained winds and a central pressure of 29.21 inches. The center moved to just a few miles west of Bradenton Beach at dawn on October 11 and continued following the coastline with much of its circulation over land in the St. Petersburg area to near Clearwater at about noon on the eleventh.

Marco was downgraded to a tropical depression in the early evening of October 11 just west of Cedar Key. The central pressure rose as the depression moved inland and it was declared extra-tropical early in the morning of October 12.

There were five aircraft reconnaissance missions into Marco with a total of 27 center fixes during the time the storm was in the Florida Straits until it moved ashore near Cedar Key. The best track peak wind measurement of 63 mph on October 11 was based on an aircraft measurement at 1500 feet. This estimate was confirmed by satellite imagery. The highest reported wind gusts from Marco were 85 mph at the Sunshine Skyway Bridge across Tampa Bay and also at Bradenton. Some selected surface observations from Tropical Storm Marco are shown in Table 14-1.

The Damage

Neither the wind gusts nor the rainfall from tropical storm Marco resulted in major agricultural damage.

1991

The 1991 season produced four hurricanes and four tropical storms, none of which made landfall in Florida. The hurricanes were Bob, Claudette, Grace and an unnamed late season storm.

TABLE 14-1. METEOROLOGICAL DATA FOR TROPICAL STORM MARCO AT SELECTED FLORIDA STATIONS, OCTOBER 11-12, 1990.

Location	Minimum pressure (inches)	Maximum sustained winds (mph)	Wind gusts (mph)	Rainfall (inches)
Key West WSO	29.49	40	46	3.08
Naples	—	23	44	2.13
Fort Myers	29.53	28	33	—
Port Charlotte	29.32	45	52	3.85
Sarasota	29.29	45	62	1.58
Sunshine Skyway Bridge	—	80	85	—
Ruskin WSO	29.48	—	—	3.80
St. Petersburg	29.49	25	37	1.40
McDill AFB	29.45	46	68	4.78
Tampa	29.50	26	45	1.78

Source: National Hurricane Center, 1990.

1992

There were only four hurricanes, Andrew, Bonnie, Charley and Frances and two tropical storms during the 1992 season with only one hurricane making landfall in the United States (Table 14-2). However, that one hurricane was Hurricane Andrew which in sheer force and ferocity as a category 4 hurricane more than made up for the low number of storms. Hurricane Andrew deserved extensive treatment and was covered in the previous chapter.

1993

The 1993 season was relatively inactive, with only four hurricanes and four tropical storms, none of which affected either the Florida peninsula or west Florida. The hurricane names were Emily, Floyd, Gert and Harvey.

The most notable feature of the season and the only hurricane to affect the United States was Hurricane Emily which began as a tropical wave near the Cape Verde Islands on August 17 and intensified to a tropical storm on August 22 near 20°N, 53°W. It then moved to the northwest before stalling about 350 miles southeast of Bermuda on August 25 where it attained hurricane status with 74 mph winds and gusts to 98 mph. Emily began to move westward on August 27, then to the northwest and finally to the north with the eye passing 20 miles east of Cape Hatteras on August 31. The western eye wall passed over Hatteras Island with strongest winds of 75 to 115 mph before turning eastward into the open Atlantic (Rappaport and Pasch, 1993).

TABLE 14-2. 1992 HURRICANES.

Name	Origin	Time span	Landfall
Andrew[1]	East Atlantic	Aug. 16-27	Florida and Louisiana
Bonnie	Mid-Atlantic	Sept. 17-Oct. 2	None
Charley	East Atlantic	Sept. 21-28	None
Frances	Mid-Atlantic	Oct. 22-29	None

[1]See Chapter 13.
Source: Neumann et al., 1993.

1994

The 1994 Atlantic Basin storm season produced three systems which attained hurricane status at some point, and four tropical storms which did not attain hurricane force. Only three systems made landfall in Florida (Table 14-3). These were tropical storm Alberto which lashed the Pensacola area on July 3, tropical storm Beryl which came ashore near Panama City on August 15, and tropical storm, but later hurricane, Gordon which came ashore in the Venice area on November 16, crossed the state, and exited the state north of Vero Beach, briefly reaching hurricane status in the open Atlantic.

Tropical Storm/Hurricane Gordon—November 16, 1994

Meteorological Considerations

The tropical depression stage of Gordon began in the western Caribbean near 12°N, 83°W on November 8. For the first two days, Gordon tracked north to north northwest along the east coast of Nicaragua, then shifted to the northeast and then east and attained tropical storm status as it shifted its course toward Jamaica on November 10 and 11. Crossing the eastern tip of Jamaica on November 13, Gordon accelerated its forward motion to pass over Guantanamo Bay Cuba the same day and then turned to the northwest passing south of the Western Bahama Islands on the fourteenth and across the Florida straits on the fifteenth. Crussing through the lower

TABLE 14-3. 1994 HURRICANES AND TROPICAL STORMS.

Name	Origin	Time span	Landfall
Alberto	Yucatan Channel	June 30-July 7	Pensacola, FL area
Beryl	Gulf of Mexico	Aug. 14-19	Panama City, FL area
Chris	East Atlantic	Aug. 18-23	None
Florence	Mid-Atlantic	Nov. 2-8	None
Gordon	Southwest Caribbean	Nov. 8-21	Florida as a tropical storm

Source: Neumann et al., 1993.

Keys, Gordon entered the Gulf of Mexico on a northwesterly course before shifting to the north and then the northeast before crossing the Florida coastline near Fort Myers with sustained winds of 45 mph at 8:00 A.M. on November 16. Continuing its northeast track, Gordon emerged into the Atlantic just north of Vero Beach at about 5:00 P.M. on the sixteenth. As Gordon moved back over water, its central pressure fell and its winds increased to hurricane force on November 17. However, hurricane force winds from Gordon were not experienced over the Florida peninsula. Meteorological data from Gordon are shown in Table 14-4.

TABLE 14-4. METEOROLOGICAL DATA DURING TROPICAL STORM GORDON AS IT CROSSED THE FLORIDA PENINSULA FROM WEST TO EAST, NOVEMBER 16, 1994.

Location	Minimum pressure (inches)	Maximum sustained winds (mph)	Wind gusts (mph)	Rainfall (inches)
Key West	29.46	26	43	0.84
Naples Airport	29.42	20	29	—
Fort Myers Airport	29.42	28	45	—
Immokalee	—	—	—	3.26
Moore Haven	—	—	—	3.39
Belle Glade	—	—	—	5.69
Fowey Rocks	29.56	46	54	—
Homestead AFB	29.57	—	—	13.15
Miami Int. Airport	29.53	26	43	6.89
Hialeah	29.64	19	30	5.19
Andytown	—	—	—	16.00
Fort Lauderdale	—	—	—	14.68
North Dade County	—	—	—	11.73
Hollywood	—	—	—	12.21
West Perrine	—	—	—	9.50
Stuart	—	—	—	7.95
Titusville	—	—	—	8.75
Daytona Beach	29.58	—	—	10.70
Orlando Airport	28.54	17	27	5.80
Sanford	—	—	—	9.14

Source: Pasch, 1995.

The Damage

Most of Gordon's damage to farms and groves was due to flooding (Pasch, 1995). This is hardly surprising with 13.15 inches of rain being recorded at Homestead Air Force Base, 16 inches at Andytown and 11.73 inches in North Dade County (Table 14-4). Vegetables and tropical fruits were the most severely affected crops. Market Watch in the December 1994 issue of Citrus and Vegetable Magazine noted that potatoes and early corn in Dade County would have to be replanted, and that the Dade County Farm Bureau predicted an 85% loss on the 6,000 acres of beans planted as of November 17, 1994. The Dade County Farm Bureau further predicted a loss of 50% of the 1,000 acres of tomatoes which had been planted and an 85% loss on an additional 1,500 acres, plus 85 to 100% loss of 2,000 acres of sweet corn which had been planted. Vegetable Digest in the January 1995 issue of Florida Grower and Rancher Magazine noted that fortunately 50 to 70% of the winter vegetables in Dade County had not been planted, but that tropical vegetables on low ground were flooded and lost. Specific losses in tropical vegetables included all of the casaba crop, 100% of the bonito crop on 1,500 acres and 60% of the bonito on the remaining 1,500 acres, 100% loss of malanga on 2,500 of a total 6,000 acres and 60% loss on the remaining 3,500 acres. Table 14-5 shows the Dade County Farm Bureau's assessment of winter vegetable losses in the county as a result of Tropical Storm Gordon.

TABLE 14-5. WINTER VEGETABLE ESTIMATED LOSSES IN DADE COUNTY FLORIDA AS A RESULT OF EXCESSIVE RAINFALL FROM TROPICAL STORM GORDON, NOVEMBER 15, 1994.

Crop	Planted percent	Planted acres	Percent loss
Squash	25-35	2,000	85
Potato	24	1,500	100
Sweet corn	33	2,000	85-100
Green bean	25-35	6,000	85
Tomato	50	2,500	50-85
Cucumber	33.3	1,000	85
Pepper	100	1,000	60
Eggplant	100	200	50

Source: Florida Grower and Rancher, 1995.

The total crop loss from Tropical Storm Gordon was estimated at $125 million (Youngblood, 1995). The range of crop losses in nine counties most affected is shown in Table 14-6.

The Florida Agricultural Statistics Service, Vegetable Summary for 1994-95 detailed the following effects from Tropical Storm Gordon by crop:

Beans: Tropical Storm Gordon, which dropped one to seven inches of rain in central and southern areas, flooded and washed out some fields. Bloom loss and scarring were also problems in many areas. The surviving crop slowly recovered from the damage.

Cabbage: Gordon buffeted very young plants and flooded some fields in the Hastings area. The fields lost to Gordon were replanted.

Carrots: Heavy rainfall caused by Tropical Storm Gordon in late November prevented germination of some seedlings around Lake Apopka with some acreage replanted.

Corn: Heavy rain generated by Tropical Storm Gordon, passing during the week ending November 20, flooded some fields around the Zellwood, Lake Okeechobee, and east coast regions, and in Dade County. Strong winds accompanying the storm whipped and laid over older plants. Most damage from the storm was the loss of recent seedings due to flooding with producers replanting most of

TABLE 14-6. AGRICULTURAL LOSSES IN NINE COUNTIES MOST AFFECTED BY TROPICAL STORM GORDON, NOVEMBER 15, 1994.

Crop	Percent loss
Tomatoes	35-85
Potatoes	30-100
Peppers	30-60
Squash	35-85
Green beans	40-85
Cucumbers	35-85
Eggplant	50-60
Sweet corn	85-100
Sugarcane	35
Citrus[1]	30-40
Tropical fruit	30-100

[1]Citrus losses are presumed to be limes in Dade County.
Source: Florida Grower and Rancher, 1995.

this lost acreage. Dade County reported at least 35% of the winter crop acreage was lost and reseeded. Most older plants recovered although Everglades growers abandoned some acreage due to reduced ear quality. Dade growers planted some corn on lost potato acreage. Zellwood producers finished picking in early December. East Coast growers began harvesting around mid-December. Producers in Dade and western Palm Beach counties started picking fields in late January that they planted after Tropical Storm Gordon passed.

Cucumbers: Strong winds generated by Tropical Storm Gordon singed leaves, and broke off leaves and stems in some east coast localities during mid to late November. Southwestern and southeastern growers marketed most storm damaged fruit from the last half of November into early December. Strong winds tossed vines in the east coast area during the last half of December.

Eggplant: The storm caused significant wind, rain and sand damage to the plants in most areas. Good quality fruit continued to be shipped, but grade out was high due to scarred and misshapen fruit.

Escarole/Endive: Escarole and endive grows mainly in two areas of muck soils, the central area which is located northeast of Lake Apopka near Zellwood, and the Everglades area located around the southeast side of Lake Okeechobee. Planting became active during September. Heavy fall rainfall from Tropical Storm Gordon caused some acreage abandonment and reduced yields.

Bell peppers: Winds accompanying Tropical Storm Gordon during November blew blooms off plants and increased fruit drop which caused a lower fall crop yield. Producers in most southern localities picked marketable fruit prior to the passage of the storm, but strong winds caused some leaves to burn and some bloom and leaves to drop from plants in fields located in southeastern and southwestern areas and in Dade County. The Palmetto-Ruskin region escaped significant storm damages.

Potatoes: The first winter potatoes were planted in the southwest area around mid-October and in Dade County in late October. Rains from Tropical Storm Gordon in late November flooded the potato fields in Dade County and killed the crop planted at that time. About one-third of Dade County potatoes were affected. All the destroyed acreage was replanted. A small acreage was damaged in the southwest and had to be replanted. Planting ended in the southwest in mid-December and in Dade County in mid-January.

Radishes: Tropical Storm Gordon flooded fields which stopped planting and harvesting during the week ending November 20. The acreage of destroyed radishes was replanted.

Squash: Tropical Storm Gordon damaged the plants, scarred the fruit, and caused heavy bloom loss to the southern crop.

Tomatoes: Prices increased in late November through most of January due to the short supplies caused by damages from Tropical Storm Gordon passing over the State on November 16.

Wind gusts from Tropical Storm Gordon, clocked at 20 to 60 miles per hour, tossed foliage, blew blooms off plants and flung sand that scarred fruit in all southern and most central localities. Heavy rain accompanying the storm flooded some fields in Dade County and the southwest. Acreage around Palmetto-Ruskin escaped the heavy rainfall, but wind damaged some foliage and fruit in unprotected areas. In the East Coast regions, the storm's wind snapped some transplants off at the first tie and the weight of fruit nearing maturity toppled stakes and plants in some blocks. Flooding killed some plants in low lying areas of Dade County. A strong market, following the passage of the storm, prompted a few Palmetto-Ruskin and northern producers to make extra picks.

Above average temperatures and mostly dry conditions during late November and early December helped plants to recover from the storm damage. The milder weather allowed plants along the southeastern coast, which snapped off during the storm, to develop sucker growth. However, some older fields in the east coast region did not have enough marketable green fruit for a third pick due to increased fruit drop and a high percentage of pink and scarred fruit caused by Tropical Storm Gordon.

During early January, east coast and southwest growers finished picking acreage affected by Tropical Storm Gordon and began harvesting plants not hurt by the storm.

1995

This was an extremely active hurricane and tropical storm season in the Atlantic Basin with a total of eleven hurricanes, beginning with Allison on June 3 and ending with Tanya on November 1

(Table 14-7), and eight tropical storms, beginning with Barry on July 6 and ending with Sebastian on October 25. Fortunately, nine of these systems remained over the open Atlantic, but four made landfall in Florida. These were Hurricane Allison which struck the Apalachicola area on June 5, Hurricane Erin which crossed peninsular Florida August 1-2 and then buffeted Pensacola on August 3, Tropical Storm Jerry which crossed the peninsula August 23-26 and Hurricane Opal which lashed the Pensacola area on October 4.

Hurricane Allison, June 5, 1995

Meteorological Considerations

Allison formed from a tropical wave initially detected as a tropical depression near 17°N, 84°W in the southwestern Caribbean Sea on June 3, and reached tropical storm status as she crossed the 20th parallel on a northerly course. Moving rapidly north to about 25°N, Allison became a minimal hurricane, turned to the northeast and then weakened slightly below hurricane strength before moving ashore in the Big Bend area early on Monday morning June 5 carrying heavy showers and thunderstorms (Pasch, 1996A). The greatest

TABLE 14-7. 1995 HURRICANES.

Name	Origin	Time span	Landfall
Allison	Western Caribbean	June 3-6	West Florida
Erin	Bahamas	July 31-Aug. 6	Florida and west Florida
Felix	West Atlantic	Aug. 8-22	None
Humberto	East Atlantic	Aug. 22-Sept. 1	None
Iris	West Atlantic	Aug. 22-Sept. 4	None
Luis	East Atlantic	Aug. 27-Sept. 11	None
Marilyn	East Atlantic	Sept. 12-22	Virgin Islands
Noel	East Atlantic	Sept. 26-Oct. 7	None
Opal	Western Caribbean	Sept. 27-Oct. 5	West Florida
Roxanne	Western Caribbean	Oct. 7-21	Yucatan
Tanya	Mid-Atlantic	Oct. 27-Nov. 1	None

Source: Neumann, et al., 1993.

precipitation was in the Big Bend and in the western peninsula where three to six inches were common (Weather and Crop News, 1996). The storm weakened as it moved inland but tropical force winds continued until 4:00 P.M. on June 5. Allison diminished to a tropical storm as it moved into south Georgia early on June 6.

The Damage

Heavy rains associated with Hurricane Allison forced growers in the southern vegetable producing areas to complete their harvests earlier than usual, and interrupted harvesting in the northern areas (Vegetable Summary 1994-95). However, only relatively minor crop damage was reported.

Hurricane Erin - August 1-3, 1995

"The wall had glowing green clouds caused by the phospho-rescence in the seawater Erin carried that night."—Peter Spyke (1998).

Meteorological Considerations

Hurricane Erin formed in the southeastern Bahama Islands near 22°N, 73°W, from a tropical wave which left the coast of Africa on July 22 and crossed the Atlantic on a west northwest course. It was estimated that the wave became Tropical Storm Erin at 7:00 P.M. on July 30 (Rappaport, 1995) and reached hurricane force near Rum Cay in the Bahamas on the evening of the thirty-first. Moving at a forward speed of about 15 mph, Erin moved first to the west north-west toward southeast Florida, then was deflected to a northwest course by an upper level low. A ragged eye was noted on satellite pic-tures on August 1 and Erin made landfall at 1:00 A.M., August 2 near Vero Beach, Florida as a category 1 hurricane on the Saffir-Simpson scale (Table 15-4). A sustained wind of 75 mph was recorded at Sebastian Inlet by a Florida Institute of Technology Anemometer, but gusts at Melbourne and Vero Beach were less than hurricane force (Table 14-8).

Erin's winds diminished to near 50 mph as the still well-orga-nized storm crossed the Florida peninsula on a west northwest course. Emerging into the Gulf of Mexico north of Tampa, Erin

TABLE 14-8. METEOROLOGICAL DATA DURING HURRICANE ERIN AS IT CROSSED THE FLORIDA PENINSULA FROM EAST TO WEST AUGUST 1-2, 1995 AND MADE LANDFALL IN WEST FLORIDA AUGUST 3, 1995.

Location	Minimum pressure (inches)	Maximum sustained winds (mph)	Winds gusts (mph)	Rainfall (inches)
Peninsular Florida				
Sebastian Inlet	29.09	75	—	2.05
Melbourne	29.11	—	66	8.81
Vero Beach	29.12	—	61	2.46
Port St. Lucie	—	45	52	—
Fort Lauderdale	29.65	—	—	6.75
West Palm Beach	29.56	22	28	3.80
Jacksonville	29.85	22	37	2.07
New Port Richey	29.33	24	39	—
Tampa Airport	—	29	38	—
Winter Haven	29.16	31	42	—
Lakeland	29.34	20	38	—
West Florida				
Panama City	—	30	45	5.40
Eglin AFB	29.29	43	58	2.78
DeFuniak Springs	—	—	—	11.00
Pensacola NAS	28.82	55	88	2.19
Hurlburt Field	29.18	70	85	4.06

Source: Rappaport, 1995.

reintensified and moved on a northwest course toward Pensacola, finally coming ashore on August 3 at Fort Walton Beach, Florida with winds in the northeast eyewall estimated at 85 mph, category 1 on the Saffir-Simpson scale (Table 15-4) (Figure 14-1).

The Damage

The headline in the Florida Citrus Mutual Triangle for August 4, 1995 read, "Florida Citrus Growers Weather Hurricane Erin with Strength." The text of the article read:

FIGURE 14-1. HURRICANE ERIN QUICKLY DIMINISHED TO A TROPICAL STORM AS IT CROSSED CENTRAL FLORIDA, AUGUST 1-2, 1995. SOURCE: RAPPAPORT, 1995.

"For most Florida citrus growers, Hurricane Erin was all bark and no bite. The tropical storm turned hurricane trekked through central Florida on the morning of August 2, forcing growers to say their prayers and keep their fingers crossed. Some of the state's $10 billion industry did see heavy winds and rain, but escaped substantial damage.

"We should all be very thankful that this storm was a level one hurricane and not anything stronger," Florida Citrus Mutual Executive Vice President/CEO Bobby F. McKown said. "Our primary concern before the storm was excessive rainfall in areas already saturated with moisture. We have heard of one isolated incident in which a grower near Melbourne received close to 16 inches of rain and has standing water in his grove—to him and any other growers like him our hearts go out."

McKown went on to emphasize that the vast majority of growers weathered the storm with only incidental damage. The following is a summary of effects throughout the citrus belt.

South Florida (Charlotte, Collier, Glades, Hendry and Lee Counties)

The groves in these areas experienced some rain, but were largely unaffected from Hurricane Erin and growers are not anticipating any future problems as a result of the storm.

South Central Florida (DeSoto, Hardee, Highlands, Hillsborough, Manatee, Polk and Sarasota Counties)

Rainfall amounts of three to six inches were reported in these areas creating some soggy ground and delaying some spraying activities. However, growers do not expect to see any further problems directly related to Erin.

East Coast (Brevard, Indian River, Martin and St. Lucie Counties)

A light concentration of growers in south Brevard County experienced the most rainfall with amounts of up to 10 inches reported. Moving south down the coast, however, growers saw rainfall levels decrease to five inches in Indian River County and one to two inches in St. Lucie County.

Evidence of high winds is more visible in the northern part of this region with a few pieces of fruit on the ground, mostly under peripheral trees. Although it is not immediately visible, some growers are concerned about seeing puncture damage as the fruit matures.

North Florida Citrus Belt (Lake, Orange, Osceola, Seminole and Volusia Counties)

Generally, these areas received three to six inches of rain which is not expected to result in future problems. Some gift fruit trees holding large sized fruit did experience very minor fruit loss, mainly in Osceola County.

Florida is seeing unusually heavy activity early in the hurricane season. "From what we understand, most of the storm activity occurs after the beginning of August, with August, September and October being the busiest months. Looks like Erin was right on time by that standard," McKown said.

The Tampa Tribune recently reported that according to Dr. William Gray, a well-known predictor of the number of tropical storms and hurricanes that will form each year, indicated that 1995 would produce 12 named storms in the Atlantic Basin, eight of which would become hurricanes. Of those eight, three, he says, will be serious.

Vegetables

In the Vegetable Summary for 1995-96, the Florida Agricultural Statistics Service (1997) noted that, "during early August, Hurricane Erin dumped five to ten inches of rain and whipped winds up to 100 mph as it moved over the southeast coast into the central peninsula and over some western Panhandle localities." As most Florida vegetable crops are not planted until later in the fall (Table 13-3), Erin was not as disruptive as it would have been if it had visited the state in September or October. The Florida Agricultural Statistics Service (FASS) made few comments regarding Erin, noting that growers of sweet corn in the northern growing areas had begun planting in July, but that virtually all acreage escaped damage from Erin. Planting of escarole and endive became very active after Hurricane Erin had passed. Tomatoes also escaped damage from Erin.

Weather and Crop News (1995), for the week ending August 3, 1995 noted under the field crops heading that Erin's rainfall interrupted corn, hay and tobacco harvesting, but that sugarcane and most cotton, peanut and soybean plantings escaped damage except for some flooded fields in the Panhandle. In the east coast area, land preparation and vegetable planting was delayed by the rains.

Tropical Storm Jerry - August 23-26, 1995

Meteorological Considerations

The system which developed into Tropical Storm Jerry formed from a tropical wave which came off the coast of Africa on August 9 (Pasch, 1996B). The wave moved across the Atlantic with some increase in convection as it approached the Lesser Antilles on August 15, but no surface pressure fall was noted. However, on August 19, evidence of a low level circulation was detected near 18°N, 75°W, but there was no intensification as the system moved northwest and interacted with the mountains of eastern Cuba. Then, having cleared the mountains, the system formed into a tropical depression southwest of Andros Island at 1:00 P.M. on August 22, and was found to have strengthened into a tropical storm at 7:00 A.M. on August 23. Moving inland near Jupiter, Florida on the afternoon of the twenty-third with 40 mph winds, Jerry quickly weakened back to a tropical depression having maintained tropical storm status for only 24 hours as it moved on a west northwest course across the state, drifted into the gulf and turned north toward Cedar Key and Tallahassee (Pasch, 1996B). As Jerry approached the Florida coast, tropical storm warnings had been posted from Flagler Beach south to Deerfield Beach, but the primary concern was rain, not wind. Rainfall totals over southeast Florida ranged from three to eight inches with even heavier amounts in Palm Beach, Martin and St. Lucie counties where nine to ten inch totals were common (Table 14-9). Golden Gate, east of Naples, measured 16.18 inches, the highest rainfall of any station in the state.

The Damage

Again, rain was the negative factor from Tropical Storm Jerry. The Triangle, August 31, 1995 reported that with a two to 12 inch rainfall common along the east and lower west coast, citrus growers and

TABLE 14-9. RAINFALL DATA FOR SELECTED STATIONS DURING THE PASSAGE OF TROPICAL STORM JERRY, AUGUST 21-28, 1995.

Station	Rain days	Rainfall (inches)
South Florida		
Miami	5	7.13
West Palm Beach	7	11.07
Fort Lauderdale	6	7.09
Fort Myers	4	9.82
Immokalee	6	7.74
Golden Gate (near Naples)	6	16.18
Belle Glade	4	5.13
Homestead AFB	4	5.24
Indian River Area		
Stuart	3	5.31
Vero Beach	5	4.84
Melbourne	5	3.55
Central and West Florida		
Avon Park	5	2.31
Ruskin	6	5.72
Lakeland	5	3.06
Tampa	5	6.18
Orlando	6	3.83
Lisbon	6	7.53
Ocala	5	2.98
North Florida		
Lake City	6	6.90
Gainesville	4	2.69
Hastings	4	3.56
Jacksonville	7	6.31
West Florida		
Quincy	4	1.08
Crestview	3	1.65
Tallahassee	4	0.19

Source: Weather and Crop News, 1995.

caretakers had pumps running around the clock to move excessive water out of the groves and away from tree root systems. Flooding was particularly severe in Lee and Charlotte counties, and damage to agriculture was estimated to be $19 million (Pasch, 1996B).

Field crops were little affected by Jerry's rains which replenished soil moisture needed in Pasco and Suwanee counties, but interrupted hay making and the harvesting of peanuts in north Florida. Tobacco harvest was essentially complete (Weather and Crop News, 1995). Vegetable crops did not suffer major damage as planting had not begun. Young tomato plants in southwest Florida were beaten down but were expected to recover. West central Florida tomato growers did not begin planting until mid-September and thus missed the storm. Field preparation and planting over southern Florida was interrupted with washouts needing repair in some southeast Florida areas.

Hurricane Opal - October 4-5, 1995

Meteorological Aspects

Hurricane Opal developed from a tropical wave which came off the west coast of Africa on September 11 and moved westward across the Atlantic Ocean to the Caribbean Sea without intensification (Mayfield, 1995). On September 23, the wave merged into a broad low pressure area centered near 15°N, 80°W and moved west northwest toward the Yucatan peninsula without much development until it formed into a tropical depression approximately 80 miles south southeast of Cozumel, Mexico on the twenty-seventh.

For the next three days, the depression moved slowly over the Yucatan peninsula and intensified into Tropical Storm Opal on September 30 as it followed a westerly course into the Bay of Campeche, where it attained hurricane strength midday on the second of October and shifted to a more northerly course.

On October 3, Opal gradually shifted to a northeasterly course and passed over warmer waters, 83 to 84°F, where it rapidly intensified to a category 4 hurricane on the Saffir-Simpson scale (Table 15-4) with a minimum pressure of 27.05 inches and a maximum wind speed estimated near 150 mph at 5:00 A.M. 250 miles south southwest of Pensacola, Florida. Fortunately, the hurricane then weakened to a marginal category 3 level before coming ashore at Pensacola Beach, Florida at 5:00 P.M. on October 4. The minimum pressure at landfall was 27.82 inches with maximum surface winds near 115 mph over a narrow area. Sustained hurricane force winds were found only in the easterly quadrant of Opal between Destin and Panama City, and rapidly diminished to category 1 or 2 on the Saffir-Simpson scale (Table 15-4). The strongest winds at a land station were 84 mph with gusts to 144 mph at Hurlburt Field. Nevertheless, extensive damage resulted from the storm surge. Table 14-10 shows the meteorological data for Hurricane Opal.

The Damage

Field Crops: The cotton harvest was just beginning in the Florida Panhandle. As a result, the crop was damaged and harvesting delayed. It most areas, the peanut harvest was almost finished and little damage occurred (Weather and Crop News, 1995).

Timber: Damage to timber in west Florida was severe, and extended as far as 70 miles into Alabama (Ward, 1998).

TABLE 14-10. METEOROLOGICAL DATA FOR WEST FLORIDA DURING HURRICANE OPAL, OCTOBER 4-5, 1995.

Station	Minimum pressure (inches)	Maximum sustained winds (mph)	Wind gusts (mph)	Rainfall (inches)
Pensacola (NPA)	28.20	60	77	6.93
Hurlburt Field	28.36	84	144	6.64
Eglin AFB	28.54[1]	80[1]	115[1]	—
Panama City	28.87	63	85	3.64
Niceville	—	—	—	9.10
Apalachicola	29.27	32	60	2.56
Crestview	—	—	—	4.16
Tallahassee	28.38	32	52	1.25
Chipley	—	—	—	5.03
Milton Expt. Sta.	—	—	—	15.52

[1]Estimated.
Source: Mayfield, 1995.

Vegetables: Strong winds from Opal laid over some tomato plants in north Florida and in the Palmetto-Ruskin area. Tomato yields were also reduced by the blowing off of blossoms, and young tomatoes and peppers suffered damage from bruising in several sections, including north Florida, the east coast, the southwest and the Palmetto-Ruskin areas. In the southwest, the development of snap beans and eggplant was slowed by the heavy rainfall which also slowed the ground preparation for potato planting in the Everglades. Standing water over roads in the southwest limited access to fields by tractors and other machinery. Additional heavy rains in mid-October led to losses of snap beans and squash in Dade County, radishes in the Everglades, and watermelons in Hardee County. Nearly all vegetables around Immokalee lost yield and some acreage in the Webster and Center Hill areas was lost due to the October flooding (Weather and Crop News, 1996).

1996

The 1996 Atlantic Basin season was active, with nine hurricanes and four tropical storms. Fortunately, only Tropical Storm Josephine, which moved inland near Perry on October 7, crossed the Florida coastline. Hurricane Lili, which crossed Cuba from the western Caribbean and moved east northeast through the Bahamas on October 19, brought moderate to heavy rains to extreme south Florida but her winds remained well offshore.

Tropical Storm Josephine - October 7, 1996

Meteorological Aspects

Josephine formed as a tropical depression near 23°N, 96°W in the southwest Gulf of Mexico on the fourth of October. The depression moved northeast for 24 hours, then due east for a day before attaining tropical storm strength late on October 6 and moving across the Gulf of Mexico on a northeasterly track striking the north Florida coast and moving through the Big Bend area on October 7 with high winds and heavy rains (Table 14-11).

TABLE 14-11. RAINFALL AMOUNTS DURING PASSAGE OF TROPICAL STORM JOSEPHINE OCTOBER 7-8, 1996.

Station	Rainfall (inches)
Northwest Florida	
Apalachicola	5.95
Monticello	5.76
Quincy	5.02
Tallahassee	8.25
North Florida	
Cross City	5.40
Gainesville	3.98
Jasper	6.44
Perry	7.69
Glen St Mary	9.91
Jacksonville	9.08
North Central Florida	
DeLand	2.90
Sanford	5.29
Orlando	1.25

Source: Climatological Data, October 1996.

The Damage

Vegetables: Heavy rains and high winds from Josephine affected crop development and slowed field work in all areas except Dade County (Weather and Crop News, 1997).

Citrus: Most citrus groves were thoroughly drenched by the tropical storm, with some groves receiving up to seven inches of rain, but flooding was not damaging.

1997

This was a relatively quiet year in the Atlantic Basin with three hurricanes and three tropical storms, only one of which, Hurricane Danny, struck the Florida coast. The low number of storms was generally attributed to the extremely strong El Niño.

Hurricane Danny, July 18-20, 1997

Meteorological Considerations

Hurricane Danny was an unusual storm in that it developed "from a weather system of non-tropical origin" (Pasch, 1997). This system consisted of an area of convection which drifted south into the Gulf of Mexico from the Mississippi River Valley and formed a weak surface low on July 14. The system remained weak until a deep convection developed and formed a tropical depression near 6:00 A.M. on the sixteenth near 28°N, 93°W. Development was slow for several hours, but the system finally reached tropical storm strength on the seventeenth and was named Tropical Storm Danny. It became a minimal hurricane on July 18 as it moved on an east northeast course toward the mouth of the Mississippi River and made landfall near the towns of Empire and Buras on the Mississippi delta. Danny then passed back over the Gulf of Mexico with a minimum pressure of 29.06 inches and a maximum wind speed of 81 mph. Moving slowly, Danny moved erratically to the mouth of Mobile Bay, eastward over the Bay and then to the north across the northwest tip of the Florida panhandle accompanied by torrential rains. As Danny moved east, a hurricane warning had been issued at 8:00 P.M. on July 19 for east of Destin, Florida to Apalachicola, Florida. The hurricane warning was replaced by a tropical storm warning 24 hours later. After 12 hours, the tropical storm warning was also discontinued. Most Florida Panhandle counties received very heavy rains from Danny. Unofficial reports indicated amounts up to eight inches in Escambia and Santa Rosa counties (Weather and Crop News, 1997). Rainfall at some select northwest Florida stations is shown in Table 14-12.

The Damage

Hurricane Danny damaged corn, soybeans, and cotton in the northwest corner of Escambia County, but other areas of Escambia County suffered only minimal crop damage. Peanut and tobacco acreage escaped significant damage (Weather and Crop News,

TABLE 14-12. RAINFALL AT SOME SELECT FLORIDA PANHANDLE STATIONS DURING THE PASSAGE OF HURRICANE DANNY, JULY 18-20, 1997.

Station	Rainfall (inches)
Chipley	1.96
Milton Expt. Station	2.40
Niceville	2.65
Panama City	5.98
Pensacola	8.48

Source: Climatological Data, July 1997.

1997). Escambia County Extension Director Lamar Christenberry (1998) indicated that losses in the northwestern tip of the county could include as much as 1,800 acres of corn, 4,250 acres of cotton and 500 acres of soybeans.

"According to the American Insurance Services Group, insured losses from Danny were about $60 million dollars. The National Hurricane Center estimated around $100 million dollars in total damages" (Pasch, 1997).

1998

El Niño shifted to La Niña in 1998, and the effect was dramatic. Whereas there had been only three hurricanes and three tropical storms in the Atlantic Basin in 1997, 1998 endured nine hurricanes and five tropical storms, and in a span of only 35 days, August 19 to September 23, ten named tropical cyclones formed with four making landfall in the United States. On one day, September 25, 1998, there were four hurricanes on the map simultaneously. These were Georges, Ivan, Jeanne and Karl. This was a historic event as it was the first time such a thing had occurred in the Atlantic Basin during the 20th century. Two significant hurricanes struck Florida, Earl which pummeled the panhandle and Georges which first lashed Key West and then did his best to wash away the western most counties of the state with incredibly heavy rains.

Hurricane Earl, August 31-September 3, 1998

"Earl just bumped us with his heine as he sashayed by. He wasn't a big deal." — Lamar T. Christenberry, Escambia County Extension Director, 1998.

Meteorological Considerations

Earl attained tropical storm status on August 31 in the Gulf of Mexico, near 22.5°N, 93.9°W, or about 575 miles south southwest of New Orleans, Louisiana. Moving north northeast, the storm attained hurricane force, as a category 1 hurricane on the Saffir-Simpson scale (Table 15-4), on September 2 as it passed through 28.8°N, 87.9°W. It continued to intensify and the winds reached the 85 mph level before losing some of its strength and moving ashore near Panama City, Florida early on the morning of September 3 as a category 1 hurricane (Figure 14-2). A few locations in west Florida received major precipitation (Table 14-13) with the greatest amount being 12.46 inches at Panama City. Gale force winds were widespread with minimal hurricane force winds estimated at some beachfront locations.

The Damage

Hurricane Earl produced some beachfront erosion and trees were blown down in several locations. However, damage to agriculture was minimal.

Hurricane Georges, September 15-29, 1998

"Hurricane Georges deposited an incredible amount of rain (38.46 inches) at Munson in the Blackwater River State Forest in western Florida."—National Weather Service, Mobile, Alabama, 1998.

Meteorological Aspects

The tropical depression which was to become Hurricane Georges formed on September 15 from a tropical wave at 9°N,

FIGURE 14-2. THE PATH OF HURRICANE EARL AS IT BROUGHT ITS WINDS AND RAIN TO PANAMA CITY, FLORIDA ON SEPTEMBER 3, 1998. SOURCE: NATIONAL WEATHER SERVICE, 1998.

25.9°W, about 400 miles south southwest of the Cape Verde Islands, much further south than the initial stages of most Atlantic Basin hurricanes. As it moved on a west northwest course, Georges attained tropical storm status on the morning of the sixteenth at 10.5°N, 32.4°W, and became a minimal hurricane, category 1 on the Saffir-Simpson scale (Table 15-4), with a developed eye, late in the afternoon of September 17 as it passed through 12.5°N, 41.1°W at a forward speed of 15 to 20 mph (Purdue, 1998).

TABLE 14-13. METEOROLOGICAL DATA FOR NORTHWEST FLORIDA DURING HURRICANE EARL, SEPTEMBER 2-3, 1998.

Station	Minimum pressure (inches)	Maximum sustained winds (mph) direction	Wind gusts (mph) direction	Rainfall (inches)
Apalachicola	29.25	70[1]	—	—
Crestview	29.40	40 NE	54 NE	6.03
Destin	29.36	34.5 NNE	47 NNE	—
Eglin AFB	29.46	—	44 —	6.31
Hurlburt Field	29.38	—	—	5.45
Marianna	29.25	37 NNE	48 NNE	5.96
Milton, Whiting Field	29.53	31 NE	43 NE	2.22
Panama City Airport	29.15	41 NE	53 NE	12.46
Pensacola NAS	29.46	37 ESE	49 E	2.81
Perry, Foley Airport	29.43	28 NE	39 ENE	4.40
Tallahassee Airport	29.16	33 SW	46 SSW	5.41

[1]Estimated by amateur radio operator.
Source: National Weather Service, Tallahassee, Florida and Mobile, Alabama, 1998.

Continuing on a west northwest course, Georges reached category 2 status with 98 mph winds at 10:00 A.M. on the eighteenth, category 3 status with 115 mph winds at 7:00 A.M. on the nineteenth, and category 4 status with 132 mph winds and a minimum central pressure of 27.87 inches at 4:00 P.M. on September 19 as it continued its west northwest surge past 15.7°N, 54.4°W. By this time, residents of the northern Leeward Islands knew that they were facing a major hurricane of disastrous proportions.

In Puerto Rico and the Virgin Islands, concerned residents crossed their fingers and prayed that Georges would swing more to the north and pass through latitude 18°N while still well east of the islands. Latitude 18°N is the magic number. If it passes north of that latitude while still well east of the islands, they can breath a sigh of relief. Unfortunately, relief was not to be forthcoming, the hurricane remained on a west northwest track and a minimum central pressure of 27.70 inches was recorded about 420 miles east of Guadeloupe on the evening of September 19 with maximum sustained winds estimated at 150 mph.

On September 20, the intensity of the hurricane decreased to the category 3 level, a still dangerous storm, and Georges made his first landfall at the island of Antigua before proceeding through the Virgin Islands and beginning his assault on Puerto Rico on the evening of the twenty-first with 115 mph winds. Raging across the island, Georges was the first hurricane since San Felipe in 1928 to cross the full length of the island from east to west. Being only 30 miles wide, Puerto Rico is a narrow target for a storm approaching from the east, but Georges scored a bulls eye traversing Puerto Rico from Humacao on the east to Mayaguez on the west pummeling the island with winds up to 150 mph along the mountain tops and heavy rains (Table 14-14). Surprisingly, the hurricane lost little strength in its passage across the island, despite encountering 3,500 to 4,000 foot peaks of the Cordillera Central. Georges emerged into the Mona Passage between Puerto Rico and the Dominican Republic with a perfectly defined eye and 120 mph winds.

Holding its west northwest track, Georges devastated the Dominican Republic and Haiti, late on September 22, where it finally lost strength as it encountered mountain peaks over 10,000 feet. Down-

TABLE 14-14. RAINFALL AT SELECTED STATIONS[1] IN PUERTO RICO FOR THE TWO-DAY PERIOD ENDING AT 7:00 A.M., SEPTEMBER 23, 1998.

Station	Rainfall (inches)
Comerio	25.68
Jayuya	18.13
Cidra	17.19
Ovocoris	14.38
San Lorenzo	14.27
Ponce	14.25
Juncas	13.18
Aibonito	12.32
Caguas	11.04
Trujillo Alto	10.92
Lares	10.27

[1]Only stations receiving over 10 inches of rain are listed.
Source: National Climatic Data Center, 1998.

graded to a minimal, category 1 hurricane, with 75 mph winds, the hurricane exited from Haiti and made its next landfall on September 23 on the shores of eastern Cuba.

The hurricane moved along the north shore of Cuba for most of the day on September 24 maintaining its 75 mph winds until the eye had completely moved away from land, after which it reintensified as it moved into the Florida Straits and stormed across the Florida Keys in the morning of September 25, lashing Key West with winds to 105 mph. Leaving the Florida Keys, Georges moved west northwest and then to the northwest pointing his might at the Florida Panhandle, Mobile Bay, and the Mississippi and Louisiana gulf coasts, creating a major scare for New Orleans as it appeared that a direct hit might be imminent. Fortunately for the Crescent City, Georges slowed down and moved slightly more to the east and came ashore instead near Biloxi, Mississippi on the twenty-seventh as a category 2 hurricane. Shortly afterward, it was downgraded to a tropical storm on the twenty-eighth and a tropical depression on the twenty-ninth as it moved northeastward into southern Alabama (Figure 14-3).

During the hurricane's approach to the gulf coast, the major rain bands were located in its northeast quadrant, soaking the westernmost Florida counties, from western Jackson and Holmes counties to Escambia, Santa Rosa, Okaloosa and Walton counties, with drenching rains exceeding 20 inches in many places (Table 14-15).

Damage in the Caribbean

Puerto Rico was a major disaster area with damaged estimated to exceed $2 billion dollars. Over 80% of the 3.8 million people on the island lost water and power, and the Federal Emergency Management Agency (FEMA) estimated that 33,133 homes were destroyed and another 50,000 suffered damage. Agriculture was devastated with 75% of the coffee crop, 95% of the plantains and 65% of the chickens lost to the winds and floods (NCDC, 1998A).

In the Virgin Islands, structural damage was surprisingly light as most buildings had been rebuilt to FEMA standards after direct hits by Hurricane Hugo in 1989 and Hurricane Marilyn in 1995. Actually, the most serious losses were to agriculture. Coconuts and mangos were stripped from the trees and livestock losses were heavy.

FIGURE 14-3. AFTER RAVISHING THE ISLANDS OF THE CARIB-
BEAN, HURRICANE GEORGES BUFFETED THE FLORIDA KEYS
THEN POINTED ITS FURY AT THE GULF COAST DELUGING THE
FLORIDA PANHANDLE EAST OF THE STORM CENTER WITH
EXTREMELY HEAVY RAINFALL. SOURCE: NATIONAL WEATHER
SERVICE, 1998.

Among the smaller islands in the Leeward chain, St. Kitts
reported 85% of all homes damaged, 3,000 residents homeless,
three deaths and an estimated $402 million in damages including
the airport terminal and control tower. Antigua reported two dead
and major structural damage.

TABLE 14-15. METEOROLOGICAL DATA FOR THE FLORIDA PAN-HANDLE DURING HURRICANE GEORGES, SEPTEMBER 28-29, 1998.

Station and county	Minimum pressure (inches)	Maximum sustained winds (mph) direction	Wind gusts (mph) direction	Rainfall (inches)[1]
Munson, Santa Rosa	—	—	—	38.46
Gulf Breeze, Santa Rosa	—	—	—	26.87
Pensacola TV station, Escambia	—	—	—	26.83
Eglin AFB, Okaloosa	29.35	48 SE	90 SE	24.24
Crestview, Okaloosa	29.50	32 WSW	49 WSW	19.98
Niceville, Okaloosa	—	—	—	19.53
Milton, Santa Rosa	29.30	44 ENE	58 E	18.41
Jay, Santa Rosa	—	—	—	18.19
Pensacola Airport, Escambia	29.50	51 SE	67 SE	15.78
Pensacola NAS, Escambia	29.50	46 SSE	70 ENE	12.84

[1]Four-day rainfall totals ending at midnight, September 28, 1998.
Source: National Weather Service, Mobile, Alabama, 1998.

The Dominican Republic and Haiti were very hard hit. Early reports were 210 dead and over 500 missing in the Dominican Republic and probably over 100 dead and 60 to 70 missing in Haiti. Crop losses totaled up to 90%. Cuba also reported major crop damage, with five deaths and 20,000 homes flooded.

Damage in the Florida Keys

In the Florida Keys, more than 150 homes were totally destroyed including 75 houseboats. In addition, 500 homes suffered major damage and over 900 suffered minor damage. All residents were without power.

Damage in the Florida Panhandle

With 18 to 30 inches of rain in the western counties of the Florida Panhandle, mandatory evacuations totaled about 225,000, including our word processor operator, Peggy Hicks and her husband Wayne of Niceville, due to severe flooding, high winds and isolated tornadoes. Nearly 700,000 residents were without power.

Damage Reports from County Extension Agents

Bruce H. Ward, Walton County: "The greatest farm damage factor in terms of dollars was the loss of 1/4 of the farm ponds and impoundments when the flood waters washed out the dams. We estimate replacement costs may be as high as $1.5 million.

"We caught the major rain bands. There were 23.6 inches of rain at my house, and some reports in the county were as high as 34 inches. The road between DeFuniak Springs and Crestview was closed for the first time in history. The rain mutilated the hay crop stored outside. With 30 inches of rain, the hay was soaked in too deep to possibly dry. Of 6,000 acres, the loss was probably 25 to 30%. We were midway through harvesting our 4,600 acre peanut crop. We lost about 30% in yield, but 40% of the economic value due to loss in quality. There are about 400 acres of pecans in the county. About 25% were lost due to wind damage. Over 4,800 acres of cotton lost both in quality and quantity and we think the loss is about 50%."

Lamar T. Christenberry, Escambia County: "The cotton acreage in Escambia County was in the order of 12,000 to 14,000 acres. It was devastated by wind and rain. The loss may total as high as 75%. We had about 5,000 acres of corn but it was lost to the drought earlier in the year. The drought also got our hay crop, the wind got half the pecans. Timber was not seriously affected, certainly not as bad as during Opal and Erin. Soybean losses were 30 to 35% of the 3,000 acres."

Mike Donahoe, Santa Rosa County: "The rain could not have come at a worse time for the cotton crop. Only 1,000 acres of the estimated 25,000 acres had been harvested. Initial estimates by the Farm Service Agency indicated a 75% loss, but this has now been reduced to between 50 and 60%. We had the same problem with peanuts. We were just starting to harvest 12,400 acres and it appears that 40 to 50% will be lost. The damage to soybeans will probably reduce the yield by 30%, there will be a 50% loss to the hay crop and a 50% loss to pecans. The Santa Rosa County corn crop was lost to the drought earlier in the year."

Gerald R. Edmondson, Okaloosa County: "The cotton bolls were beaten into the ground by the wind and rain. We are looking at a 40 to 50% reduction in yield on 4,600 acres. The corn crop was mostly lost to the drought last spring so Georges was of no consequence. The peanut crop was just maturing and we could lose as much as 40%. Peanuts were hit pretty bad with 40 to 50% on the ground. These were green pecans prior to shuck split. Timber came through well with only tropical storm force winds. Trees had been thinned out earlier by Hurricane Opal in October 1995. Okaloosa County is not a big soybean producer so damage to that crop was not major."

David M. Solger, Washington County: "Most of the heavy rains were to the west of us. We only had six to seven inches. There was a 1 to 2 week delay in harvesting peanuts and we may lose a few, but not too serious."

Logan B. Barbee, Calhoun County: "There was some water damage to the cotton, but we are waiting to see. The corn had been lost to the drought. Damage to peanuts, soybeans and timber was light."

Henry E. Jowers, Jackson County: "We were on the eastern edge of the storm and we didn't get as much damage as the counties to the west of us. Only the western edge of the county was affected. We are a major cotton producing county, but the crop was not too good before the hurricane. The storm did not help it any."

Personal Observations

Even the completion of the manuscript for this book was adversely affected by Hurricane Georges. Peggy Hicks (1998), our word processor operator par excellence in Niceville, Florida, was forced to move twice by the torrential rains and work on the manuscript was suspended for the duration and then some. Peggy writes:

"The high water forced us to evacuate to a hotel, but the first two floors of the hotel had soggy, wet carpet. When we finally got back home, we found two inches of standing water downstairs. We then rented a cottage for three nights while the carpet at home was being replaced. Guess what! The carpet at the cottage was also wet and soggy. I have never in all my life been so tired of soggy, wet carpet!"

Peggy Hicks is a former staff member of the Citrus Research and Education Center at Lake Alfred, and contributed to the preparation of my earlier book, *A History of Florida Citrus Freezes.* She now resides with her husband, Wayne, in Niceville in the Florida Panhandle.

Steve Howe (1998) writes to the Tropical Fruit News from Big Pine Key:

" . . . My heart feels storm-tossed and crushed like our wonderful trees. They are indeed in sad shape since Hurricane Georges put its 110 mph northeast eyewall over us for six hours.

"The smell of dying vegetation is all in and about the still, hot airless atmosphere . . . All of the southern half of the Keys from Marathon to Key West looks like someone took a giant egg beater to it."

Hurricane Mitch, October 26-November 4, 1998

Hurricane Mitch was not a major event as far as Florida agriculture was concerned as it was only a tropical storm when it crossed the southern part of the state. However, in the early days of this hurricane, it attained category 5 status on the Saffir-Simpson scale (Table 15-4) with 180 mph sustained winds and gusts to 200 mph, and its heavy rains caused catastrophic floods in Central America. For that reason, a description of Hurricane Mitch is appropriate for this book.

Meteorological Considerations

The first advisory from the National Hurricane Center reported a tropical depression in the southern Caribbean Sea near 12.8°N, 77.90°W at 10:00 P.M., October 21. The depression reached tropical storm force at 4:00 P.M. on October 22 and was assigned the name Mitch, a name which will be long remembered for the death and destruction it would bring to the countries of Central America.

Moving on a course almost due north, Mitch attained minimal hurricane status on October 24. Continuing to the north, the hurricane reached the category 3 level on the Saffir-Simpson scale (Table 15-4) on October 25 with maximum winds reaching 110 mph and a central pressure of 27.91 inches (945 mb). With the hurricane still on a course generally to the north, Florida agricultural interests began to pay close attention to advisories from the National Hurricane Center. On the twenty-sixth, Mitch's winds reached 125 mph, category 4, with the central pressure at 27.43 inches (929 mb), and on the twenty-seventh, the winds attained category 5 level with 155 mph sustained and gusts possibly as high as 200 mph, with a central pressure of 26.75 inches (906 mb). Mitch remained a category 5 level for a continuous period of 33 hours, the longest continuous period for a category 5 storm since the 36 continuous hours of Hurricane David in 1979 (National Climatic Data Center, 1998). However, fortunately for Florida, the hurricane had shifted first to the northwest and then due west on a course which threatened the north coasts of Nicaragua and Honduras. Losing much of its intensity, the storm moved inland over Honduras on the twenty-eighth and twenty-ninth bringing torrential rains and catastrophic floods. Precipitation in Honduras and Nicaragua was estimated at 50 to 75 inches, reportedly producing 25 inches of rain in six hours at one location (National Climatic Data Center, 1998).

After drifting slowly to the west through the mountains of Honduras as a weak tropical depression, the remnants of the hurricane moved into the Bay of Campeche on November 2 and regained its tropical storm status. Rushing to the northeast, Mitch brushed the western coast of the Yucatan Peninsula, briefly losing its tropical storm winds, then reintensified again in the Gulf of Mexico, lashed Key West, Florida with gale winds and heavy rains on November 4 and 5 aggravating the damages caused by Hurricane Georges in September. In a final swan song, Mitch crossed south Florida on a line roughly from Naples to Stuart with wind gusts to 70 mph and six to eight inch rains before becoming extra tropical in the Atlantic (Figure 14-4).

The Damage in Central America

The first word of the horrible situation in Honduras was delivered to the Salvation Army by amateur radio (Shaver, 1999). The initial

FIGURE 14-4. HURRICANE MITCH HAD DISINTEGRATED TO A TROPICAL STORM WHEN IT CROSSED SOUTH FLORIDA, BUT LEFT HEAVY RAINS IN ITS WAKE. SOURCE: NATIONAL CLIMATIC DATA CENTER, 1998.

reaction was one of skepticism, as it didn't seem possible that one storm could leave so many dead and homeless. However, when CNN and the other major news agencies reached the scene, they confirmed what the Salvation Army already knew through amateur radio.

It is estimated that the loss of life in Honduras and Nicaragua made Hurricane Mitch the most deadly Caribbean hurricane since

1780 when approximately 22,000 people died in the eastern Caribbean (National Climatic Data Center, 1998). Preliminary casualty reports by country indicated 6,500 deaths in Honduras and 3,800 in Nicaragua which coupled with adjoining countries projected a total death toll of up to 11,000. However, two months after the hurricane, Honduran authorities had significantly reduced the death toll in that country by 25% or more with an exact count almost impossible due to lack of an accurate body count, many having been washed away in floods or buried in mudslides. Initial assessments had assumed that many reported missing were included among the dead. Efforts were underway to produce a more accurate figure.

Agricultural damage in Central America was catastrophic with an estimated 70% of crops destroyed in Honduras and massive crop damage also in Nicaragua, El Salvador and Guatemala. Particularly hard hit were the banana, coffee and sugar crops. Honduran authorities projected a loss of as much as 90% of the banana crop, and the Costa Rican Coffee Institute reported the loss of 112,000 130 pound bags from the 1998-99 crop. The Guatemalan Coffee Association estimated a 25% loss of that country's crop. To complicate the already serious situation, many roads and bridges needed to move crops to market were washed out preventing timely delivery of coffee and sugar (USA Today, 1998). As a result, coffee and sugar future's contracts moved sharply upward on the New York commodity exchange.

Florida dodged the bullet where agricultural damage was concerned as Mitch never regained hurricane strength before crossing the peninsula. Some flooding occurred in the Okeechobee mucklands and in the Delray area, but losses were minor compared to losses suffered in Central America.

Chapter 15

Hurricane Warnings, The National Hurricane Center

As a result of the catastrophic damage from hurricanes to the Caribbean islands, and the vulnerability of island residents and shipping operations in Caribbean waters, there was great interest in hurricanes in the West Indies from the time of Columbus forward. Therefore, it is not surprising that the first attempts to study these damaging storms and predict their course were begun in 1831 in the West Indies by Lt. Col. William Reid of the Royal Engineers of England who was stationed in Barbados (Calvert, 1935; Sheets, 1990). Reid published a book in 1838 in which he proposed a "law of storms" to help ships avoid sailing into the centers of tropical storms and hurricanes (Dunn and Miller, 1964).

The next major contribution to tropical storm forecasting in the West Indies was made by Father Benito Vines at Belen College in Havana, Cuba. Fr. Vines began organizing a hurricane warning system in 1870, and issued a hurricane warning for Cuba on September 11, 1875 (Dunn, 1971). This was probably the first instance of a hurricane warning being issued in the western hemisphere. Father Vines' objectives were to demonstrate the existence of a storm as far away from the observer as possible, to determine its movement, and to forecast its size, intensity and the velocity of its winds.

Also in 1870, the United States Congress made its first appropriation to establish a national meteorological service which would be operated by the Signal Corps of the U.S. Army. This service began on August 6, 1873, initially using weather reports sent by cable from

Havana. Later in the year, these reports were augmented by daily observations from Kingston, Jamaica and Santiago de Cuba. The first warning by the Signal Corps was issued from Washington on August 23, 1873 and regarded "a storm of tropical origin" (Calvert, 1935) which affected the middle Atlantic coast. The printed weather map on September 28, 1874 was the first to show a hurricane which was centered off the Atlantic coast between Jacksonville and Savannah on that date.

The Signal Corp's hurricane warning service suffered a reverse in 1876 due to lack of funding for the reports received from the islands by cable, and again in 1881 by a view that it was illegal to spend funds on activities outside the borders of the United States. However, these problems were quickly overcome and reports began to be received from additional stations in Barbados, Guadeloupe, and St. Thomas, as well as Havana, Kingston and Santiago, and by 1884, reports became available through the Naval Observatory of Cuba, from Puerto Rico, Santa Cruz and Antigua. In 1890, the U. S. Weather Bureau was created as a branch of the U.S. Department of Agriculture (Sheets, 1990).

The next major milestone in hurricane forecasting occurred in 1898 with the outbreak of the Spanish-American War. President McKinley is said "to have feared a hurricane more than he feared the Spanish Navy" (Dunn, 1971), and it was immediately recognized that the existing system was not adequate to protect naval and merchant vessels operating in Caribbean waters. Consequently, a bill was sent to Congress on June 16, 1898 and approved July 7, 1898 which would provide additional warning stations throughout the area. The bill was implemented July 22, 1898 and in August official U.S. trained observers were stationed in Kingston, Jamaica; Port-of-Spain, Trinidad; Willemstad, Curacao; Santo Domingo, and Santiago de Cuba with the forecast center at Kingston. Further stations were established in mid-September at Basse-Terre, St. Kitts and Bridgetown, Barbados.

Although the war ended quickly, expansion of the service continued with the addition of new stations at Roseau, Dominica and San Juan, Puerto Rico, and on February 1, 1899 the headquarters were moved from Kingston to Havana and provided a system for timely warnings to ships of all nationalities. When the U.S. occupation of

Cuba ended in 1902, the forecast headquarters were moved from Havana to Washington, D.C. Later, in 1919, a second major forecast center was established in Puerto Rico.

The next milestone was the widespread use of radio and on August 26, 1909, the U.S. Cartago, near the coast of the Yucatan peninsula, made the first ship report of a hurricane using the new medium of communication. Ship reports by radio immediately provided a much wider range of observations than could be carried out solely by cable from land-based stations. During the 1934 hurricane season, over 21,000 reports were received from ships in the Atlantic Ocean, Caribbean Sea and the Gulf of Mexico. Colonel Joseph P. Duckworth made the first aircraft reconnaissance flight into a hurricane center.

When criticism of the hurricane warning system reached Congress and the President, the Presidential Science Advisory Board recommended the appropriation of funds to the Department of Agriculture in 1935 to reorganize the hurricane warning service into three centers: San Juan, Puerto Rico to cover the islands east of longitude 75°W and south of latitude 20°N; New Orleans, LA to cover the Gulf of Mexico west of longitude 85°W and Jacksonville, FL to cover the remainder of the Atlantic, Caribbean and gulf south of latitude 35°N. There would be no warnings issued from Washington, D.C. until a tropical storm or hurricane moved north of latitude 35°. A more sophisticated scheme for receipt of radio reports was also implemented. It included the use of designated ships to transmit reports twice daily, 0000 and 1200 GMT (7:00 P.M. and 7:00 A.M. EST) which would include latitude and longitude, wind direction, weather observed, wind force, barometric pressure, visibility and temperature, except that during the hurricane season period from June 16 to November 15, the daily observations would be expanded to four with the addition of reports at 0600 and 1800 GMT (11:00 A.M. and 11:00 P.M. EST). During this period, the data transmitted would also include swell, clouds, temperature difference between air and water, ships course, barometer change and past weather. And, when a hurricane was actually in progress, the forecast area was divided into 5° squares and efforts were made to contact ships in specified squares for reports which might otherwise not be received. Cooperation with the U.S. Coast Guard was also put on a more systematic basis and a teletype system was organized to provide more rapid receipt and dissemination of observations. The teletype system connected the Jacksonville and New Orleans centers with offices in Tampa, Miami, Key West, Pensacola, Mobile, Port Arthur, Houston, Galveston, Corpus Christie and Brownsville, and operated 24 hours a day, 7 days a week to keep newspapers and the public promptly apprised of serious weather developments.

In 1937, a radiosonde network was established which enabled hurricane forecasters to analyze a hurricane's steering currents for the first time (Dunn, 1971), and in 1944, regular aircraft reconnaissance was begun. In 1943, the hurricane forecast center was moved from Washington to Miami and put under the direction of Mr. Grady Norton, an outstanding forecaster. Mr. Norton, Dean of hurricane forecasters, died in 1954 during Hurricane Hazel, a tragic loss for the service. Miami was designated as the National Hurricane Center in the mid-1950s and in 1955, Dr. Robert H. Simpson began the National Hurricane Research Project.

By 1960, a "radar force" (Dunn, 1971) had been put into use along the Atlantic and Gulf coasts; and on April 1, 1961, the first weather satellite was put into orbit. In 1965, the Hurricane Warning Service was concentrated in one office at the National Hurricane Center with Gordon Dunn as Director of the National Hurricane Warning Service for the Atlantic Basin (Sheets, 1990). This was followed by the designation of the Tropical Analysis Center to prepare analyses for the area from 40° north latitude to 40° south latitude from Central Africa west to the eastern Pacific Ocean. The name was later changed to the Regional Center for Tropical Meteorology under the World Meteorological Organization (WMO). The net effect was that all hurricane-track and intensity forecasts for the Atlantic basin, including the Caribbean Sea and the Gulf of Mexico, became the responsibility of the National Hurricane Center.

Gordon Dunn retired in 1967, at which point the staff of the Miami office had grown to 83. He was succeeded by Robert H. Simpson who held the post from 1968 to 1973. Simpson promoted applications of satellite technology and the staff at the National Hurricane Center increased to 100. The National Oceanic and

Atmospheric Administration (NOAA) was created in October 1970 and the Weather Bureau became the National Weather Service. Neil Frank became Director in 1973 and stressed preparedness and the use of radio and television to motivate residents to take the necessary precautions. During the 1970s, weather satellites became the primary forecasting tool and the Tropical Satellite Analysis Center was created at the National Hurricane Center at Coral Gables, Florida near Miami. As a result, many activities were automated and the staff was reduced from 100 to 37. Neil Frank retired in 1987 and was succeeded by Bob Sheets.

In 1989, the National Hurricane Center was given responsibility for that area from the equator to 30° north latitude and from Africa to 140° west longitude. This included the Atlantic Basin plus the eastern Pacific Basin and included functions previously performed by Weather Service offices in Miami and San Francisco and the staff was increased to 40 people (Sheets, 1990).

After Hurricane Andrew severely damaged Coral Gables in 1992, the National Hurricane Center was moved to the campus of Florida International University in Miami where it functions as the forecast arm of the Tropical Prediction Center. With the use of satellite images, Doppler radar and other technological advances, the tools used by the hurricane forecaster have changed mightily since the days of Grady Norton.

Florida Agricultural Areas Most Subject to Hurricane Damage

Looking at the map of Florida, the agricultural areas most vulnerable to hurricane damage are the southernmost counties, particularly those along the Atlantic and Gulf Coasts. The most battered region is Dade County, particularly the Homestead area, with its valuable lime, avocado, mango and other tropical fruits and its bountiful vegetable crops such as tomatoes, beans, cabbage, squash, sweet corn, cucumbers, potatoes and strawberries. Hurricanes moving inland from the Atlantic along latitude 25.5°N to 26.5°N will first devastate Dade County and then move across the citrus growing areas of Collier, Hendry and Lee counties which have moved to the forefront of orange production during the past 30 years.

Florida Hurricanes—Where Do They Originate?

With regard to vulnerability to hurricane damage, Florida is unique among the lower 48 states both for the length of its coastline, 1,200 miles, and the fact that hurricanes may strike the state from three different directions, east, south or west. The number of hurricanes and tropical storms striking peninsula Florida and the Florida panhandle by decade are shown in Table 15-1. Many major hurricanes have struck peninsular Florida from the east. Some examples are the major hurricanes in 1926, 1928, 1935, 1947, 1949, Hurricane Betsy in 1965 and, of course, extremely destructive Hurricane Andrew in 1992. These are included in Table 15-2, years in which major 20th century hurricanes have followed paths which would seriously damage today's Florida agriculture.

Hurricanes from the south may turn north in the Fort Myers area and follow a path directly up the peninsula, as did the destructive October 1944 hurricane and destructive Hurricane Donna in 1960,

TABLE 15-1. NUMBER OF HURRICANES (H)[1] AND TROPICAL STORMS (T) AFFECTING THE WEST FLORIDA PANHANDLE OR PENINSULAR FLORIDA BY DECADE.

Decade	Panhandle	Peninsular Florida
1871-80	8 (H or T)[2]	15 (H or T)[2]
1881-90	9 (H or T)[2]	13 (H or T)[2]
1891-1900	5H, 2T	5H, 8T
1900-10	2H, 7T	5H, 3T
1911-20	3H, 1T	1H, 3T
1921-30	3H, 3T	4H, 2T
1931-40	2H, 6T	5H, 3T
1941-50	2H, 2T	12H, 5T
1951-60	2H, 5T	1H, 6T
1961-70	1H, 3T	6H, 4T
1971-80	2H, 0T	1H, 1T
1981-90	1H, 1T	1H, 6T
1991-96	2H, 3T	2H, 2T

[1]Hurricanes varied significantly in intensity, some minimal storms had winds of hurricane force at sea but lost strength rapidly upon moving inland.
[2]Hurricanes were first distinguished from tropical storms in 1887.
Source: National Climatic Data Center, 1998.

TABLE 15-2. YEARS IN WHICH MAJOR 20TH CENTURY FLORIDA HURRICANES, CATEGORY 3 OR GREATER, HAVE FOLLOWED A PATH WHICH WOULD CAUSE MAJOR DAMAGE TO PENINSULAR FLORIDA AGRICULTURE TODAY.

Year	Month	Category[1]	Origin	Name
1910	October	3	W. Caribbean	—
1921	October	3	W. Caribbean	—
1926	September	4	Atlantic	—
1928	September	4	Atlantic	—
1933	September	3	Atlantic	—
1935	September	5	Atlantic	—
1944	October	3	W. Caribbean	—
1945	September	3	Atlantic	—
1947	September	4	Atlantic	—
1948	September	3	W. Caribbean	—
1949	August	3	Atlantic	—
1950	October	3	W. Caribbean	King
1960	September	4	Atlantic	Donna
1965	September	3	Atlantic	Betsy
1992	August	4	Atlantic	Andrew

[1]Saffir-Simpson scale, Table 15-4.
Source: National Climatic Data Center, 1998.

or a hurricane from the south may track north through the Gulf of Mexico and smash into the Florida Panhandle between Pensacola and Apalachicola as have many, many hurricanes throughout Florida history (Table 15-3). Finally, hurricanes forming in the western Caribbean or the Gulf of Mexico may move east or east northeast and cross the state from the Gulf of Mexico to the Atlantic Ocean. Examples of hurricanes following such a course are the September 21-22, 1948 hurricane and Hurricane Isbell in 1968.

Major hurricanes striking Florida have tended to cluster in certain ten-year periods (Table 15-1). The decade of the 1920s produced two very severe hurricanes, September 1926 and September 1928; the 1930s only one severe hurricane which did catastrophic damage to the Florida Keys, then the 1940s, particularly the period from 1944 to 1950, produced one or two hurricanes annually. Among the most destructive of these were in 1944, 1947, 1949 and 1950.

TABLE 15-3. YEAR IN WHICH MAJOR 20TH CENTURY FLORIDA HURRICANES, CATEGORY 3 OR GREATER, HAVE STRUCK THE WEST FLORIDA PANHANDLE.

Year	Month	Category[1]	Origin	Name
1917	September	3	Atlantic	—
1926	September	3[2]	Atlantic	—
1936	July	3	Atlantic	—
1950	September	3	Atlantic	Easy
1975	September	3	Atlantic	Eloise

[1]Saffer-Simpson scale, Table 15-4.
[2]Hurricane was category 4 when it struck southeast Florida.
Source: National Climatic Data Center, 1998.

The decade of the 1950s was most kind to the "Sunshine State" regarding the passage of hurricanes. During these years, a number of hurricanes formed in the Atlantic Basin, but they either moved north, striking the Atlantic Coast of the Carolinas or points even further north, or else they invaded the Gulf Coast from Alabama to Texas. However, this idyllic period ended with Hurricane Donna, which swept north over the entire peninsula from Fort Myers to Jacksonville. The 1960s continued to be busy with Hurricanes Cleo, Dora and Isbell in 1964, Hurricane Betsy in 1965, Hurricane Alma and Inez in 1966 and Hurricane Gladys in 1968.

Florida was fortunate again in experiencing no severe hurricanes in the 1970s and 1980s, although Florida agriculture was repeatedly devastated by a series of very severe freezes in the 1980s (Attaway, 1997). Then, in 1992, south Florida was devastated by Hurricane Andrew which, with the Miami Hurricane of September 1926, was one of the most damaging hurricanes to ever strike the state.

Florida hurricanes originate in two principal areas. Most of those which strike the state from August to mid or late September form in the eastern Atlantic Ocean in an area south and west of the Cape Verde Islands, between 10° and 15° north latitude, and have become known as Cape Verde hurricanes. These storms are usually the largest and most violent and destructive of the hurricanes which strike the Florida peninsula. Hurricanes which occur from late September through October, occasionally into November, but very rarely in December normally form in the western Caribbean Sea and the Gulf of Mexico. While these storms are usually smaller and less intense than the Cape Verde hurricanes, due to their proximity to the southeast coast, they are more likely to hit Florida than Cape Verde storms which frequently curve to the north and follow a track through the north Atlantic Ocean.

The history of a typical Cape Verde storm might be as follows: A low pressure area forms over Central Africa in late August and drifts into the Atlantic in an area called the intertropical convergence zone. The National Hurricane Center will take note of this system and newscasters will describe it as a tropical disturbance, a tropical wave, or an easterly wave as it moves in a westerly direction toward the islands of the eastern Caribbean at about 20 miles per hour. As it crosses 40° west longitude, a clearly defined low pressure center is noted with winds of 35 miles per hour, and the system is classified as a tropical depression. A few hours later, the wind speed is found to be in excess of 39 miles per hour and the system is upgraded to become a named tropical storm and a hurricane hunter plane may be sent to investigate. As the storm crosses 50° west longitude, its forward speed slows, winds of 75 miles per hour are reported and the system is then a full-fledged hurricane. With high pressure above the storm, it then may rapidly intensity further and become a major hurricane with wind speeds of 120-140 miles per hour. At that time, the storm is in a critical area. Typically, it will turn from a west to a west-northwest course passing near Puerto Rico and the Virgin Islands into the southern Bahamas. If it holds a west-northwest course, it will threaten Florida agriculture. Many of our most destructive hurricanes have followed this course. If it turns to the northwest or north-northwest, it may threaten the mid-Atlantic coast or New England or eventually turn to the northeast and remain at sea.

The history of a hypothetical October hurricane might be as follows: On October 5, a westward moving tropical wave crosses Jamaica with winds of 35 miles per hour and the formation of a tropical depression is noted by the National Hurricane Center. As the depression moves south of the Cayman Islands, its forward speed slows, its wind speed increases and it becomes a named tropical storm. The storm then turns slowly to the northwest and on October 8 it reaches hurricane force as it approaches 20° north latitude, 85° west longitude. On October 10, the hurricane passes through the Yucatan Channel and a hurricane hunter plane notes that the wind speed has increased to 90 miles per hour as the storm begins a slow curve to the north and passes over the western tip of Cuba. The hurricane is then in a critical area regarding its potential threat to Florida agriculture. A turn to the east-northeast will bring it into the Naples/Fort Myers area and damage citrus and vegetables in Lee, Hendry, and Collier counties and later the Everglades and Indian River areas as it emerges into the Atlantic. A turn to the northeast toward the Sarasota/Bradenton and Tampa areas will damage vegetables and citrus in the west coast area and citrus in the ridge district, while a continued move to the north will threaten agricultural interests in the Florida panhandle.

The Saffir-Simpson Scale for Describing Hurricane Intensity

The Saffir-Simpson scale is used by the National Hurricane Center to characterize hurricanes by strength. The scale ranges from 1 to 5, with 1 being the least intense and 5 being the most violent. Numbers on the scale are determined by measuring the central pressure. The scale is shown in Table 15-4 with approximate description of damage which might be expected to tree crops.

TABLE 15-4. SAFFIR-SIMPSON SCALE.

Scale number	Wind speed (mph)	Expected damage
1	74-95	Some loss of leaves and fruit, heaviest in exposed areas
2	96-110	Considerable loss of leaves and fruit with some trees blown down
3	111-130	Heavy loss of foliage and fruit, many trees blown over
4	131-155	Trees stripped of all foliage and fruit, many trees blown over and away from property
5	over 155	Damage would be almost indescribable, groves and orchards completely destroyed

Source: Simpson and Riehl, 1981.

Chapter 16

The Commodity Futures Markets

The Orange Juice Futures Market

Since trading began in the orange juice futures market in 1966, destructive freezes have resulted in massive upswings on many occasions, particularly during the 1980s (Attaway, 1997). Freezes in January 1981, January 1982, December 1983, January 1985 and December 1989 figuratively pinned the needle to the top of the chart. However, to date, there has not been a hurricane with enough impact on orange juice production to significantly affect the market.

Historically, there were severe hurricanes in October 1944, August 1949 and September 1960 which might have impacted the market had it been in existence at the time. In 1944, losses of all varieties were estimated at 23 million boxes from a total crop of 69 million boxes. In 1949, the loss was estimated at 15 million boxes from a total crop of 88 million boxes, and in 1960 the loss from Hurricane Donna was estimated at 22 to 24 million boxes from a total crop of 124 million boxes (Citrus Industry, 1960).

At first glance, one would conclude that these hurricanes would have very severely affected the supply of orange juice. This is not really so, because by far the largest losses were in grapefruit production (Table 16-1). However, it should be noted that fruit blown from the trees in August or September will be a total loss as it will be too immature to salvage for juice.

Table 16-1 shows that in the three most destructive hurricanes to strike the Florida citrus industry, the loss to the orange crop varied from only 3.6 million boxes in 1949, to 8 million boxes in Hurricane Donna in 1960, to 9.2 million boxes in the 1944 hurricane, enough to affect the market but not the way it was affected by the freezes of the 1980s. This does not mean that these hurricanes did not cause catastrophic damage to orange groves in the coastal counties. However, as Florida oranges are grown in many different sections of the state, devastating damage from a hurricane is usually limited to a few miles from the storm center and is most severe near the coast. In contrast, a freeze may totally wipe out the groves in many different areas (Attaway, 1997). In 1944, the most severe damage was in the southwest Florida counties as the hurricane passed ashore west of Fort Myers (Chapter 7). In 1949, the Indian River area was devastated as the hurricane made landfall at Jupiter (Chapter 7), and in 1960, Hurricane Donna was most destructive in Lee, Charlotte and DeSoto counties (Chapter 9).

Other noteworthy hurricanes and the total loss in boxes of citrus of all varieties are shown in Table 16-2. Historically, there were other damaging hurricanes to strike Florida which did not have an important affect on the Florida citrus industry statewide. All of these were covered in previous chapters.

An example of a hurricane which occurred since the orange juice futures market was initiated in 1966 was minimal Hurricane David in September 1979 (Table 16-3). David seemed to produce an uptick in the market as it closed up 1.20 to 110.20 basis the September contract on Friday, August 31, 1979. This was Friday before the

TABLE 16-1. ESTIMATED LOSSES OF GRAPEFRUIT AND ORANGES DURING THE PASSAGE OF A SEVERE HURRICANE THROUGH FLORIDA'S CITRUS PRODUCING COUNTIES.

Hurricane year	Grapefruit	Oranges
	--------------- million boxes ---------------	
1944	13.7 million	9.2 million
1949	11.5 million	3.6 million
1960	14 million	7 million

Source: Citrus Industry, 1960.

TABLE 16-2. ADDITIONAL HURRICANES WHICH AFFECTED ALL VARIETIES OF FLORIDA CITRUS PRODUCTION AND THE COUNTIES MOST AFFECTED.

Date and year	Total crop	Estimated loss	Counties
	- - - - million boxes - - - -		
Sept. 16-18, 1928	27.8	1.0	DeSoto, Hardee & Polk
Sept. 4-5, 1933 (Labor Day Hurricane)	28.8	4.5	Indian River, St. Lucie, Martin & Palm Beach
Sept. 15-16, 1945	86.2	4.0	Highlands
Oct. 6-8, 1948	83.3	2.0	Lee, Charlotte & Sarasota
Sept. 17-18, 1947	91.3	6.0	Lee & DeSoto
Oct. 4-5, 1948	93.1	2.0	Lee, Charlotte & Sarasota
Oct. 17-18, 1950 (Hurricane King)	105.4	2.5	St. Lucie & Indian River
Aug. 27-28, 1964 (Hurricane Cleo)	123.7	3.0	Martin & St. Lucie
Sept. 3-4, 1979 (Hurricane David)	283.6	5.0	Martin, St. Lucie & Indian River

Source: Climatological Data, 1928-1979 and Florida Agricultural Statistics Service, 1997.

long Labor Day weekend and the hurricane was causing massive crop damage in Haiti and the Dominican Republic as its winds peaked to 125 mph. However, over the weekend, David dropped to only tropical storm strength as it moved through the Bahamas, and on Tuesday, September 4, 1979 as it brushed Florida with winds of barely hurricane force, the market closed down 2.60 at 107.60. Whether this small market reaction prior to the weekend was due to Hurricane David is not certain. It could have resulted from other factors. In later trading, the market made a contract low of 103.00 on September 7 and a contract high of 111.00 on September 14.

The Cotton Futures Market

The three major cotton-producing areas of the United States are the southeast including west Florida, Texas and the central valley of

TABLE 16-3. ACTION OF ORANGE JUICE FUTURES MARKET (SEP-TEMBER 1979 CONTRACT) DURING APPROACH OF HURRICANE DAVID TO FLORIDA.

Date	High-low	Close	Status of hurricane
August 27	110.00-109.50	109.50	Attained hurricane force near 50° west
August 28	111.25-109.55	109.80	Approaching Lesser Antilles
August 29	109.75-108.00	108.30	Passing through Lesser Antilles with winds to 94 mph
August 30	110.20-108.75	109.00	Passing south of Puerto Rico
August 31	110.20-108.00	110.20	Winds to 125 mph cause severe crop damage in Dominican Republic and Haiti
September 1-2	Weekend		Diminished to tropical storm winds ENE 64 mph at Nassau
September 3	Labor Day Holiday		Moves inland over Indian River area as minimal hurricane
September 4	109.00-107.00	107.60	Moves offshore with course toward Savannah Beach, Georgia

Source: Barry, 1999 and Herbert, 1980.

California (Barry, 1999). With cotton production in west Florida totaling almost 110,000 bales in 1997 (Florida Agricultural Statistics Service, 1997), and Hurricane Georges bringing gale force winds with occasional gusts to hurricane force, and torrential rains from western Jackson County on the east to Holmes and Escambia counties on the west and then moving into the cotton-producing counties of south Alabama and southwest Georgia, it would be reasonable to speculate that the cotton future's market might strengthen substantially with the approach of a hurricane to west Florida, much as the orange juice future's market moves sharply upward with the approach of freezing weather. This conclusion was strengthened by estimates of 40 to 75% losses in cotton production reported by Florida Agricultural Statistics Service (1999), Bruce Ward (1998), L. T. Christenberry (1998), Mike Donahoe (1998) and G. R. Edmondson (1998). However, an examination of the movement of the cotton future's market during the approach of Hurricane Georges, and the immediate aftermath of the hurricane, indicate that damage from Hurricane Georges to west Florida, Mississippi, south Alabama and south Georgia was not sufficient to move the market upward. In fact, there was a slight down trend in the market during this period (Barry, 1999).

Appendix One

Glossary of Terms

Anemometer—An instrument for measuring wind speed.

Atmospheric Pressure—The force exerted by a column of air over a specific area. It is usually measured by a barometer and expressed as inches of mercury or millibars (mb). The standard pressure at sea level is 29.92 inches of mercury or 1013.35 mb.

Barometer —An instrument for measuring atmospheric pressure.

Beaufort Wind Scale—A scale used historically by seamen to describe or estimate wind speed. See Table 1-2.

Cyclone—A low pressure system whose winds blow counterclockwise in the northern hemisphere.

Easterly Wave—A low pressure trough moving in a westerly direction toward the Caribbean Islands or the southern coast of the United States. It may also be called a tropical wave.

Eye—The relatively calm center of a hurricane. The average diameter of a hurricane eye is 14 miles. However, hurricane eyes have been known to range from only 4 miles in a developing hurricane to 20-25 miles in a large, mature hurricane (Dunn and Miller, 1960).

Gale—A wind equal to or exceeding 39 mph.

Hurricane—A tropical cyclone with maximum sustained winds of 74 mph or greater.

Hurricane Force Winds—Winds of 74 mph or greater.

Hurricane Warning—A warning that hurricane force winds may be felt in a specified area with 24 hours.

Hurricane Watch—An indication that hurricane force winds may be felt in a specified area within 36 hours.

Intertropical Convergence Zone—Zone near the equator along which the opposing trade winds of the northern and southern hemispheres meet.

Maximum Sustained Winds—Maximum wind speed sustained over a one minute period.

Millibar (mb)—A unit of pressure.

Peak Wind Gust—Highest wind speed. According to Dunn and Miller (1960), wind speeds in hurricanes fluctuate so rapidly that peak gusts in hurricanes are difficult to record accurately. Peak wind gusts may exceed the highest sustained wind speed by 30 to 50%. Therefore, gusts of over 200 mph could occur in hurricanes with maximum sustained winds of 150 mph. Such gusts are responsible for the massive destruction in severe hurricanes.

Recurvature—The change in direction of an Atlantic Basin hurricane or tropical storm from a west or northwest course to a north or northeasterly course.

Quadrant—A quarter-circle (90%) of the wind field of a hurricane. In an Atlantic Basin hurricane, the strongest winds are found in the northeast quadrant.

Tropical Cyclone—A cyclone which has developed over tropical waters.

Tropical Depression—A tropical cyclone with winds up to 38 mph.

Tropical Disturbance—An area of thunderstorms over tropical waters. May move as part of a tropical wave.

Tropical Storm—A tropical cyclone with maximum sustained winds ranging from 39-73 mph (34-63 knots).

Tropical Storm Warning—A warning that tropical force winds may be felt in a specified area within 24 hours.

Tropical Storm Watch—An indication that tropical force winds may be felt in a specified area within 36 hours.

Tropical Wave—A low pressure trough moving from east to west. Also called an easterly wave. In the Atlantic Basin, the wave will usually be moving toward the Caribbean islands.

Saffir-Simpson Scale—A scale for measuring the relative destructive potential of hurricanes. See Table 15-4.

Wind Shear—Winds aloft which may shear top of a developing storm thus preventing hurricane formation.

Appendix Two

This appendix shows Florida counties with crops vulnerable to hurricane damage. Crops include tropical fruits, citrus, vegetable crops and field crops including cotton.

TABLE A2-1. COUNTIES PRODUCING TROPICAL FRUITS. ALL THREE COUNTIES ARE VERY VULNERABLE TO HURRICANE DAMAGE.

County	Production area	Fruit type
Dade	Southeast coast	Limes, Lemons, Avocadoes, Mangoes and minor varieties
Collier	Southwest	Limes, Lemons, et al.
Hendry	Southwest	Limes, Lemons, et al.

Source: Florida Agricultural Statistics Service, 1996-97.

TABLE A2-2. MAJOR CITRUS PRODUCING COUNTIES MOST VULNERABLE TO HURRICANE DAMAGE.

County	Production area	Boxes of citrus produced annually (thousands)
Hendry	Southwest	34,931
St. Lucie	Indian River	34,234
Indian River	Indian River	21,768
Martin	Indian River	15,180
Collier	Southwest	12,402
Charlotte	Southwest	6,177
Palm Beach	Southeast	4,360
Okeechobee	Interior	3,748
Lee	Southwest	3,743
Glades	Southwest	3,616
Brevard	Indian River	2,769

Source: Florida Agricultural Statistics Service, Citrus Summary 1996-97.

TABLE A2-3. MAJOR CITRUS PRODUCING COUNTIES MODERATELY VULNERABLE TO HURRICANE DAMAGE.

County	Production area	Boxes of citrus produced annually (thousands)
Polk	Ridge	36,250
DeSoto	Peace River	27,541
Highlands	Ridge	27,137
Hardee	Peace River	22,324
Hillsborough	West coast	11,073
Manatee	West coast	9,097
Osceola	Interior	6,151
Lake	Ridge	4,961
Pasco	West coast	2,818
Orange	Interior	2,357

Source: Florida Agricultural Statistics Service, Citrus Summary 1996-97.

TABLE A2-4. VEGETABLE CROPS IN HURRICANE PRONE AREAS NORMALLY UNHARVESTED DURING PART OR ALL OF THE HURRICANE SEASON.

Crop	Date that harvesting begins	Period of peak harvest	Counties at highest risk
Beans	Oct. 15	Nov. 1-May 1	Dade, Broward, Palm Beach
Cabbage	Oct. 25	Jan. 1-April 15	Dade, Palm Beach, Martin
Celery	Oct. 25	Dec. 15-June 1	Palm Beach
Corn (sweet)	Sept. 25	Nov. 15-Jan. 15	Dade, Palm Beach, Lee, Hendry and Collier
Cucumbers	Sept. 20	Nov. 1-Dec. 15	Dade, Palm Beach, Lee, Hendry and Collier
Eggplant	Oct. 1	Nov. 15-July 1	Palm Beach, Lee, Hendry and Collier
Escarole & Endive	Oct. 20	Nov. 15-May 25	Palm Beach
Lettuce & Romaine	Oct. 20	Dec. 1-May 1	Palm Beach
Peppers (green)	Oct. 20	Nov. 15-June 15	Palm Beach, Lee, Hendry and Collier
Squash	Sept. 1	Nov. 15-May 15	Dade, Palm Beach, Lee, Hendry and Collier
Tomatoes	Oct. 15	Nov. 15-June 1	Dade, Palm Beach, Martin, Lee, Hendry, Collier and St. Lucie

Source: Florida Agricultural Statistics Service, Vegetable Summary 1995-96.

TABLE A2-5. FIELD CROP PRODUCTION IN FLORIDA PANHANDLE COUNTIES WHICH MAY BE SUBJECT TO HURRICANE DAMAGE.

County	Corn	Peanuts	Soybeans	Cotton
			acres	
Calhoun	1,600	3,600	4,600	6,300
Escambia	7,000	—	3,600	17,300
Gadsden	1,400	700	600	2,000
Holmes	2,400	5,500	1,600	3,100
Jackson	16,900	34,700	7,400	30,700
Jefferson	4,100	1,100	1,100	2,700
Leon	1,200	400	—	—
Okaloosa	1,000	900	400	4,700
Santa Rosa	1,100	12,400	1,400	26,900
Walton	2,100	5,500	1,100	5,000
Washington	3,500	2,300	1,200	—
Other	1,500	500	300	1,400
TOTAL	43,800	67,600	23,200	100,100

Source: Florida Agricultural Statistics Service, 1995.

Appendix Three

This appendix uses figures developed by Pielke and Landsea (1998) to compare the most costly hurricanes in Florida history.

TABLE A3-1. THE MOST COSTLY HURRICANES IN FLORIDA HISTORY, NORMALIZED TO 1995 DOLLARS BY INFLATION, PERSONAL PROPERTY INCREASES AND COASTAL COUNTY POPULATION CHANGES (1925-1995).

Rank	Hurricane-year	Category	Damage (billions of U.S. dollars)
1	Miami-1926	4	72.303[1]
2	Andrew-1992	4	33.094[1]
3	S.W. Florida-1944	3	16.864
4	S. Florida-1938	4	13.795
5	Betsy-1965	3	12.434[1]
6	Donna-1960	4	12.048[2]
7[5]	S. Florida-1947	4	8.308[3]
8	S. Florida-1945	3	6.313
9	S. Florida-1949	3	5.838
10	Dora-1964	2	3.108
11	Opal-1995	3	3.000[4]
12	Cleo-1964	2	2.435
13	King-1950	3	2.266
14	Florida Keys-1935	—	2.191

[1]Includes damages in Louisiana.
[2]Includes damages in other east coast states.
[3]Includes damages in Alabama and Louisiana.
[4]Includes damages in Alabama.
[5]Hurricane Agnes was omitted as heavy dollar damages included floods in northeastern states.
Source: Pielke and Landsea, 1998.

404

TABLE A3-2. YEARS IN WHICH 20TH CENTURY HURRICANES, OF CATEGORY 2 OR BETTER, HAVE FOLLOWED A COURSE WHICH WOULD BE DAMAGING TO THE FLORIDA CITRUS INDUSTRY AS PRESENTLY CONSTITUTED.

Year	Month	Category according to Saffir/Simpson scale
1903	September	2
1906	October	2
1910	October	3
1921	October	3
1926	September	4
1928	August	2
1928	September	4
1933	September	3
1941	October	2
1944	October	3
1945	September	3
1947	September	4
1947	October	2
1948	September	3
1948	October	2
1949	August	3
1950	October (King)	3
1960	September (Donna)	4
1964	August (Cleo)	2
1964	September (Dora)	2
1979	September (David)	1
1992	August (Andrew)	4

Source: National Climatic Data Center.

TABLE A3-3. TROPICAL STORMS OR HURRICANES, BEGINNING BY MONTH, FROM 1886-1992.

Month	Number of tropical storms or hurricanes
May	14
June	56
July	68
August	217
September	308
October	189
November	42
December	6

Source: *Tropical Cyclones of the North Atlantic Ocean, 1871-1992* Historical Climatology Series 6-2. National Climatic Data Center, Asheville, NC in cooperation with the National Hurricane Center, Miami, Florida.

Appendix Four

Conversions

Over the years, Weather Bureau reports may record wind speeds in nautical miles (knots), miles per hour and meters per second. Pressures may be recorded in either inches of mercury or millibars. For consistency, all wind speeds have been converted to miles per hour and all pressure readings to inches of mercury using the Smithsonian Meteorological Tables, Fifth Revised Edition, Smithsonian Institution, Washington, D.C., 1939. Tables used were:

Table 11. Barometric inches (mercury) into millibars,
pages 36-37

Table 17. Interconversion of nautical and statute miles,
page 48

Table 35. Meters per second into miles per hour,
pages 66-67

References

Adams, A. 1998. Personal Communication. Fort Pierce, FL.

Alexander, T. R. 1967. Effect of Hurricane Betsy on the Southeastern Everglades. Quarterly J. Fla. Acad. Sci. 30(1):1-24.

Allison, E. 1945. The Lyonizer. Citrus Ind. Mag. 26(10):18.

Allison, R. V. 1929. Fla. Agr. Expt. Sta. Ann. Rpt. :85-88.

Alsheimer, F. W. and R. F. Morales, Jr. 1997. A Climatological Comparison Between Rainfall, Freezes and Tropical Activity During El Nino/La Nina Years for West Central and Southwest Florida. Nat. Weather Serv. Tech. Attachment :22.

Anderson, R. L. 1951. Climatological Data, Fla. Section.: 182.

Annual Agricultural Statistical Summaries. 1921 through 1958. Fla. State Mktg. Bur., Jacksonville, FL.

Ashenberger, A. 1912. The Tropical Storm of September 13-14, 1912. Climatological Summary, District No. 2, Sept. 1912:5 and Monthly Weather Rev. 16:1307.

Attaway, J. A. 1997. *A History of Florida Citrus Freezes.* Fla. Sci. Source, Lake Alfred, FL.

Avila, L. A. and R. J. Pasch. 1991. *North Atlantic Hurricanes - 1991.* Natl. Hurricane Ctr., Miami, FL.

Barbee, L. B. 1998. Personal Communication. Calhoun County Extension Service, Blountstown, FL.

Barnes, J. 1998. *Florida's Hurricane History.* Univ. North Carolina Press, Chapel Hill, NC.

Barry, T. 1999. Personal Communication, Citrus Expo, Fort Myers, FL. Citrus Associates, New York Cotton Exchange.

Bartram, W. 1943. Travels in Georgia and Florida, 1773-74. A Report to Dr. John Fothergill. Trans. Amer. Philosophical Soc. Vol. XXXIII, Part II, Philadelphia, PA.

Bowditch, N. 1938. *American Practical Navigator, An Epitome of Navigation and Nautical Astronomy.* U.S. Govt. Printing Office :275-297.

Bowie, E. H. 1921. The Hurricane of October 25, 1921 at Tampa, Florida. Monthly Weather Rev. 49:567-570.

Box, M. 1998. Personal Communication. Okeechobee, FL.

Boyer, H. B. 1926. Destructive Gust at Jupiter, Fla., Following the Miami Hurricane. Monthly Weather Rev. 54:410.

Brooks, J. R. 1946. Hurricane Damage to Commercial Fruit Trees in Dade County. Proc. Fla. State Hort. Soc. 59:149-151.

Brown, C., Jr. 1991. *Florida's Peace River Frontier.* Univ. Central Fla. Press, Orlando, FL.

Brown, E. 1998. Personal Communication. Immokalee, FL.

Brown, L. G. 1993. *Totch, A Life in the Everglades.* Univ. Press of Fla., Gainesville, FL.

Burpee, R. W. 1988. Grady Norton: Hurricane Forecaster and Communicator Extraordinaire. Weather Forecasting 3:247-254.

Butson, K. 1959. Climatological Data, October 1959.

Butson, K. 1960. Special Weather Summary, Hurricane Donna. Climatological Data, Fla. Section 64(9):132-133.

Butson, K. 1964A. Special Weather Summary, Hurricane Cleo. Climatological Data, Fla. Section 68(8):94-95.

Butson, K. 1964B. Special Weather Summary, Hurricane Dora. Climatological Data, Fla. Section 68(9):110-111.

Butson, K. 1965. Special Weather Summary, Hurricane Betsy. Climatological Data, Fla. Section 69(9):118-119.

Butson, K. 1966. Special Weather Summary, Hurricane Alma. Climatological Data, Fla. Section 70(6):63.

Byers, H. R. 1935. On the Meteorological History of the Hurricane of November 1935. Monthly Weather Rev. 63:318-322.

Calvert, E. B. 1935. The Hurricane Warning Service and Its Reorganization. Monthly Weather Rev. 63:84-88.

Camp, A. F. 1949. Care of Storm Injured Groves. Citrus Mag. 12(2):19-20.

Campbell, R. J., C. W. Campbell, J. Crane, C. Balerdi, and S. Goldweber. 1993. Hurricane Andrew Damages Tropical Fruit Crops in South Florida. Fruit Var. J. 47:218-225.

Carlton, R. 1998. Personal Communication. Fort Pierce, FL.

Case, R. A. 1987. *North Atlantic Tropical Cyclones - 1987.* National Hurricane Ctr., Miami, FL.

Case, R. A. and M. Mayfield. 1990. Atlantic Hurricane Season of 1989. Monthly Weather Rev. 118:1165.

Christenberry, L. T. 1998. Personal Communication. Escambia County Ext. Serv., Cantonment, FL.

Citrus Industry. 1927. Clearing House Purposes Explained. 8(11):28.

Citrus Industry. 1928. 9(12):23.

Citrus Industry. 1929. Fla. State Marketing Bureau Report. 10(8):9.

Citrus Industry. 1935. Impressions. 16(10):12.

Citrus Industry. 1944A. Citrus Growers Sustain Heavy Losses. 25(11):5-13.

Citrus Industry. 1944B. The Lyonizer. 25(11):14.

Citrus Industry. 1948. Slight Hurricane Damage. 29(2):10.

Citrus Industry. 1949A. Citrus Groves Hit Hard. 30(9):10.

Citrus Industry. 1949B. The Lyonizer. 30(10):18.

Citrus Industry. 1949C. The Lyonizer. 30(12):19.

Citrus Industry. 1950A. The Lyonizer. 31(10):16.

Citrus Industry. 1950B. The Citrus Situation. 31(11):10.

Citrus Industry. 1960. Federal Crop Estimate Indicates $300 Million Crop. 41(11):14.

Citrus Magazine. 1944A. Pepper-Stewart Amendment. 7(5):3.

Citrus Magazine. 1944B. October Storm Straddled the State. 7(5):7.

Citrus Magazine. 1945. September Storm Traversed the State. 8(2):13-15.

Citrus Magazine. 1947. Estimates Loss. 10(2):7.

Citrus Magazine. 1949A. Hurricane Loss Estimated to be 35% on Grapefruit, 6% on Oranges. 12(1):7.

Citrus Magazine. 1949B. Advertising Fund Cut by Hurricane Loss. 12(1):7.

Citrus Magazine. 1949C. Pope Summer Oranges Average $7.45 Per Box. 12(1):13.

Citrus Magazine. 1949D. Minton Predicts Citrus Shortage on His Grove. 12(2):9.

Citrus and Vegetable Magazine. 1960. 23(2):12, 14, 21 and 35.

Citrus and Vegetable Magazine. 1992A. Tropical Research Center Damaged by Storm. 56(2):16-18.

Citrus and Vegetable Magazine. 1992B. Dade Farmers Working to Get Back in Business. 56(2):20-24.

Citrus and Vegetable Magazine. 1992C. Dade County's Agriculture Heavily Hit. 56(2):24.

Citrus and Vegetable Magazine. 1992D. More Support for Hurricane Victims. 56(2):26.

Citrus and Vegetable Magazine. 1992E. Hurricane Relief. 56(2):48.

Citrus and Vegetable Magazine. 1992F. USDA Lab Sustains Damage. 56(3):74.

Citrus and Vegetable Magazine. 1994. Market Watch 58(4):9.

Clemmons, H. L. 1997. Personal Communication. Ocala, FL

Climate and Crop Service, Fla. Section. 1897-1909.

Climatological Data, U.S. Weather Bur., Fla. Section, 1897-1998. Vols. 1-102.

Climatological Summary. 1910. District No. 2, South Atlantic and East Gulf States.

Commerce, Department of. NOAA. 1977. *Some Devastating Hurricanes of the 20th Century.* Obtained from the Natl. Climatic Data Ctr., Asheville, NC.

Covington, J. W. 1957. *The Story of Southwestern Florida.* Lewis Historical Publishing, Inc., NY.

Craighead, F. C. and V. C. Gilbert. 1962. The Effects of Hurricane Donna on the Vegetation of Southern Florida. Quarterly J. Fla. Acad. Sci. 25:1-28.

Crane, J. H. and C. Balerdi. 1996. Effect of Hurricane Andrew on Mango Trees in Florida and Their Recovery. Fifth Intl. Mango Symp., Israel.

Crane, J. H., C. Balerdi, R. Campbell, C. Campbell, and S. Goldweber. 1993. Hurricane Damage Update. Fla. Grower and Rancher 86(2):25-27.

Crane, J. H., C. Balerdi, R. Campbell, C. Campbell, and S. Goldweber. 1994. Managing Fruit Orchards to Minimize Hurricane Damage. HortTechnology 4(1):21-26.

Crane, J. H., R. J. Campbell, and C. F. Balerdi. 1993. Effect of Hurricane Andrew on Tropical Fruit Trees. Proc. Fla. State Hort. Soc. 106:139-144.

Crane, J. H., A. J. Dorsey, R. C. Ploetz, and C. W. Weekley, Jr. 1994. Post Hurricane Andrew Effects on Young Carambola Trees. Proc. Fla. State Hort. Soc. 107:338-339.

Cry, G. W. 1965. Tropical Cyclones of the North Atlantic Ocean. U.S. Weather Bur. Tech. Paper No. 55.

Cry, G. W., W. Haggard, and M. White. 1959. North Atlantic Tropical Cyclones. U.S. Weather Bur. Tech. Paper No. 36.

Cutler, H. G. 1923. *History of Florida, Past and Present.* Vol. 1. Lewis Publishing, NY.

Davis, R. A., S. C. Knowles, and M. J. Bland. 1989. Role of Hurricanes in the Holocene Stratigraphy of Estuaries: Examples from the Gulf Coast of Florida. J. Sedimentary Petrology 59:1052-1061.

Day, H. 1998. Personal Communication. St. Petersburg, FL.

Day, P. C. 1929. The Weather Elements. Monthly Weather Rev. 57:391.

DeAngelis, R. M. and W. T. Hodge. 1972. Preliminary Climatic Data Rpt., Hurricane Agnes June 14-23, 1972. NOAA Technical Memorandum EDS NCC-1. Natl. Climatic Data Ctr., Asheville, NC.

DeFoor, J. A., II. 1994. Odet Philippe at Tampa Bay. Tampa Bay History. 16 (Fall/Winter 1994):5.

DeFoor, J. A., II. 1997A. Odet Philippe: Peninsular Pioneer. Safety Harbor Museum of Regional History, Safety Harbor, FL.

DeFoor, J. A., II. 1997B. Personal Communication. Miami, FL.

Division of Water Survey and Research, State of Fla., State Board of Conservation. 1948. *Observed Rainfall in Florida, Monthly Totals from Beginning of Records to 31 December 1947.* Tallahassee, FL.

Donahoe, M. 1998. Personal Communication. Santa Rosa County Ext. Serv., Milton, FL.

Dorn, H. W. 1928. The Avocado Today in Dade County. Proc. Fla. State Hort. Soc. 41:161-170.

Douglas, M. S. 1947. *The Everglades, River of Grass.* Hurricane House Publications, Inc., Coconut Grove, FL.

Dovell, J. E. 1947. A History of the Everglades of Florida. PhD Thesis, Dept. of History, Univ. of North Carolina.

Dunn, G. E. 1961. The Hurricane Season of 1960. Monthly Weather Rev. 88:99-108.

Dunn, G. E. 1971. A Brief History of the United States Hurricane Warning Service. Muse News 3:140-143.

Dunn, G. E. and B. I. Miller. 1964. *Atlantic Hurricanes.* Louisiana State Univ. Press, Baton Rouge, LA.

Dunn, G. E. and Staff. 1965. The Hurricane Season of 1964. Monthly Weather Rev. 93:176-187.

Edmondson, G. R. 1998. Personal Communication. Okaloosa County Ext. Serv., Crestview, FL.

El Mundo. 1996. San Juan, PR, Sept. 29.

English, H. M. 1997. Personal Communication. A. Duda & Sons, Inc., Ft. Myers, FL.

Everglades Drainage District. 1947.

Fassig, O. L. 1913. Hurricanes of the West Indies. U.S. Weather Bur. Bul. X.

Fassig, O. L. 1928. San Felipe - The Hurricane of September 13, 1928 at San Juan, PR. Monthly Weather Rev. 56:350.

FEMA Interagency Hazard Mitigation Team Report. 1992. FEMA 955-DR-FL.

Fielding, E. 1998. Personal Communication. Winter Haven, FL.

Florida Agricultural Statistics Service. Field Crops Summaries. 1920-1997. Fla. Agr. Stat. Serv., Orlando, Fl.

Florida Dispatch, Farmer and Fruit Grower. 1892. Jacksonville, FL, June 16.

Florida Grower. 1926A. Storm Reduces Florida Citrus Crop 15 Per Cent. 34(13):11 and 16.

Florida Grower. 1926B. Weekly Citrus Market Summary. 34(14):14.

Florida Grower. 1926C. In the Wake of the Storm. 34(15):8.

Florida Grower. 1926D. Florida Crop Report for October. 34(17):18.

Florida Grower. 1927A. Second Freeze Boosts Citrus Loss. 35(4):14, (5):8 and (6):8

Florida Grower. 1927B. Developments in the Citrus Clearing House Plan. 35(15):3.

Florida Grower. 1928. High Winds Reduce Citrus Fruit Crop. 36(10):17.

Florida Grower. 1944. Florida News of the Month. 52(12):8.

Florida Grower. 1946. Florida News of the Month. 54(12):7.

Florida Grower. 1947. Florida News of the Month. 55(10A):6.

Florida Grower. 1948. Florida News of the Month. 56(10):7.

Florida Grower. 1949. Florida News of the Month. 57(10):23.

Florida Grower. 1950. Florida News of the Month. 58(12):11.

Florida Grower and Rancher. 1960. 68(10):36.

Florida Grower and Rancher. 1992A. Hurricane Update – Growers Hear Chiles Report on Hurricane—Governor Addresses Indian River Citrus League. 85(10):30.

Florida Grower and Rancher. 1992B. Agricultural Impact of Hurricane Andrew. 85(10):8-10.

Florida Grower and Rancher. 1992C. Florida Agricultural Industry Assists Hurricane Victims. 85(10):11.

Florida Grower and Rancher. 1992D. Federal Help Available for Victims of Hurricanes. 85(10):13.

Florida Grower and Rancher. 1993. Bonus from Hurricane - Storm Debris Mulched for Use in Fields. 86(4):27, 43.

Florida Grower and Rancher. 1995. Florida Farms Assess Storm Damage. 88(1):35.

Florida Hurricane Report Concerning Hurricane Donna. 1961. Trustees of the Internal Improvement Fund of the State of Fla., Tallahassee, FL.

Florida Hurricane Survey Report. 1965. Cabinet of the State of Fla., Tallahassee, FL.

FOCUS. 1992. University of Florida/IFAS. 3(4): July/August.

Forget, L. 1998. Personal Communication. Fort Pierce, FL.

Fort Myers News Press. 1947, 1960. Fort Myers, FL.

Frankenfield, H. C. 1917. The Tropical Storm of September 21-29, 1917. Monthly Weather Rev. 45:457-459.

Friday, E. W., Jr. 1983. Modernizing the Nation's Weather Communications Systems. Bul. Amer. Meteorological Soc. 64:355-358.

Frisbie, S. L. 1949. Citrus Groves Hit Hard. Citrus Ind. Mag. 30(9):10.

Gallenne, J. H. 1940. Tropical Disturbances of August 1940. Monthly Weather Rev. 67:217-218.

Garriott, E. B. 1901. Forecasts and Warnings. Monthly Weather Rev. 29:341-347.

Garriott, E. B. 1903. Forecasts and Warnings. Monthly Weather Rev. 31:407-410.

Garriott, E. B. 1906. West Indian Hurricanes. U.S. Weather Bur. Bul. H.

Garriott, E. B. 1909. Weather, Forecasts and Warnings for the Month. Monthly Weather Rev. 37:538-539.

George, P. S. 1995. *A Journal Through Time, A Pictorial History of South Dade.* The Donning Co., Homestead, FL.

Glades County Florida History. 1985. Rainbow Books, Moore Haven, FL.

Goldweber, S. 1997. Personal Communication. Homestead, FL.

Goodwin, G. 1926. The Hurricane at Turks Island, September 16, 1926. Monthly Weather Rev. 54:416-417.

Graham, J. H. and T. R. Gottwald. 1991. Research Perspectives on Eradication of Citrus Bacterial Diseases in Florida. Plant Dis. 75:1193-1200.

Gray, R. W. 1933. Florida Hurricanes. Monthly Weather Rev. 61:11-13.

Gray, W. M. 1984. Atlantic Seasonal Hurricane Frequency: Part I: El Nino and 30 MB Quasi-biennial Oscillation Influences. Monthly Weather Rev. 112:1649-1668.

Green, P. M., D. M. Legler, C. J. Miranda V, and J. J. O'Brien. 1997. The North American Climate Patterns Associated with the El Nino-Southern Oscillation. COAPS Project Rpt. Series 97-1:8.

Grierson, W. 1998. Personal Communication. Winter Haven, FL.

Grismer, K. H. 1949. *The Story of Fort Myers.* St. Petersburg Printing Co., St. Petersburg, FL.

Hall, W. H. 1927. Marketing Florida Citrus, Summary of 1926-27 Season and Florida Grower 35(12):18.

Handbook on Disaster Preparedness and Recovery. 1997. Univ. Fla., IFAS, Gainesville, FL.

Hanna, A. J. and K. A. Hanna. 1948. *Lake Okeechobee, Wellspring of the Everglades.* The Bobbs-Merrill Co., NY.

Hardy, N. G. 1992. Help When its Needed. Citrus Ind. 73(10):12.

Hartwell, F. E. 1930. The Santo Domingo Hurricane of September 1-5, 1930. Monthly Weather Rev. 58:362-364.

Harvey, B. 1998. Personal Communication. Okeechobee, FL.

Hendrix, C. 1998. Personal Communication. Clermont, FL.

Henry, A. J. 1896. Notes Concerning the West Indian Hurricane of September 29-30, 1896. Monthly Weather Rev. 24:368-369.

Herbert, P. J. 1976. Atlantic Hurricane Season of 1975. Monthly Weather Rev. 104:453-465.

Herbert, P. J. 1980. Atlantic Hurricane Season of 1979. Monthly Weather Rev. 108:973-990.

Herbert, P. J., J. D. Jarrell, and M. Mayfield. 1995. *The Deadliest, Costliest and Most Intense United State Hurricanes of this Century (and*

Other Frequently Requested Hurricane Facts). Natl. Hurricane Ctr., Miami, FL. Updated April 7, 1997 by B. Maher and J. Bevan.

Hicks, P. 1998. Personal Communication. Niceville, FL.

Hobgood, J. S. and R. S. Cerveny. 1988. Ice-Age Hurricanes and Tropical Storms. Nature 333:243-245.

Howard, R. A. and L. Schokman. 1995. Recovery Responses of Tropical Trees after Hurricane Andrew. Harvard Paper in Bot. 6:37-74.

Howe, S. 1998. Letter from Big Pine Key. Trop. Fruit News 32:2-3.

Huffman, H. 1997. Personal Communication. Redlands, FL.

Hughes, P. 1987. Hurricanes Haunt Our History. Weatherwise 40:134-148.

Huie, S. 1997. Personal Communication. Haines City, FL.

Hull, D. L. and A. W. Hodges. 1993. Impact of Hurricane Andrew on Ornamental Nursery Profitability in Dade County Florida. Proc. Fla. State Hort. Soc. 106:307-311.

Hunt, Z. 1998. Personal Communication. Okeechobee, FL.

Hurricane Loss Study for East Central Florida. 1986. Final Report. East Central Fla. Regional Planning Council, Winter Park, FL.

Hutchinson, J. 1975. *History of Martin County.* Gilbert's Bar Press, Hutchinson Island, FL.

Johnson, W. O. 1944. October Hurricane Straddled the State. Citrus Mag. 7(5):7.

Johnson, W. O. 1945. September Storm Traversed the State. Citrus Mag. 8(2):13.

Jordan, C. L. and Lt. F. J. Schatzle, USN. 1961. The Double Eye of Hurricane Donna. Monthly Weather Rev. 89:354-356.

Jowers, H. E. 1998. Personal Communication. Jackson County Ext. Serv., Marianna, FL.

Kaczor, B. 1995. Erin Rake's Florida's Panhandle. The Ledger, Lakeland, FL.

Kadel, B. C. 1926. An Interpretation of the Wind Velocity Record at Miami Beach, Florida September 17-18, 1926. Monthly Weather Rev. 54:414-416.

Kender, W. J. 1999. Personal Communication. Lake Alfred, FL.

King, J. 1998. Personal Communication. Fort Pierce, FL.

King, M. L. 1972. History of Santa Rosa County. Privately Published.

Krause, S. 1992A. Hurricane Recovery Could Take Years. Citrus and Veg. Mag. 56(2):10-12.

Krause, S. 1992B. Citrus, Vegetables Mostly Spared from Andrew's Wrath. Citrus and Veg. Mag. 56(2):12.

Krome, P. 1997. Personal Communication. Homestead, FL.

Krome, W. H. 1992. Notes and Observations Following Hurricane Andrew. An Unpublished Report, Homestead, FL.

Krome, W. H. 1997. Personal Communication. Homestead, FL.

Krome, W. H. and S. Goldweber. 1987. Commercial Fruit Production in Dade County 1900-1987. Proc. Fla. State Hort. Soc. 100:268-272.

Lamberts, M. and H. Bryan. 1993. Non-Cultural Factors Affecting Dade County Vegetable Production After Hurricane Andrew. Proc. Fla. State Hort. Soc. 106:175-176.

Lawrence, F. P. 1964. The Citrus Scene. Fla. Field Rpt., Sept. 24, 1964 :10.

Lawrence, F. P. and W. H. Mathews. 1960. Treatment of Hurricane Damaged Trees. Ext. Serv. FC Memo No. 2-60.

Lawrence, M. B. 1977. Atlantic Hurricane Season of 1976. Monthly Weather Rev. 105:505.

Lawrence, M. B. 1979. Atlantic Hurricane Season of 1978. Monthly Weather Rev. 107:477.

Lawrence, M. B. 1982. Atlantic Hurricane Season of 1981. Monthly Weather Rev. 110:858.

Lawrence, M. B. 1985. Preliminary Report, Hurricane Bob. Natl. Hurricane Ctr., Coral Gables, FL.

Lawrence, M. B. 1989. Hurricane Hugo Report. Natl. Hurricane Ctr., Coral Gables, FL.

Lawrence, M. B. 1996. Preliminary Report, Hurricane Lili, 14-27 October 1996, Natl. Hurricane Ctr., Miami, FL.

Lesley, J. T. 1944. Personal Communication to Forrest F. Attaway, Sr., Haines City, FL.

Liu, Kam-biu and M. L. Fearn. 1993. Lake-Sediment Record of Late Holocene Hurricane Activities from Coastal Alabama. Geology 21:793-796.

Longboat; Yesterday, Today and Tomorrow. 1984. First Edition, Lindsay Curtis Publishing Co., Sarasota, FL.

Loomis, E. Contributions to Meteorology, 1874-1889. Amer. J. Sci. :14-129.

Loomis, H. F. 1946. Hurricane Damage to Tropical Plants. Proc. Fla. State Hort. Soc. 59:146-148.

Loope, L., et al. 1994. Hurricane Impact on Uplands and Freshwater Swamp Forest. Bioscience 44:238-246.

Ludlum, D. M. 1963. *Early American Hurricanes, 1492-1870.* Amer. Meteorol. Soc., Boston, MA.

Maddox, W. T. 1998. Personal Communication. LaBelle, FL.

Maher, B. and J. Beven. 1997. The Costliest Hurricanes in the United States 1900-1996. Natl. Hurricane Ctr., Miami, FL.

Maloney, W. C. 1876. A Sketch of the History of Key West, Florida. Facsimile Reproduction, Univ. Fla. Press, Gainesville, FL, 1968.

Mann, G. W., Jr. 1998. Personal Communication. Polk County, FL.

Mayfield, M. 1990. North Atlantic Hurricanes - 1990. Reprinted 1991 Mariners Weather Log, 35. Natl. Hurricane Ctr., Miami, FL.

Mayfield, M. 1992. North Atlantic Hurricanes - 1992. Natl. Hurricane Ctr., Miami, FL.

Mayfield, M. 1994. Preliminary Report, Tropical Storm Beryl, August 19, 1994. Natl. Hurricane Ctr., Miami, FL.

Mayfield, M. 1995. Preliminary Report, Hurricane Opal, September 27-October 6, 1995. Natl. Hurricane Ctr., Miami, FL.

Mayfield, M. and L. Avila. 1992. North Atlantic Hurricanes - 1992. Natl. Hurricane Ctr., Miami, FL.

Mayfield, M. and L. Avila. 1993. Atlantic Hurricanes – Hurricane Andrew, August 25, 1992 20:20 GMT. Weatherwise 46:18-25.

Mayfield, M., L. Avila, and E. N. Rappaport. 1994. Atlantic Hurricane Season of 1992. Monthly Weather Rev. 122:517-531.

Mayfield, M. and B. Case. 1989. North Atlantic Tropical Cyclones, 1989. Reprinted 1990 Mariners Weather Log, 34. Natl. Hurricane Ctr., Miami, FL.

McDonald, W. F. 1931. Tropical Storms of September 1931 in North American Waters. Monthly Weather Rev. 59:364.

McDonald, W. F. 1935. The Hurricane of August 31 to September 6, 1935. Monthly Weather Rev. 63:269-271.

McKay, D. B. 1959. *Pioneer Florida.* Southern Publishing Co., Tampa, FL.

McLean, W. Foundation. 1998. Recordings of *Hold Back the Waters* and other Will McLean music are available from the Foundation at P. O. Box 77, Holder, FL 34445.

Mickelson, J. E. 1968. Special Weather Summary. Climatological Data, Fla. Section 72(10):117-118.

Mitchell, A. J. 1920. Tropical Storm, September 29-30, 1920. Monthly Weather Rev. 48:524.

Mitchell, A. J. 1928. General Summary. Climatological Data, Sept.

Mitchell, A. J. and W. P. Hardin. 1906. The West Indian Hurricane of the Second Decade of October 1906. Monthly Weather Rev. 34:478-479.

Mitchell, C. L. 1924. Notes on the West Indian Hurricane of October 14-23, 1924. Monthly Weather Rev. 52:497-498.

Mitchell, C. L. 1926. The West Indian Hurricane of September 14-22, 1926. Monthly Weather Rev. 54:409-415.

Mitchell, C. L. 1928. The West Indian Hurricane of September 10-20, 1928. Monthly Weather Rev. 56:347-352.

Mitchell, C. L. 1933. Tropical Disturbances of 1933. Monthly Weather Rev. 61:274-275.

Monthly Weather Review. 1882-1928.

Morgan, C. 1992. Andrew Produced Botanical Wonders. Miami Herald, Sept. 13 :6J.

National Climatic Data Center. 1998. Asheville, N.C.

National Hurricane Center, Miami, FL. 1981. Preliminary Report, Hurricane Dennis.

National Hurricane Center, Miami, FL. 1983. Preliminary Report, Hurricane Barry.

National Hurricane Center, Miami, FL. 1984. Preliminary Report, Tropical Storm Isidore.

National Hurricane Center, Miami, FL. 1987. Preliminary Report, Hurricane Floyd.

National Hurricane Center, Miami, FL. 1988. Preliminary Report, Tropical Storm Keith.

National Hurricane Center, Miami, FL. 1990. Preliminary Report, Tropical Storm Marco.

National Weather Service, Mobile, AL. 1998.

NCDC. 1998A. Georges Pummels Caribbean, Florida Keys and U.S. Gulf Coast.

NCDC. 1998B. Mitch: The Deadliest Atlantic Hurricane Since 1780.

Neumann, C. J., B. R. Jarvinen, C. J. McAdie, and J. D. Elms. 1993. *Tropical Cyclones of the North Atlantic Ocean, 1871-1992.* Historical Climatology Series 6-2. Natl. Climatic Data Ctr., Asheville, NC and Natl. Hurricane Ctr., Coral Gables, FL.

NOAA. 1982. *Some Devastating North Atlantic Hurricanes of the 20th Century.* U.S. Department of Commerce, Natl. Oceanic and Atmospheric Admin., Washington, DC.

Norton, G. 1948. Climatological Data, Fla. Section: 73.

Norton, G. 1949. Hurricanes of August, 1949. Climatological Data, Fla. Section. 53:128-131.

Norton, G. 1951. Hurricanes of the 1950 Season. Monthly Weather Rev. 79:8-15.

O'Brien, J. J., T. S. Richards, and A. G. Davis. 1995. The Effect of El Nino on U.S. Landfalling Hurricanes. Submitted to the Bulletin of the American Meteorological Society. Ctr. for Ocean Atmospheric Prediction Studies, Fla. State Univ., Tallahassee, FL 32306-3041.

Ogden, J. C. 1992. The Impact of Hurricane Andrew on the Ecosystems of South Florida. Conservation Biol. 6:488-491.

Padgett, R. L. 1946. The Lyonizer. Citrus Industry Mag. 27(11):12.

Padgett, R. L. 1947. The Lyonizer. Citrus Industry Mag. 28(10):16.

Parrish, J. R., et al. 1982. Rainfall Patterns Observed by Digitized Radar During the Landfall of Hurricane Frederic (1979). Monthly Weather Rev. 110:1933-1944.

Pasch, R. J. 1995. Preliminary Report, Hurricane Gordon, November 8-21, 1994. Natl. Hurricane Ctr., Miami, FL.

Pasch, R. J. 1996A. Preliminary Report, Hurricane Allison. Natl. Hurricane Ctr., Miami, FL.

Pasch, R. J. 1996B. Preliminary Report, Tropical Storm Jerry. Natl. Hurricane Ctr., Miami, FL.

Pasch, R. J. 1996C Preliminary Report, Tropical Storm Josephine, 4-8 October 1996. Natl. Hurricane Ctr., Miami, FL.

Pasch, R. J. 1997. Preliminary Report, Hurricane Danny, 16-26 July 1997. Natl. Hurricane Ctr., Tropical Prediction Ctr., Miami, FL.

Pielke, R. A., Jr. and C. W. Landsea. 1998. Normalized Hurricane Damages in the United States: 1925-95. Weather and Forecasting 13:621-631.

Pimm, S. L., G. E. Davis, L. Loope, C. T. Roman, T. J. Smith, III, and J. T. Tilmant. 1994. Hurricane Andrew. Bioscience 44:224-229.

Pratt, A. M. 1926. Weekly Market Summary. Fla. Grower. 34(18):22.

Purdue University. 1999. http:\\weather.unisys.com/hurricane/index.html/1998.

Purvis, J. C., S. F. Sidlow, D. J. Smith, W. Tyler, and I. Turner. 1990. Hurricane Hugo, Climate Report G-37. South Carolina State Climatology Office, Columbia, SC.

Putnam, D. 1998. Personal Communication. Bartow, FL.

Raley, T. 1997. Personal Communication. Winter Haven, FL.

Rappaport, E. N. 1992. Preliminary Report Hurricane Andrew. Natl. Hurricane Ctr., Miami, FL.

Rappaport, E. N. 1994. Preliminary Report Tropical Storm Alberto, July 3-4, 1994. Natl. Hurricane Ctr., Miami, FL.

Rappaport, E. N. 1995. Preliminary Report Hurricane Erin. Natl. Hurricane Ctr., Miami, FL.

Rappaport, E. N. and J. Fernandez-Partagas. 1995. *The Deadliest Atlantic Tropical Cyclones, 1492-Present.* Natl. Hurricane Ctr., Miami, FL. (Updated 1997 by Jack Bevin).

Rappaport, E. N. and R. J. Pasch. 1993. North Atlantic Hurricanes - 1993. Natl. Hurricane Ctr., Miami, FL.

Redfield, W. C. 1836. On the Gales and Hurricanes of the Western Atlantic. The Naval Mag. 1(39):301-319.

Reep, R. 1998. Personal Communication. Gainesville, FL.

Reid, W. 1838. The Law of Storms. London.

Reitz, H. J. 1960. Florida Citrus Mutual Triangle, Sept. 19.

Reitz, H. J. 1997. Personal Communication. Lake Alfred, FL.

Reese, J. H. 1926. *Florida's Great Hurricane.* Lysle E. Fessler Publishing, Miami, FL.

Ritter, S. 1993. The Relandscaping of Dade County Following Hurricane Andrew. Proc. Fla. State Hort. Soc. 106:306-307.

Robinson, J. H. 1898. The Telegraph Service with the West Indies. Monthly Weather Bur., Sept. :410.

Roman, C. J. 1994. Hurricane Andrew's Impact on Freshwater Resources. Bioscience 43:247-255.

Romans, B. 1775. A Concise Natural History of East and West Florida. Aitken, NY. Reprinted 1961, Pelican Publishing Co., New Orleans, LA.

Rose, G. N. 1973. Tomato Production in Florida—A Historic Data Series. Econ. Rpt. 48. Food and Resource Econ. Dept., Univ. Fla., Gainesville, FL.

Rose, G. N. 1974. Pepper Production in Florida—A Historic Data Series. Econ. Rpt. 64. Food and Resource Econ. Dept., Univ. Fla., Gainesville, FL.

Rose, G. N. 1975A. Celery Production in Florida—A Historic Data Series. Econ. Rpt. 69. Food and Resource Econ. Dept., Univ. Fla., Gainesville, FL.

Rose, G. N. 1975B. Snap Bean Production in Florida—A Historic Data Series. Econ. Rpt. 74. Food and Resource Econ. Dept., Univ. Fla., Gainesville, FL.

Rouse, R. E. 1992. Hurricane Andrew . . . A Near Miss for Southwest Florida Citrus. Citrus and Veg. Mag. 56(2): 8.

Roy, M., C. O. Andrew, and T. H. Spreen. 1996. *Persian Limes in North America, An Economic Analysis of Production and Marketing Channels.* Fla. Sci. Source, Lake Alfred, FL.

Schokman, L. M. 1993A. Impact of Hurricane Andrew at the Kampong. Kampong Notes 20(3):5.

Schokman, L. M. 1993B. Hurricane Andrew. Kampong Notes 20(4):3.

Scott, J. E. W. 1998. Personal Communication. Westminster, CA.

Scruggs, F. H. 1944. Hurricane Crop Damage Reported. Fla. Grower 52(11):19-22.

Shaver, A. S. 1999. SATERH Assists Hurricane Mitch Recovery. QST 83(2):80.

Sheets, R. C. 1990. The National Hurricane Center - Past, Present and Future. Weather Forecasting 5:185-232.

Simpson, R. H. and H. Riehl. 1981. The Hurricane and Its Impact. Louisiana State Univ. Press, Baton Rouge, LA.

Smith, D. J., M. E. Brown, P. Robinson, H. Poteat, C. M. Myers, and I. Nwabude. 1993. Hurricane Emily, Special Climate Summary. Southeast Regional Climate Ctr., Columbia, SC.

Solger, D. M. 1998. Personal Communication. Washington County Ext. Serv., Chipley, FL.

South Dade News Leader. Sept. 2, 1987. Dade County, FL.

Sprague, J. T. 1848. The Origin, Progress and Conclusion of the Florida War. Facsimile Reproduction 1964 Univ. Fla. Press, Gainesville, FL.

Spyke, P. 1998. Personal Communication. Fort Pierce, FL.

Stewart, A. 1951. Personal Communication. Haines City, FL.

Strano, R. 1997. Personal Communication. Florida City, FL.

Sugg, A. L. 1966. The Hurricane Season of 1965. Monthly Weather Rev. 94:184-190.

Sugg, A. L. 1968. Beneficial Aspects of the Tropical Cyclone. J. Applied Meteorol. 7:39-45.

Sugg, A. L. and P. J. Hebert. 1969. The Hurricane Season of 1968. Monthly Weather Rev. 97:226-238.

Summary of the 1965-66 Season. 1966. Fla. Agr. Stat. Serv., Orlando, FL.

Sumner, H. C. 1941A. North Atlantic Tropical Disturbances of 1941. Monthly Weather Rev. 68:363.

Sumner, H. C. 1941B. Hurricanes of October 3-12 and Tropical Disturbance of October 18-21, 1941. Monthly Weather Rev. 68:303-304.

Sumner, H. C. 1943A. North Atlantic Hurricanes and Tropical Disturbances of 1942. Monthly Weather Rev. 70:49-51.

Sumner, H. C. 1943B. North Atlantic Hurricanes and Tropical Disturbances of 1943. Monthly Weather Rev. 71:179-183.

Sumner, H. C. 1944A. The North Atlantic Hurricane of September 8-16, 1944. Monthly Weather Rev. 72:187-189.

Sumner, H. C. 1944B. The North Atlantic Hurricane of October 13-21, 1944. Monthly Weather Rev. 72:221-223.

Sumner, H. C. 1944C. North Atlantic Hurricanes and Tropical Disturbances of 1944. Monthly Weather Rev. 72:237-240.

Sumner, H. C. 1946A. North Atlantic Hurricanes and Tropical Disturbances of 1945. Monthly Weather Rev. 74:1-5.

Sumner, H. C. 1946B. North Atlantic Hurricanes and Tropical Disturbances of 1946. Monthly Weather Rev. 74:215-217.

Sumner, H. C. 1947. North Atlantic Hurricanes and Tropical Disturbances of 1947. Monthly Weather Rev. 75:251-255.

Sumner, H. C. 1948. North Atlantic Hurricanes and Tropical Disturbances of 1948. Monthly Weather Rev. 76:277-280.

Tannehill, I. R. 1952. *Hurricanes, Their Nature and History. Particularly those of the West Indies and the Southern Coasts of the United States.* 8th ed. Princeton Univ. Press, Princeton, NJ.

Tebeau, C. W. 1971. *A History of Florida.* Univ. Miami Press, Coral Gables, FL.

Terry, R. R. 1993. Impact of Hurricane Andrew on Tropical Fruit Acreage in Dade County. Proc. Fla. State Hort. Soc. 106:116-117.

Thullbery, H. A. 1960. Post Storm Impressions. Citrus and Veg. Mag. 23(3):8.

Todd, J. W. 1965. Memo to P.E. Shaw dated October 4, 1965. Fla. Agr. Stat. Serv., Orlando, FL.

Todd, N. 1998. Personal Communication. LaBelle, FL.

Tomson, F. 1998. Personal Communication. Gainesville, FL.

Townsend, J. C., Jr. 1949. Hurricane Smashes Citrus Crop. Fla. Grower 57(10):4.

Triangle. 1960-1995. Florida Citrus Mutual, Lakeland, FL.

Tropical Storm Marco. 1990. Preliminary Report. Natl. Hurricane Ctr., Miami, FL.

Trovillion, F. 1997. Personal Communication. Lakeland, FL.

USA Today. 1998. Mitch Devastates Agricultural Economics. www.usatoday.com/weather/ news/1998/wmecono.htm

Valentive, T. 1997. Lake Okeechobee 'Lady of Mistery,' Yesterday-Today-Tomorrow Supplement to Okeechobee News, Clewiston News, The Sun, Glades County Democrat and Caloosa Belle.

Van Landingham, K. S. 1976. Pioneer Families of the Kissimmee River Valley. Privately Published, Okeechobee, FL.

Van Landingham, K. S. and A. Hetherington. 1978. *History of Okeechobee County.* Daniels Publishers, Orlando, FL.

Vegetable Summaries. 1917 to 1997-98. Fla. Agr. Stat. Serv., Orlando, FL and Fla. State Mktg. Bur., Jacksonville, FL.

Walden, B. 1997. Personal Communication. Homestead, FL.

Walker, M. H. 1926. Florida Loses 10% Orange and 20% Grapefruit Crop. California Citrograph 12(1):3.

Ward, B. H. 1998. Personal Communication. Walton County Ext. Serv., DeFuniak Springs, FL.

Weather and Crop News. 1964-1997. Fla. Agr. Stat. Serv., Orlando, FL.

White, C. 1998. Personal Communication. Okeechobee, FL.

Wilder, A. 1998. Personal Communication. Fort Pierce, FL.

Will, L. E. 1968. *Swamp to Sugar Bowl; Pioneer Days in Belle Glade.* Great Outdoor Publishing, St. Petersburg, FL.

Williams, Ada Coates. 1998. Personal Communication. Fort Pierce, FL.

Williams, J. L. 1837. *The Territory of Florida . . . From the First Discovery to the Present Time.* A. J. Goodrich, New York. Facsimile Edition, 1962, Univ. Fla. Press, Gainesville, FL.

Williams, J. M. and I. W. Duedall. 1997. Florida Hurricanes and Tropical Storms. Univ. Fla. Press, Gainesville, FL.

Wilson, H. F. 1930-1961. Marketing Florida Citrus. Fla. State Mktg. Bur.

Young, F. D. 1910. Climatological Summary, District No. 2, South Atlantic and East Gulf States.

Youngblood, E. G. 1995. Growers Keep Nervous Eye on Erin. The Ledger, Lakeland, FL.

Zoch, R. T. 1949. North Atlantic Hurricanes and Tropical Disturbances of 1949. Monthly Weather Rev. 79:339-341.

Index

A

C

I

M